MECHANISMS OF LYMPHOCYTE ACTIVATION AND IMMUNE REGULATION VI

Cell Cycle and Programmed Cell Death in the Immune System

ADVANCES IN EXPERIMENTAL MEDICINE AND BIOLOGY

Recent Volumes in this Series

Volume 400A
EICOSANOIDS AND OTHER BIOACTIVE LIPIDS IN CANCER, INFLAMMATION, AND RADIATION INJURY, Part A
Edited by Kenneth V. Honn, Santosh Nigam, and Lawrence J. Marnett

Volume 400B
EICOSANOIDS AND OTHER BIOACTIVE LIPIDS IN CANCER, INFLAMMATION, AND RADIATION INJURY, Part B
Edited by Kenneth V. Honn, Santosh Nigam, and Lawrence J. Marnett

Volume 401
DIETARY PHYTOCHEMICALS IN CANCER PREVENTION AND TREATMENT
Edited under the auspices of the American Institute for Cancer Research

Volume 402
AIDS, DRUGS OF ABUSE, AND THE NEUROIMMUNE AXIS
Edited by Herman Friedman, Toby K. Eisenstein, John Madden, and Burt M. Sharp

Volume 403
TAURINE 2: Basic and Clinical Aspects
Edited by Ryan J. Huxtable, Junichi Azuma, Kinya Kuriyama, and Masao Nakagawa

Volume 404
SAPONINS USED IN TRADITIONAL AND MODERN MEDICINE
Edited by George R. Waller and Kazuo Yamasaki

Volume 405
SAPONINS USED IN FOOD AND AGRICULTURE
Edited by George R. Waller and Kazuo Yamasaki

Volume 406
MECHANISMS OF LYMPHOCYTE ACTIVATION AND IMMUNE REGULATION VI: Cell Cycle and Programmed Cell Death in the Immune System
Edited by Sudhir Gupta and J. John Cohen

Volume 407
EICOSANOIDS AND OTHER BIOACTIVE LIPIDS IN CANCER, INFLAMMATION, AND RADIATION INJURY, Volume 2
Edited by Kenneth V. Honn, Lawrence J. Marnett, Santosh Nigam, Robert L. Jones, and Patrick Y-K. Wong

Volume 408
TOWARD ANTI-ADHESION THERAPY FOR MICROBIAL DISEASES
Edited by Itzhak Kahane and Itzhak Ofek

MECHANISMS OF LYMPHOCYTE ACTIVATION AND IMMUNE REGULATION VI

Cell Cycle and Programmed Cell Death in the Immune System

Edited by

Sudhir Gupta

University of California, Irvine
Irvine, California

and

J. John Cohen

University of Colorado Health Sciences Center
Denver, Colorado

SPRINGER SCIENCE+BUSINESS MEDIA, LLC

Library of Congress Cataloging-in-Publication Data

Mechanisms of lymphocyte activation and immune regulation VI : cell cycle and programmed cell death in the immune system / edited by Sudhir Gupta and J. John Cohen.
p. cm. -- (Advances in experimental medicine and biology ; v. 406)
"Proceedings of the Sixth International Conference on Lymphocyte Activation and Immune Regulation ... held February 2-4, 1996, in Newport Beach, California"--T.p. verso.
Includes bibliographical references and index.
Other title: Cell cycle and programmed cell death in the immune system.

DOI 10.1007/978-1-4899-0274-0
1. Lymphocytes--Congresses. 2. Cell cycle--Congresses. 3. Apoptosis--Congresses. I. Gupta, Sudhir. II. Cohen, J. John. III. International Conference on Lymphocyte Activation and Immune Regulation (6th : 1996 : Newport Beach, Calif.) IV. Series.
[DNLM: 1. Apoptosis--physiology--congresses. 2. Cell Cycle--congresses. 3. Lymphocytes--physiology--congresses. W1 AD559 v.406 1996 / QH 671 M486 1996]
QR185.8.L9M427 1996
616.07'9--dc20
DNLM/DLC
for Library of Congress 96-29392
CIP

Proceedings of the Sixth International Conference on Lymphocyte Activation and Immune Regulation: Cell Cycle and Programmed Cell Death in the Immune System,
held February 2 – 4, 1996, in Newport Beach, California

Originally published by Plenum Press, New York in 1996
MyCopy version of the original edition 1996

10 9 8 7 6 5 4 3 2 1

PREFACE

Since *programmed cell death* was first described in insects in 1964 and *apoptosis* was described in 1972, rapid progress has been made in understanding the basic mechanisms and genes regulating programmed cell death and apoptosis. In addition, defects in various genes regulating programmed cell death have been delineated in several experimental models of human diseases. This volume surveys various aspects of these rapidly developing areas of research in programmed cell death/apoptosis.

This volume should be of interest to basic immunologists and molecular biologists. The volume begins with a historical perspective of cell death. The remainder of the volume is divided into four different parts. Part I deals with the signaling pathways in apoptosis, including cell cycle control of apoptosis, role of ceramide in apoptosis, role of antibody signaling, and biochemical regulation of apoptosis. The mechanisms for recognition of apoptotic lymphocytes by macrophages are also reviewed. Part II examines the role of various genes that regulate apoptosis, including the role of Fas, FasL, and other TNF family members in apoptosis and homeostatic regulation of immune response. Recently described splice variants and their influence on apoptosis are also reviewed, and the role of the members of the Bcl-2 family in apoptosis is discussed in detail. Part III reviews various aspects of apoptosis in B lymphocytes, including mechanisms that regulate apoptosis/survival of B lymphocytes and the regulation of Fas-mediated apoptosis in B lymphocytes. One chapter deals with apoptotic cell death in the avian bursa of Fabricius, and B cell apoptosis in relation to autoreactivity is also discussed. Part IV adopts a similar approach to the analysis of apoptosis in T cells, including the mechanisms of apoptosis in thymocytes and the genes that regulate apoptosis in thymocytes. The regulators of programmed cell death and their role in controlling lymphocyte activation and proliferation are reviewed, and programmed cell death in human peripheral blood T cells and defects in diseases of the immune system are discussed in detail. The role of *fas* and *nur 77* in experimental models of autoimmune diseases is also reviewed, as are the mechanism and role of granzyme-induced apoptosis.

The editors wish to thank Miss Nancy Doman for her excellent secretarial assistance.

Sudhir Gupta
J. John Cohen

CONTENTS

I. Cell Cycle and Signaling Pathways in Apoptosis

II. Apoptosis-Regulating Genes

III. Apoptosis in B Cells

IV. Apoptosis in T Cells

1

APOPTOSIS/PROGRAMMED CELL DEATH

A Historical Perspective

Sudhir Gupta

Division of Basic and Clinical Immunology
University of California
Irvine, California 92717

1. A SIMPLE CELL DEATH

The concept of cell death dates back to 1858, when cell death at the gross level was discussed in Virchow's Cellular Pathology as degeneration, mortification, and necrosis equivalent to the term "gangrene"[1]. Carl Weigart in 1877 described coagulation necrosis in which he observed that necrotic cells lose their nuclei [2]. When stains became available in 1885, Walther Flemming described spontaneous cell death as a physiological event[3] . This was perhaps the first morphological description of apoptosis. He observed that the epithelial lining of regressive ovarian follicles was full of cells whose nuclei were breaking up. The broken nuclei ultimately disappeared. He described half-moons of pyknotic chromatin and loose chromatin in the follicular cavity. He termed the entire process "chromatolysis". Soon after the description of Walther Flemming, Franz Nissen, a German medical student, described chromatolysis in the lactating mammary gland[4]. Chromatin margination was described in 1890[5]. In 1892, Strobe gave a detailed account of chromatolysis and nuclear pathology in breast cancer cells, which would be a modern description of apoptosis[6] . Ludwig Grapher in 1914 published a paper in German under the title "Eine neue Anschauung uber physiologische Zellausschaltung", whose English translation is, "A new point of view regarding the elimination of cells"[7]. His hypothesis was that there must be an amitotic mechanism to counterbalance mitosis. Based upon his studies, he concluded that physiological elimination of cells occurs by chromatolysis during the shrinkage of organs; a sister cell engulfs a neighboring cell that breaks down. Glucksmann understood the significance of chromatolysis in morphogenetic mechanism. In a landmark paper, he described physiological cell death in the embryo[8]. In 1960 Majno et al , using a model of ischemic cell death, established a role of protein denaturation in cell death[9]. The concept of cell suicide was proposed by De Duve, who postulated that the cells might be killed from within by an explosion of their lysosomes acting as "suicide bags"[10].

Mechanisms of Lymphocyte Activation and Immune Regulation VI, Edited by Gupta and Cohen
Plenum Press, New York, 1996

2. PURPOSEFUL CELL DEATH

2.1 Apoptosis

In 1965, Kerr performed the critical experiment that led to the definition of purposeful suicide-apoptosis[11]. He ligated the large branch of the portal vein that supplies the left and median lobes of rat liver. Within hours of ligation, he observed patches of confluent necrosis around the terminal hepatic vein. Initially, the periportal parenchyma was normal; however, over a period of weeks, liver parenchyma subsequently shrank. During this period of parenchymal shrinkage, he observed that individual hepatocytes were continually converted into small round cytoplasmic masses, some of which also contained specks of pyknotic chromatin (a manifestation of cell death). However, their histology was different from that of necrotic cells, in that these cells did not exhibit inflammation (a common feature with necrotic cells). Staining for acid phosphatase showed that lysosomes in necrotic cells rupture, whereas lysosomes in round cytoplasmic masses were intact. Identical round cytoplasmic masses were also observed in very small numbers in the livers of healthy rats, suggesting the physiological nature of this process. In 1971, Kerr used electron microscopy to study the development of the round cytoplasmic masses[12]. They were comprised of membrane enclosed cellular fragments containing crowded well-preserved organelles with occasional pieces of condensed chromatin. He originally termed these changes as "shrinkage necrosis"[13] . However, realizing that the term "necrosis" for a physiological phenomenon is inappropriate, Kerr and his colleagues named the morphological changes "apoptosis"[14].

2.2 Programmed Cell Death

The first indirect evidence of possible programmed cell death in immune cells came from the observations of Cole and Ellis in 1957, who showed a dose-dependent accumulation of soluble polydeoxyribonucleotides in the spleen and bone marrow of irradiated animals, and suggested its possible importance in interphase death[15]. In the 1960's and 70's, a number of investigators showed that polydeoxyribonucleotides are products of chromatin degradation[16–17]. They also observed chromatin degradation in lymphocytes following exposure to glucocorticoids. The kinetics of the appearance of chromatin fragmentation was similar to that of cell death measured by the appearance of pyknosis. The earliest direct evidence of programmed cell death was demonstrated in chick embryo[18] and in the metamorphosis of Pernyi silkworm[19,20]. In 1964, Lockshin and Williams[20] coined the term "programmed cell death". In the 970's, Skalka et al[21], Yamada et al[22] and Zhivotosky et al[23] described the electrophoretic changes in chromatin of lymphocytes that were exposed to irradiation. They showed that the DNA of polydeoxyribonucleotides was similar in length to DNA of nucleosomes and their oligomers. They observed a ladder-like pattern, suggesting that the fragments were multiples of nucleosome. In 1984, Wyllie et al[24] linked the ladder pattern with the morphological changes of apoptosis. In 1986, application of flow cytometry was introduced to analyze DNA fragmentation of individual cells[25, 26].

Although the terms are commonly used interchangeably, programmed cell death (PCD) and apoptosis are two distinct features of cell death. The PCD is a biochemical characteristic, whereas apoptosis is the morphological characteristic. Although these two characteristics are often observed together, programmed cell death can be observed without apoptosis. Programmed cell death (PCD)/apoptosis is perhaps one of the most fundamental and physiological suicide mechanisms for maintaining cellular homeostasis in a variety of cells, including cells of the immune system, by removing undesired cells.

3. GENES REGULATING APOPTOSIS/PROGRAMMED CELL DEATH

3.1 *ced* Genes

It is now evident that the final pathway of apoptosis is the result of activation of endonuclease leading to DNA fragmentation at the internucleosomal junction (approximately 180–200 bp each). The process of apoptosis is controlled by a number of genes that either promote or inhibit apoptosis. The family or families of these genes is rapidly expanding. The genes regulating programmed cell death were first discovered in the nematode *Caenorhabditis elegans*. During *C. elegans* development, the generation of 959 somatic nuclei of the hermaphrodite is accompanied by the generation and subsequent death of 131 cells by PCD[27]. This process of PCD was positively controlled by two genes, the *ced-3* and *ced-4*[28, 29] The cloning of *ced*-3 gene showed a strong homology to human IL-1β converting enzyme (ICE), a protease enzyme that cleaves the inactive precursor of IL-1β to yield active IL-1β[30]. This observation also provided evidence for a possible connection of ICE family proteases to PCD[31]. Studies of inhibitors of ICE-like proteases support their role in core apoptotic pathways (reviewed in [32]). The members of the ICE protease family are rapidly increasing (see chapter by Dixit et al, in this volume). Recently, Nicholson et al[33] have reported another member of ICE/*ced-3* family, apopain/CPP32, which appears to be important in the initiation of mammalian apoptotic cell death by cleaving poly (ADP-ribose) polymerase (PARP). More recently *ced-9* has been shown to protect cells from PCD in *C. elegans*[34]. Furthermore, it has been shown that *ced-9* is homologous to mammalian *bcl*-2 gene[35].

3.2 Bcl-2 Gene

The *bcl-2* proto-oncogene was originally discovered in the immunoglobulin locus as a result of a chromosomal translocation t(14;18) in human follicular lymphoma/ leukemia[36,37]. The cloning of normal *bcl*-2 gene revealed that *bcl*-2 oncogene was identical to normal *bcl*-2 gene [38]. The *bcl-2* gene encodes for a 26 kDa membrane-associated protein that is localized to the mitochondrial membrane and perinuclear membrane[39,40]. Transfection and transgenic experiments have demonstrated that the product of *bcl-2* gene plays a crucial role in the survival of hematopoietic cells[41–43]. A number of investigators, using transgenic and SCID mice, have presented evidence in favor of the role of *bcl-2* in the selection and development of B cells and the maintenance of B cell memory [44, 45]. Takayama et al[46] cloned a novel bcl-2 binding protein, the BAG-1, with anti-apoptotic activity. They demonstrated that BAG-1 cooperate with bcl-2 to inhibit Fas-induced killing in Jurkat T cells. The *bcl-2* gene family is expanding rapidly and now consists of several members, which may function as promoters of PCD or inhibitors of PCD.

3.2 Bcl-x Genes

The *bcl-x* gene was identified by Boise et al[47] in an attempt to identify an avian homologue of *bcl-2*. The highest level of *bcl-x* mRNA is expressed in thymus, bursa, and the central nervous system (CNS). They identified two forms of *bcl-x* cDNAs that differ only in open reading frames (ORF). The longer cDNA ($bcl\text{-}x_L$) contains ORF with 233 amino acids, whereas the shorter cDNA ($bcl\text{-}x_S$) encodes a 170 amino acid protein. $bcl\text{-}x_L$ overexpression is associated with inhibition of apoptosis. In contrast, overexpression of $bcl\text{-}x_S$ prevents *bcl-2* induced protection to PCD.

3.3 Bax Gene

Bax (Bcl-2-associated x-protein) is a 21 kDa protein that was identified by coprecipitation with human Bcl-2[48]. Bax appears to non-covalently bind to Bcl-2. *Bax* mRNA has a wide tissue distribution and is not restricted to cells of lymphoid lineage. *Bax* is alternatively spliced to form a 1.0 kb and 1.5 kb RNA transcript. The functional significance of *bax* has been assigned to 1.0 kb species. Bax accelerates PCD and its overexpression reverses the protection to PCD conferred by Bcl-2. However, the latter property of Bax critically depends upon the ratio of Bcl-2-Bax. When Bcl-2 is in excess, Bax/Bcl-2 heterodimers are formed and cells are protected from PCD; however, in Bax excess Bax homodimers are formed and cells are susceptible to PCD.

3.4 MCL1/A1 Genes

In 1993, two additional *bcl*-2 homologues were cloned. Kozopas et al[49] cloned *MCL1* that appears to have sequence homology to *bcl-2* in the carboxyl-terminal of the protein, including the boxes of high homology and the membrane-binding domain of COOH-terminal. The amino-terminal half contains sequences that are important in protein-protein interactions. The second homologue of *bcl-2 (A1)* was cloned by Lin et al[50]. A1 is expressed in cells of myeloid and T cell lineage and can be induced in macrophages with lipopolysachharide and superinduced by cyclohexamide. A1 shows homology with *Bcl-2* in the central portion of *bcl-2*. The function of these two genes remains to be determined.

3.5 p53

The tumor suppressor protein p53 is one of the most important regulators of cell cycle. Although p53 is considered a "guardian of the genome", there is evidence to suggest that it also plays a role in PCD. Yonish-Rouach et al[51] demonstrated that wild-type p53 is necessary for induction of PCD in a myeloid cell line as demonstrated by transfection of cell line with a temperature sensitive mutant of p53. A role of p53 in apoptosis was demonstrated by Barbara Osborne's group[52] and Andrew Wyllie's group[53]. Using p53 knock-out mice, Clarke et al[53] established that p53 was required for radiation-induced apoptosis in thymocytes. The bcl-2 is capable of inhibiting p53-induced cell death[54]. It has also been shown that p53 is a regulator of *bcl-2* and *bax* gene expression *in vitro* and *in vivo*[55]. p53 appears to be a transcriptional silencer for the *bcl-2* gene[56]. A 195 bp fragment of the 5' untranslated region of the *bcl-2* gene, called p53-dependent negative response element (PNRE), appears to be necessary for the down-regulation of *bcl-2* by p53. Because *bcl-2* lacks a consensus positive response element for wild-type p53, it is also plausible that p53 regulates the production of another transcriptional regulator, the *bax*.

3.6 *c-myc* Gene

Recently, *c-myc* proto-oncogene has been shown to cause PCD. Similar to p53 gene, *c-myc* is also cell type and stimulus specific. Shi et al[57] showed that *c-myc* antisense oligonucleotides inhibit the activation-induced cell death of T cell hybridoma. Evan et al[58] showed that the presence of *c-myc* under conditions of growth arrest (e.g., deprivation of growth factor) can induce apoptosis. Several investigators have demonstrated that *bcl-2* is capable of preventing *c-myc*-induced cell death [59,60].

3.7 Fas & FasL

In 1989, two independent groups of investigators developed monoclonal antibodies that induced cell death in human lymphocytes. Peter Krammer and his colleagues developed monoclonal antibodies against the human B-lymphoblastoid cell line SKW6.4[61]. One of these antibodies, anti-APO-1, reacted with an antigen (APO-1) of approximately 50 kDa on activated human lymphocytes, malignant lymphoid cell line, and leukemic cells. *In vitro* anti-APO-1 antibody induced apoptosis[61]. Later these investigators purified and characterized APO-1 antigen . Yonehara and his colleague developed a monoclonal antibody (anti-Fas) which had a cytolytic activity against various human cells[62]. The sequence analysis of Fas antigen showed that the human Fas antigen is a type I membrane protein consisting of 325 amino acids with a single transmembrane domain. The mouse fibroblasts transfected with Fas antigen underwent cell death within hours following incubation with anti-Fas antibody[63]. Oehm et al[64] isolated APO-1 cDNA and observed that APO-1 and Fas antigens are identical. Amino acid sequence of Fas/APO-1 (CD95) indicated that CD95 is a member of the TNF/NGF receptor family[65]. Mutational studies of the Fas antigen[66] and of type I TNF receptor[67] demonstrated the homology between their cytoplasmic domain. Furthermore, it was demonstrated that the cytoplasmic domain is responsible for PCD signal transduction mediated by both CD95 and type I TNF receptor. Enari et al[68] demonstrated that a sequential activation of ICE-like and CPP32-like proteases is required in Fas-mediated apoptosis. Recently a number of gene products (e.g. FADD, FAP-1, RIP etc.) that interact with the cytoplasmic domains of Fas, have been identified[69–72]. These proteins could be variably expressed among cells of different lineages and affect downstream Fas-induced signaling /apoptotic pathways.

3.8. Disorders of Fas/FasL

3.8.1 Experimental Models. A role of Fas antigen in the development and maintenance of the immune system was supported by the observation of two types of Fas mutation in mice with *lpr* (lymphoproliferative) phenotype[73]. These mice either express significantly reduced Fas transcript levels (*lpr* mutation) because of the presence of retroviral transposon in the *Fas* gene[74–76] or they express a normally abundant but dysfunctional form of *Fas* (*lpr*^{cg} mutation) which cannot effectively couple to the PCD pathway due to an amino acid substitution (I to N) in the cytoplasmic domain[77]. These observations supported the view that *lpr* gene (which is associated with lymphoproliration of CD4-CD8- T cells) encodes the structural gene for the mouse Fas antigen that transduces the apoptotic signal into cells. The most definitive evidence of the cause and effect relationship of *Fas* mutation in *lpr* disorder was presented by Wu et al[78], who showed that T cell specific expression of the non-mutated Fas protein in *lpr* mice abrogated the onset of the autoimmune syndrome. Soon thereafter, ligand for Fas (Fas-L) was cloned[79,80] and it was demonstrated that interaction between Fas-Fas-L resulted in an induction of apoptosis[81]. Furthermore, Suda et al[79] demonstrated that FasL belongs to TNF family. Takahashi et al[82] reported another mutation, the *gld* mutation in mice, that is associated with lymphoproliferation similar to that in *lpr* mice. They demonstrated that *gld* gene encodes a functional FasL. Bone marrow chimeric experiments have shown that the transfer of normal stem cells along with the *gld* stem cells into *gld* mice ameliorates the *gld* lymphoproliferative manifestations, while normal and *lp*r stem cells did not abrogate or reduce the *lpr* stem cell transfer pathogenecity.

3.8.2 Human Diseases. Recently, a mutation of Fas has also been reported in human disorders associated with lymphoproliferation and autoimmunity[83, 84]. Fisher et al[84] described 5 children with a rare autoimmune lymphoproliferative disorder associated with an expanded population of TCR-CD3+CD4-CD8- lymphocytes. Each child had defective Fas-mediated T cell apoptosis associated with a unique *fas* gene mutation. More recently, Kashahara et al[85] reported a defect in Fas-induced apoptosis in two siblings with a lymphoproliferative disorder characterized by pancytopenia, adenopathy, hepatosplenomegaly and increase of $TCR\alpha\beta^+CD4^-CD8^-$ T cells in the peripheral blood. Although these patients had normal expression of Fas antigen, Fas-induced apoptosis was defective in activated T cells, activated or EBV-transformed B cells, and freshly isolated neutrophils. Molecular analysis of the PCR products from EBV transformed B cells from both patients showed 4 bp insertion between exon 7 and 8 in mutated Fas mRNA, resulting in the deletion of cytoplasmic death domain. The gene appears to have dominant inheritance. This study demonstrates that Fas-mediated apoptosis plays a role not only in clonal deletion of T cells but in deletion of the other hematopoietic cells as well.

3.9 NGF1-B/nur77/TINUR Genes

The most recently discovered genes regulating apoptosis belong to a family of orphan receptors, the NGF1-B/nur 77 family. Liu et al[86] and Woronicz et al[87] simultaneously and independently described a role of *nur77* gene and Nur 77 receptor in apoptosis of T cell hybridoma. Okabe et al[88] have cloned a transcriptionally inducible nuclear receptor (*TINUR*) from an apoptotic human T cell line. The *TINUR* gene is induced within 1 hour after cross-linking of the T cell antigen receptor complex. TINUR binds to the same DNA sequence as NGF1-B/nur 77 receptor and belongs to the same family of orphan receptors.

4. CONCLUDING REMARKS

It is anticipated that more genes (in the existing families) regulating apoptosis, as well as new families of genes, will be discovered in the near future. The downstream signaling pathways and their components leading to apoptosis/PCD will be better defined. The most difficult challenge will be to establish the functional relationship between the various genes regulating apoptosis/PCD. Perhaps, studies of the "experiments of nature", the human diseases and experimental models of human diseases, will be instrumental in deciphering these complex relationships among various genes regulating apoptosis/PCD.

REFERENCES

1. Virchow R. Cellular Pathology as Based Upon Physiological and Pathological Histology. Ed. 2 (translated from German by B. Chance, 1859); reproduced by Dover Publications, New York pp 356 (1971).
2. Weigert C. Uber Croup und Diphtheritis. Ein experimeteller und anatomischer Beitrag zur Pathologie der specifischen Entzundungsformen. Virchows Arch. Pathol. Anat. 72: 461 (1877–1878)
3. Flemming W. Uber die Bildung von Richtungsfiguren in Saugethiereiern beim Untergang Graaf'scher Follikel. Arch Anat EntwGesh 221 (1885).
4. Nissen F. Uber das Verhalten der Kerne in den Milchdrusenzellen bei der Absonderung. Arch Mikroskop Anat. 26: 337 (1886).
5. Arnheim G. Coagulationsnekrose und Kernschwund. Virchows Arch Path. Anat. 120: 367 (1890).
6. Stobe H. Zur Kenntnis verschiedener cellularer Vorgange und Erscheinungen in Geschwulsten. Beitr. Pathol. Anat. 11: 1 (1914).
7. Grapher L. Eine neue Anschauung uber physiologische Zellausschaltung. Arch. Zellforsch 12: 373 (1914).
8. Glucksmann A. Cell death in normal vertebrate ontogeny. Biol. Rev. Camb. Philos. Soc. 26: 59 (1951).

9. Majno G., La Gattuta M., Thompson T.E. Cellular death and necrosis: chemical, physical and morphologic changes in rat liver. Virchows Arch. Pathol. Anat. 333: 421 (1960).
10. Majno G. and Joris I. Cells, Tissues and Disease: Principles of General Pathology. Blackwell Science, Inc. (in press)
11. Kerr J.F.R. A histochemical study of hypertrophy and ischemic injury of rat liver with special reference to changes in lysosomes. J. Pathol. Bactriol. 90: 557 (1965).
12. Kerr J.F.R. An electron-microscope study of liver cell necrosis due to heliotrine. J. Pathol. 97: 557 (1969).
13. Kerr J.F.R. Shrinkage necrosis: a distinct mode of cellular death. J. Pathol. 105: 13 (1971).
14. Kerr J.F.R., Wyllie A.H., and Currie A.R. Apoptosis: a basic biological phenomenon with wide-ranging implications in tissue kinetics. Brit. J. Cancer 26: 239 (1972).
15. Cohen L.J. and Ellis M.E. Radiation-induced changes in tissue nucleic acids: release of soluble deoxynucleotides in the spleen. Radiat. Res. 7: 508 (1957).
16. Skalka M. and Matyasova J. The effect of low radiation doses on the release of deoxyribonucleotides in hematopoietic and lymphatic tissues. Int. J. Radiat. Biol. 7: 41 (1963).
17. Vodolazskaya N.A. and Vermolaeva N.V. Investigation of DNP degradation produced by γ-irradiation, hydrocortisone and degranol in rat thymus. Radiobiologiya 11: 335 (1971).
18. Saunders J.W. Jr. Death in embryonic systems. Science 154: 604 (1966).
19. Lockshin R.A. and Williams C.W. Programmed cell death. I. Cytology of generation in the intersegmental muscles of the Pernyi Silkmoth. J. Insect Physiol. 11: 123 (1965).
20. Lockshin R.A. & Williams C.M. Programmed cell death II. Endocrine potentiation of the breakdown of the intersegmental muscles of silkworm. J. Insect Physiol. 10: 643 (1964).
21. Skalka M., Matyasova J., Cejkova M. DNA in chromatin of irradiated lymphoid tissues degrades in vivo into regular fragments. FEBS Lett. 72: 271 (1976).
22. Yamada T., Ohyama H., Kinjo Y., Watanabe M. Evidence for internucleosomal breakage of chromatin in rat thymocytes irradiated in vitro. Radiat. Res. 85: 544 (1981).
23. Zhivotovsky B.D., Zvonareva N.B., Hanson K.P. Characteristics of rat thymus chromatin degradation products after whole body irradiation. Int. J. Rad. Biol. 39: 437 (1981).
24. Wyllie A.H., Morris R.G., Smith A.L., Dunlop D. Chromatin cleavage in apoptosis: association with condensed chromatin morphology and dependence on macromolecular synthesis. J. Pathol. 142: 67–77, 1984
25. Pachatnikov V.A., Afanasyev V.N., Korol B.A., Korneev V.N., Rochev Y.A., and Umansky S.R. Flow cytometry analysis of DNA degradation in thymocytes of γ-irradiation or hydrocortisone treated rats. Gen. Physiol. Biophys. 5: 273 (1986).
26. Afanasyev V.N., Korol B.A., Mantsygin Y.A., Nelipovich P.A., Pechatnikov V.A. and Umansky S.R. Flow cytometry and biochemical analysis of DNA degradation characteristics of two types of cell death. FEBS Lett. 194: 347 (1986).
27. Sulston J.E. and Horvitz H.R. Post embryonic lineages of the nematode *Caenorhabditis elegans.* Dev. Biol. 82: 110 (197 7).
28. Ellis H.M. and Horvitz H.R. Genetic control of programmed cell death in the nematode C. elegans. Cell 44: 817 (1986).
29. Yuan J. and Horvitz H.R. The *Caenorhabditis elegans* gene *Ced-3* and *Ced-4* act autonomously to cause programmed cell death. Dev. Biol. 138: 33 (1990).
30. Yuan J., Shaham S., Ledoux S., Ellis J.M. and Horvitz H.R. The *C. elegans* cell death gene *Ced-3* encodes a protein similar to mammalian interleukin-1β converting enzyme. Cell 75: 641 (1993).
31. Miura M., Zhu H., Rotello R., Hartwieg E.A. & Yuan J. Induction of apoptosis in fibroblasts by IL-1 beta-converting enzyme, a mammalian homolog of the *C.*elegans cell death gene *ced-3.* Cell 78: 653–660, 1993.
32. Henkart P.A. ICE family proteases: mediators of all apoptotic cell death. Immunity 4: 195 (1996).
33. Nicholson D.W., Ali A., Thornberry N.A., Vaillancourt J.P, Ding C.K., Gallant M., Gareau Y., Griffin P.R., Labelle M., Lazebnik Y.A., Munday N.A., Raju S.M., Smulson M.E., Yamin T-T., Yu V.L. and Miller D.K. Identification and inhibition of the ICE/CED-3 protease necessary for mammalian apoptosis. Nature 376: 37 (1995)
34. Hengartner M.O., Ellis R.E. and Horvitz H.R. *Caenorhabditis elegans* gene *ced-9* protects cells from programmed cell death. Nature 356: 494 (1995).
35. Hengartner M.O., & Horvitz H.R. *C. elegans* cell survival gene *Ced-9* encodes a functional homologue of the mammalian proto-oncogene bcl-2. Cell 76: 665 (1994).
36. Tsujimoto Y., Finger L.R., Yunis J., Nowell P.C. and Croce C.M. Cloning of chromosomes break point of neoplastic B cells with the t(14:18) chromosome translocation. Science 226: 1097 (1989)

37. Tsujimoto Y., & Croce C.M. Analysis of structure, transcripts, and protein products of *bcl-2*, the gene involved in human follicular lymphoma. Proc. Natl. Acad. Sci. (USA) 83: 5214 (1985).
38. Chen-Levy Z., Nourse J., and Cleary M.L. The bcl-2 candidate proto- oncogene product is a 24-kilodalton integral membrane protein highly expressed in lymphoid cell lines and lymphomas carrying the (14:18) translocation. Mol. Cell Biol. 9: 701 (1989).
39. Hockenbery M.O., Nunez Z., Milliman C., Schreiber R.D., and Korsmeyer S.J. Bcl-2 is an inner mitochondrial membrane protein that blocks programmed cell death. Nature 348: 334 (1990).
40. Chen-Levy Z. and Cleary M.L. Membrane topology of the Bcl-2 proto- oncogene protein demonstrated in vitro. J. Biol. Chem. 265: 4929 (1990).
41. Siegal R.M., Katsumata M., Miyashita T., Louie D.C., Greene M.I., and Reed J.C. Inhibition of thymocyte apoptosis and negative antigenic selection in *bcl-2* transgene mice. Proc. Natl. Acad. Sci. (USA) 89: 7003 (1992).
42. Strasser A., Harris A.W., & Cory S. bcl-2 transgene inhibits T cell death and perturbs thymic self censorship. Cell 67: 889 (1991).
43. Vox D.L., Cory S., and Adams J. *Bcl-2* gene promotes haematopoietic cell survival and cooperates with*c-myc* to immortalize pre-B cells. Nature 335: 440 (1988).
44. Nunez G., Hockenbery D., McDonnell T.J., Sorensen C.M., and Korsmeyer S.J. Bcl-2 maintains B cell memory. Nature 353: 71 (1991).
45. Strasser A., Harris A.W., Corcoran L.M., and Cory S. Bcl-2 expression promotes B but not T-lymphoid development in SCID mice. Nature 368: 457 (1994).
46. Takayama S., Sato T., Krajewski S., Kochel K., Irie S., Millan J. and Reed J. Cloning and functional analysis of BAG-1: a novel Bcl-2 binding protein with anti-cell death activity. Cell 80: 279 (1995).
47. Boise L.H., Gonzalez-Garcia M., Postema C.E., Ding L., Lindsten T., Turka L.A., Mao X., Nunez G., and Thompson C. bcl-x, a bcl-2 related gene that functions as a dominant regulator of apoptotic cell death. Cell 74: 597 (1993).
48. Oltvai Z.N., Miliman C., and Korsmeyer S.J. Bcl-2 heterodimerizes in vivo with a conserved homologue, Bax, that accelerates programmed cell death. Cell 74: 609 (1993).
49. Kozopas K.M., Yang T., Buchan H.L., Zhou P., and Craig R.W. MCL 1, a gene expressed in programmed myeloid cell differentiation, has sequence similarity to Bcl-2. Proc. Natl. Acad. Sci. (USA) 90: 3516 (1993).
50 Lin E.Y., Orlofsky A., Berger M.S., Prystowsky. Characterization of A1, a novel hematopoetic-specific early-response gene with sequence similarity to bcl-2. J. Immunol. 151: 1979 (1993).
51. Yonish-Rouach E., Resnizky D., Lotem J., Sachs L., Kimchi A., and Oren M. Wild-type p53 induces apoptosis in myeloid leukemic cells that is inhibited by interleukin-6. Nature 352: 345 (1991).
52. Lowe S.W., Schmitt E.M., Smith S.W., Osborne B.A., and Jacks T. p53 is required for irradiation-induced apoptosis in murine thymocytes. Nature 362: 847 (1993).
53. Clarke A.R., Purdie C.A., Harrison D.J., Morris R.G., Bird C.C., Hooper M.L., and Wyllie A.H. Thymocyte apoptosis induced by p63 dependent and independent pathways. Nature 362: 849 (1993).
54. Chiou S.K., Rao L., and White E. Bcl-2 blocks p53-dependent apoptosis. Mol. Cell. Biol. 14: 2556 (1994).
55. Miyashita T., Krajewski S., Krajewska M., Wang H.G., Lin H.K., Liebermann D.A., Hoffman B., and Reed J.C. Tumor suppressor p53 is a regulator of *bcl-2* and *bax* gene expression *in vitro* and *in vivo*. Oncogene 9: 1799 (1994).
56. Miyashita T., Harigai M., Hanada M., and Reed J.C. Identification of a p53- dependent negative response element in the *bcl-2* gene. Cancer Res. 54: 3131 (1994).
57. Shi Y., Glynn J., Guilbert L.J., Cotter T.G., Bisonette R.P., and Green D.R. Role of *c*-myc in activation-induced apoptotic cell death in T-cell hybridoma. Science 257: 212 (1992).
58. Evan G.I., Wyllie A.H., Gilbert C.S., Lttlewood T.D., Land H., Brooks M., Waters C.M., Penn L.Z. and Hancook D.C. Induction of apoptosis in fibroblasts by c-myc protein. Cell 69: 119 (1992).
59. Bissonnette R.P., Echeverri F., Mahboubi A., and Green D.R. Apoptotic cell death induced by *c-myc* is inhibited by *bcl-2*. Nature 359: 352 (1992).
60. Fanidi A., Harrington E.A., and Evan G.I. Cooperative interaction between *c-myc* and *bcl-2* proto-oncogenes. Nature 359: 554 (1992).
61. Trauth B.C., Klas C., Peters A.M.J., Matzku S., Moller P., Falk W., Debatin K.M., and Krammer P.H. Monoclonal antibody-mediated tumor regression by induction of apoptosis. Science 245: 301 (1989).
62. Yonehara S., Ishi A., and Yonehara M. A cell-killing monoclonal antibody (anti-Fas) to a cell surface antigen co-downregulated with the receptor of tumor necrosis factor. J. Exp. Med. 169: 1747 (1989).
63. N., Yonehara S., Ishii A., Yonehara M., Mizushima S.I., Sameshima M., Hase A., Seto Y. & Nagata S. The polypeptide encoded by the cDNA for human cell surface antigen Fas can mediate apoptosis. Cell 66: 233 (1991).

64. Oehm A., Behrmann I., Falk W., Pawlita M., Maier G., Klas C., Li-Weber M., Richards S., Dhein J., Trauth B.C., Ponsting H., and Krammer P.H. Purification and molecular cloning of the APO-1 cell surface antigen, a member of the tumor necrosis factor/nerve growth factor receptor superfamily. Sequence identity with the Fas antigen. J. Biol. Chem. 67: 10709 (1992).
65. Watanabe-Fukunaga R., Brennan C.I., Itoh N., Yonehara S., Copeland N.G., Jenkins N.A., and Nagata S. The cDNA structure, expression, and chromosomal assignment of the mouse Fas antigen. J. Immunol.148: 1274 (1992).
66. Itoh N. and Nagata S. A novel protein domain required for apoptosis.J. Biol. Chem. 268: 10932 (1993).
67. Tartaglia L.A., Ayres T.M., Wong G.H.W., and Goeddel D.V. A novel domain within the 55 kd TNF receptor signals cell death. Cell 74: 845 (1993)..
68. Enari M., Hug H. and Nagata S. Involvement of an ICE-like protease in Fas- mediated apoptosis. Nature 375: 78 (1995).
69. Chennaiyan A.M., O'Rourke K., Tiwari M. and Dixit V. FADD, a novel death-containing protein, interacts with the death domain of Fas and initiates apoptosis. Cell 61: 505 (1995).
70. Stanger B.Z., Leder P., Lee T., Kim E., Seed B. RIP: a novel protein containing a death domain that interacts with Fas/APO-1 (CD95) in yeast and causes cell death. Cell 81: 515 (1995).
71. Sato T., Irie S., Kitada S. and Reed J. FAP-1: a protein tyrosine phosphatase that associates with Fas. Science 268: 411 (1995).
72. Boldin M.P., Varfoloneef E.E., Pancer Z., Mett I., Camonis J. and Wallach D. A novel protein that interacts with the death domain of Fas/APO-1 contains a sequence motif related to the death domain. J. Biol. Chem. 270: 7795 (1995).
73. Cohen P.L., and Eisenberg R.A. The lpr and gld genes in systemic autoimmunity: Life and death in the Fas lane. Immunol. Today 13: 427 (1992).
74. Wu J., Zhou T., He J. and Mountz J.D. Autoimmune disease in mice due to integration of an endogenous retrovirus in an apoptosis gene. J. Exp. Med. 178: 461 (1993).
75. Chu J-L., Drappa J., Parnnasa A., and Elkon K.B. The defect in *fas* mRNA expressed in MRL/*lpr* is associated with insertion of the retrotransposon, Etn. J. Exp. Med. 178: 723 (1994).
76. Adachi M., Watanabe-Fukunaga R. & Nagata S. Aberrant transcription caused by the insertion of an early transposable element in an intron of the Fas antigen gene of *lpr* mice. Proc. Natl . Acad. Sci. (USA) 90: 1756 (1993).
77. Watanabe-Fukunaga R., Brannan C.I., Copeland N.G., Jenkins N.A., and Nagata S. Lymphopoliferation disorder in mice explained by defects in Fas antigen that mediate apoptosis. Nature 356: 314 (1992).
78. Wu J., Zhou T., Zhang J., He J., Gause W.C. and Mountz J.D. Correction of accelerated autoimmune disease by early replacement of the mutated *lpr* gene with the normal *fas* apoptosis gene in the T cells of transgenic MRL-*lpr/lpr* mice. Proc. Natl. Acad. Sci. (USA) 91: 2344 (1994).
79. Suda T., Takahashi T., Golstein P., & Nagata S. Molecular cloning and expression of Fas ligand, a novel member of the tumor necrosis factor family. Cell 75: 1169 (1992).
80. Takahashi T., Tanaka M., Inazawa J., Abe T., Suda T., & Nagata S. Human Fas ligand: gene structure, chromosome location and species specificity. Int. Immunol. 6: 1567 (1994).
81. Suda T. and Nagata S. Purification and characterization of the Fas ligand that induces apoptosis. J. Exp. Med. 179: 873 (1994).
82. Takahashi T., Tanaka M., Brannan C.I., Jenkin N.A., Copeland N.G., Suda T., and Nagata S. Generalized lymphoprolireative disease in mice, caused by a point mutation in the Fas ligand. Cell 76: 969 (1994).
83. Rieux-Laucat F., Le Diest F., Hivroz C., Roberts I.A., Debatin K.M., Fischer A., and de Villartay J.P. Mutation in Fas associated with human lymphoproliferative syndrome and autoimmunity. Science 268: 1347 (1995).
84. Fisher G.H., Rosenberg F.J., Straus S., Dale J., Middelton L.A., Lin A.Y., Strober W., Lenardo M., and Puck J.M. Dominant interfering Fas gene mutations impair apoptosis in a human autoimmune lymphoproliferative syndrome. Cell 81: 935 (1995).
85. Kashahara T., Wada Y., Niida I., Ohno A., Yachie H., Seki T., Miyawaki N., and Taniguchi N. Human lymphoproliferative disease associated with FAS mutation. The FASEB J. 10 (6): A1063, 1996
86. Liu Z.G., Smith S.W., McLaughlin K.A., Schwartz L.M. and Osborne B.A. Apoptotic signals delivered through the T-cell receptor of a T-cell hybrid require the immediate-early gene *nur 77*. Nature 367: 281 (1994).
87. Woronicz J.D., Calnan B., Ngo V., and Winoto A. Requirement for the orphan steroid receptor Nur77 in apoptosis of T cell hybridomas. Nature 367: 277 (1994).
88. Okabe T., Takayanagi R., Imasaki K., Haji M., Nawata H., and Watnabe T. cDNA cloning of a NGF1-B/nur77-related transcription factor from an apoptotic human T cell line. J. Immunol. 154: 3871–3879, 1995.

2

APOPTOSIS AND ITS REGULATION

J. John Cohen

Department of Immunology
University of Colorado Medical School
Denver, Colorado 80262

1. INTRODUCTION

There is only one popular way to be conceived, but a myriad of ways to die. We die from some external force, such as an accident or foul play, or else from "natural causes". It is interesting how good the parallel is with the cells out of which we are made. All our somatic cells arise by the process of mitosis, regardless of their location in the body or their ultimate destiny. And these cells die by either of two processes, roughly equivalent to accidents and natural causes. Just as a pathologist will determine cause of death by examining the body, we distinguish the two forms of cell death by morphology[1].

Necrosis is the morphology most often seen when cells die from severe and sudden injury, such as anoxia, hyperthermia, physical trauma, or chemical damage. This could be referred to as accidental cell death, as it almost always results from some unanticipated pathological process. In necrosis there are early changes in mitochondrial shape and function, and the cell rapidly becomes unable to maintain homeostasis. The plasma membrane may be the major site of damage; it loses its ability to regulate osmotic pressure, and the cell swells and bursts. Cellular contents are spilled into the surrounding tissue space where they provoke an inflammatory response. Because injury of this type is often extensive, involving millions of cells simultaneously, the inflammatory response is desirable, as it helps clear away debris so that repair can begin.

Apoptosis is a form of cell death in which the process is more subtle. It refers to a series of morphological changes during cell death that are clearly different from those seen in necrosis, and because it is associated with normal processes of tissue regulation it is of general interest to biologists. It is often equated with *programmed cell death*, but it is useful to keep the two concepts separate. In vertebrates cell death may be acceptable, usual and even highly predictable, but it is not programmed as precisely as the death of individual cells in invertebrates, which in some cases can be predicted with perfect accuracy[2,3]. One must accept the confusing idea that cell death may be both programmed and random. For example, as B lymphocyte populations develop, about 95 cells out of 100 will die[4] for any of a variety of reasons (faulty gene rearrangement, expression of antiself receptors, lack of stimulation, etc.) but it is impossible to predict beforehand which ones they will be.

Mechanisms of Lymphocyte Activation and Immune Regulation VI, Edited by Gupta and Cohen
Plenum Press, New York, 1996

Apoptosis seems to be the most common morphology when cell death is physiologically determined. Examples include the death of cells with short half-lives (such as the neutrophil)[5]; the elimination of self-reactive T cells[6]; involution of cells deprived of necessary growth factors[7]; morphogenetic death of cells during embryonic and early postembryonic development[8]; and killing of cells which serve as targets for T, NK, or antibody-dependent cell-mediated cytotoxic mechanisms[9]. The reason the cell must die is different from cell type to cell type, as is the initiating or triggering mechanism. Often the signal to die comes from the environment, as in exposure to or withdrawal of a hormone or growth factor. This is conceptually interesting, as it means that the fate of one cell is dependent upon the activity of another; the death may be programmed, but the locus of the program is outside the cell that must die. Because normal apoptosis of a cell usually confers some survival advantage on the organism, one might call this "obligatory altruistic suicide", a term actually coined in reference to human societies by the social anthropologist Durkheim[10]. In other cases, for example neutrophil turnover, it seems as if the cell has an autonomous internal clock controlling its death program (true "cell autonomous" programmed death has been identified in the nematode, and some of the genes ordering it have been elucidated[11].) The pathways to activation of apoptosis will be different in different cells, but the mechanism of death itself may always be the same, that is, there may be a final common pathway in apoptosis. In studying this problem it is therefore important to try to distinguish processes which are cell-specific (of potential therapeutic interest) from ones which are death-specific (of great theoretical interest).

There may be many triggers for a suicide pathway, even in a single cell type. Not all of these will seem to be "programmed" in any simple sense. For example, thymocytes may be signalled to activate the apoptosis pathway by glucocorticoids[12] or by antigen receptor crosslinking[13,14], and these both seem to reflect physiological events. But thymocytes will undergo an identical-appearing apoptotic process after low-dose ionizing irradiation[15], or exposure to certain poisons[16]. Since neither of these agents are physiological, this death cannot be strictly defined as programmed. But perhaps some cells, when nonspecifically damaged, physiologically activate a "death program". This implies that the damage must not be too severe for the program to be expressed: thymocytes heated to 43°C will die by necrosis if kept at this temperature, but if returned soon enough to 37°C, die by apoptosis[17]. Teleologically, it has been suggested that it may be too dangerous for lymphocytes to attempt repair of certain kinds of damage, as faulty repair could lead to autoimmunity or leukemia; instead, they activate their apoptotic mechanism. The somatic cell's altruistic suicide increases the chances of survival of the body as a whole. It will be interesting to see what cells besides lymphocytes have programmed this sort of reaction to injury. Possible examples might be found in the gut and the developing brain, and it may in fact be a widespread phenomenon. Other, less dangerous cells, may invest more energy in damage repair (Fig. 1.)

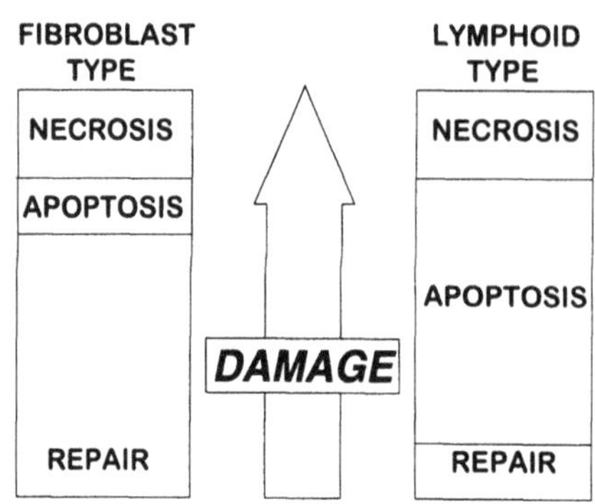

Figure 1. At low levels of damage, all cell types are capable of repair. As damage becomes more severe the apoptotic pathway is engaged; this happens more readily in the lymphoid class of cells, of which faulty repair might have dire consequences for the organism. Extreme damage kills the cell directly.

2. MORPHOLOGY AND BIOCHEMISTRY OF APOPTOSIS

The cellular changes in apoptosis are numerous, and it is still not clear which of them are directly associated with death, and which are of the greatest physiological importance. The plasma membrane in apoptosis becomes ruffled and blebbed, in a way that is more pronounced and much more active than is seen in necrosis; the phenomenon has been called "zeiosis"[18,19]. As might be predicted from this appearance, the cell may break up into what are called apoptotic bodies, but these remain sealed with intact osmotic gradients. There is no spilling of intracellular contents, and no provocation of inflammation. It is possible that the apoptotic cell strengthens its membranes against the risk of lysis by the activation of protein cross-linking enzymes such as tissue transglutaminase[20]. One might speculate from its appearance that the cell undergoes profound alterations in its cytoskeleton. Although up to now this question has not been studied in detail, there are examples in which disruption of the cytoskeleton leads to apoptosis[21], while conversely, cytoskeletal stabilization inhibits apoptosis[22]. The calcium-dependent protease calpain has been implicated in several physiological models of apoptosis[23], and it is of interest that certain cytoskeletal components, known to be calpain substrates, are degraded during apoptosis[24].

An early step in apoptosis *in vivo* is the recognition of the cell and its fragments by phagocytic cells. One interesting way in which this may happen is by a loss of normal membrane phospholipid asymmetry, with the appearance of phosphatidylserine on the outer leaflet of the plasma membrane[25]; some macrophages and many normal cells may have receptors for this normally-hidden phospholipid[26]. Other macrophages use a combination of a vitronectin receptor and CD36 to bind thrombospondin, which in turn binds apoptotic cells[27]. The result is that the cell or apoptotic body is phagocytosed while still alive; no intracellular contents are released to excite inflammation. Furthermore, the phagocyte ingesting an apoptotic cell is not activated[28].

The cell undergoing apoptosis shrinks, losing up to a third of its volume, and electron micrographs show an extremely condensed cytoplasm with normal-appearing organelles[29]. The most likely explanation is that the cell has lost water and, as it does not immediately swell, it has probably lost ions isosmotically. How this might happen is unknown. No specific lesions in cytoplasmic organelles have been described, although they may not have been looked for adequately. Swelling of the cisternae of endoplasmic reticulum has been observed, which may lead to cytoplasmic vacuolization. The cell rapidly decreases its synthesis of RNA and proteins and degrades them[30,31], but the timing of these events in relation to morphologic changes is not well characterized. Although it was originally thought that the transition from normal to apoptotically-shrunken was so rapid that an intermediate form could not be observed, the flow cytometer may help characterize the change[32–37], and in combination with cell sorting has begun to provide fresh insights into the order of events.

The nucleus is the locus of the most impressive action in apoptosis. The changes are most clearly observed by transmission electron microscopy, but are also readily seen in live cells using fluorescent cell-permeant DNA stains such as Hoechst 33342 or acridine orange[32]. Although the exact pattern is different from cell to cell, in general the nucleus shrinks and its chromatin becomes very dense, collapsing into patches, then into crescents in tight apposition to the nuclear envelope, and finally, in many cells, into one or several dense spheres. This change is often accompanied by fragmentation of the DNA into a ladder of regular subunits, the result of apparently random double-stranded DNA breaks in the linker regions between nucleosomal cores[12]. There may be over a million such breaks in each cell, a situation which could never be repaired and in itself must result in the cessation of transcription. But in other cell types typical apoptosis, including the nuclear changes, is observed microscopically in the absence of internucleosomal DNA cleavage[38]. In a few systems in which it has been studied, some form of DNA damage (extensive single-stranded nicking or rarer double-stranded breaks) *is* observed[39], and may be necessary for the morphological changes. In general, the production of large fragments, 50 to 300 kbp, takes place before nucleosomal cleavage, and may involve a different nuclease[40]. There is no consensus on the role DNA damage plays in the cell's death. Although transcription stops, that in itself cannot kill the cell as fast as it is observed to die during apoptosis. Several workers have reported that aurin tricarboxylic acid, which inhibits nucleases, also inhibits all aspects of apoptosis[41–43]; but it is unsafe to conclude that therefore the destruction of DNA is the critical step in apoptosis, as aurin tricarboxylic acid inhibits many or most other cellular metabolic reactions. It should also be noted that in spite of all the focus on the characteristic nuclear changes of apoptosis, it has been claimed that they are completely unnecessary for the process[44].

3. GENETIC REGULATION

Before discussing the genetic regulation of apoptosis it is necessary to summarize briefly some relevant observations about the activation of the process in different cell lineages[45]. Some of the examples described above depend on new gene expression after exposure to the apoptotic stimulus, and are blocked if mRNA or protein synthesis are inhibited. These are collectively referred to as *induction* mechanisms, and this mechanism seems to be the most important in true physiological cell death. There are other models, for example the cell line HL-60, in which apoptosis is, in contrast, triggered by the inhibition of mRNA or protein synthesis[46]. Because these cells behave as though the suicide program is constitutively present, but cannot be expressed until inhibitory factors are removed, the mechanism is referred to as *release*. These two mechanisms are clearly very closely related biochemically. Finally, there are systems -- most notably the activation of apoptosis in target cells by cytotoxic T cells -- in which inhibitors of macromolecular synthesis have no effect, positive or negative[9]. These have been called *transduction* mechanisms, and their existence means that some, perhaps all, cells may have all the necessary molecules for apoptosis within them at all times. Transduction is the mode of apoptotic cell death in the targets of killer cells. Damage can invoke any of these pathways: the pathological activation of a physiological process. The implication of this for gene cloning strategies is uncertain. Regardless of the triggering mechanism, all forms of apoptosis are morphologically similar, leading to the suggestion, at present still a hypothesis, that different triggers lead into a "final common pathway" of apoptosis (Fig.2.)

As mentioned above, in induction models of apoptosis there is a requirement for new gene expression for both the morphologic changes and death itself to occur[45]. These observations led to the conclusion that the cell itself committed suicide in response to the signal; it was not murdered. This is most plainly illustrated in comparing immature thymo-

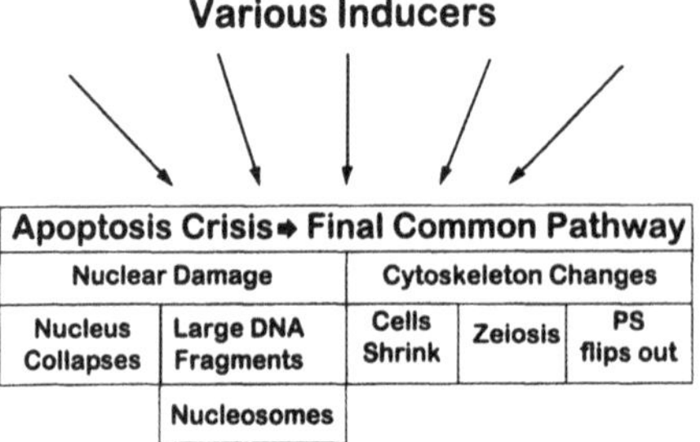

Figure 2. Because all forms of apoptosis have certain features in common, the concept of a final common pathway, elicited by any of a host of inducers, has emerged. The nature of the "apoptotic crisis" is still eluding investigators.

cytes to T cells; most of the former die on exposure to glucocorticoids, while the latter, with the same number of glucocorticoid receptors, are resistant[47]. As both cell types have the same complement of genes, clearly they must have different patterns of gene expression on exposure to the steroid. The thymocyte expresses "suicide genes" in response to this signal, while the T cell no longer does so. Are the glucocorticoid-responsive suicide genes methylated in T cells? There has been some recent progress in identifying putative genes that regulate apoptosis.

Individual genes have been associated with apoptosis in two ways: either they are activated in cells undergoing apoptosis, or their modulation affects the process. Among the latter, the *c-myc* protooncogene may play a part in regulating the choice between proliferation and apoptosis. Fibroblasts which express *c-myc* do not undergo growth arrest in low serum concentrations as do wild-type fibroblasts, but readily undergo apoptosis[48]. It may be that *c-myc* programs a cell to grow, and if this is thwarted (by lack of growth factors or a second oncogene) the cell may commit suicide in frustration. Mechanistically, one might imagine that the expression of one signal in the absence of a normally cognate second signal leads to a chaotic breakdown of cytoplasmic biochemical order, and at some point an "alarm" reaction (the apoptotic crisis) triggers the apoptotic cascade. In T cell hybridomas induced to die by antigen receptor crosslinking, *c-myc* antisense oligonucleotides were reported to block apoptosis[49]. Again, it may be that *c-myc* expression is necessary for apoptosis; more likely, the expression of *c-myc* in the absence of other mitogenic stimuli is abnormal and the cell, experiencing chaos, activates its suicide program. The adenovirus gene *E1A* may behave in a manner analogous to *c-myc*[50].

In the nematode worm *Caenorrhabditis elegans* a group of cell death genes has been identified[51,52]. *ced-4* is necessary for cell death; it encodes a protein with possible Ca^{++}-binding motifs. *ced-3* codes for a cysteine protease, with homology to a growing family of eukaryotic proteases related to ICE, the interleukin-1ß converting enzyme. There is much excitement about the possible role these proteases might play in apoptosis, and also considerable confusion. Although the earlier evidence was that ICE itself is involved in apoptosis, more recently it appears that ICE is responsible for IL-1ß processing but not apoptosis[53]. This should not diminish the importance of ICE in inflammatory conditions, but the search is on for ICE-like proteases that are directly involved in apoptosis. Perhaps the ICE homolog CPP32, which was recently given the caconym apopain, will turn out to be a major player. Another cysteine protease, calpain, which is not a member of the ICE family, has also been shown to be required for activation of apoptosis in several important models[54].

The antioncogene *p53* has also been associated with apoptosis. This gene's product arrests cell proliferation[55] and may instead switch the cell to a differentiation mode. In many cell lineages, differentiation equals death, and p53 will cause apoptosis when expressed in myeloid or epithelial cell lines[56,57]. In one case this apoptosis could be prevented by the addition of a growth factor[56]. The true role of p53 may be to hold a damaged cell in G1 while the damage is repaired; cells which try to oppose the G1 block may end up by activating the suicide pathway[58]. The interactions between *p53* and *c-myc* will be fascinating to unravel. In tumors, activated p53 may induce apoptosis in the relatively hypoxic center; cells which undergo spontaneous inactivation mutations in p53 therefore have a growth advantage under hypoxic conditions, and eventually predominate[59].

Fas is a gene whose product is a membrane-spanning protein homologous to tumor necrosis factor and nerve growth factor receptors[60,61]. In a cell which expresses Fas, cross-linking by antibody to Fas induces apoptosis. Fas is identical to the human cell surface molecule identified as APO-1, and is now called CD95. Micc bcaring the *lpr* mutation are defective in Fas[62], and it is possible that their characteristic lymphoproliferation is more correctly referred to as "lymphoaccumulation", because their lymph nodes fill up with cells that do not die as they would eventually do in a normal animal. The Fas/APO-1 system provides an exciting model to study not only normal tissue turnover, but also the possibility of activating apoptosis as a therapeutic modality in many conditions including malignancies[63]. It is especially relevant that Fas has been shown to be a target molecule for cytotoxic T cells[64] which bear a newly-identified Fas ligand[65]. This Fas/Fas-ligand interaction may be primarily important in the regulation of T cell responses, as activated T cells can kill each other by this mechanism at the end of an immune response[66]. This requires an exquisite level of regulation. Although Fas is expressed on the surfaces of activated T cells early after stimulation, it does not become effectively coupled to the lethal mechanism for several days; this presumably allows an immune response to proceed but prepares it for termination. One might safely predict that abnormalities of the Fas system will be identified and prove to be significant causes of disease. In fact, a group of children with autoimmunity and lymphoproliferation have been described, in each of which there is a dominant-interfering mutation of the Fas gene[67]. Whether Fas abnormalities are involved in other, more common autoimmune conditions like systemic lupus erythematosus remains to be seen[68,69]. A number of nonlymphoid tissues in the body normally express the Fas ligand, and seem able to protect themselves from activated T cells (which express Fas.) This may explain the phenomenon of immunologically privileged sites[70].

The putative oncogene *bcl-2* is commonly overexpressed in human follicular B cell lymphomas. Transfected into an IL-3 dependent B lymphoblastoid cell line, it does not make these cells factor independent; instead, it allows them to survive factor removal, whereas such cells otherwise commit suicide[71,72]. Thus it has properties suggestive of an "anti-apoptosis" gene. It does not prevent apoptosis under all circumstances; for example, it will not protect target cells from cytotoxic T cells[73], it does not seem to work in every cell lineage, and it also can protect against necrosis[74]. Nevertheless, its role in the determination of cell fate is intriguing. There is a growing family of gene products that are related to or interact with Bcl-2[75,76]. Bax is a pro-apoptotic relative which dimerizes with Bcl-2; the balance between Bcl-2/Bax and Bax/Bax dimers may determine a cell's fate[77]. How Bcl-2 and its related species delay apoptosis is not known, but some evidence suggests they plays a role in protection against oxidant or free radical damage. They may thus synergize with mutated p53 during carcinogenesis[59].

There are some genes whose expression increases in apoptotic cells, although their role in the process, if any, has yet to be determined. *TRPM-2*, whose protein product is known by many names including clusterin and SGP-2, is expressed in a number of tissues --primarily of the urogenital tract-- during apoptosis[78]. It is not seen in tissues undergoing morphogenetic death in the embryo[79]. This gene product may play a role in secretion and lipid transport, and could be involved in the response to damage. *RP-2* and *RP-8* are two of a family of genes whose messages increase in abundance after the induction of apoptosis in thymocytes[80]. They were isolated by subtractive hybridization. *RP-8*, at least, seems to be expressed in a number of interesting examples of programmed cell death, including morphogenesis of the limb bud and central nervous system, as well as in lymphoid tissues[81]. *RP-2* codes for an ATP-gated calcium channel[82]. The subtractive and similar approaches to suicide gene isolation are promising, although they have potential pitfalls[83].

4. CONCLUSIONS

Twenty years after it was first described, apoptosis has become a favorite topic in many branches of cell biology. It seems to be an inherent property of vertebrate body cells, and it may be that death by necrosis is actually a rarity under all but the most harsh circumstances. As discussed earlier, lymphocytes respond to damage, if they can, by apoptosis. If a drug induces apoptosis in a lymphocyte, it usually tells us more about the cell than the drug, and such experiments need to be interpreted with caution. In other cases, however, knowledge that cell death is by apoptosis offers a promise of intervention. Suppose we find, for example, that the model of CD4 cell loss in AIDS proposed by several groups[84–87] is correct; this model suggested that crosslinking of CD4 by a complex of gp120 and antibody might change an uninfected CD4 cell's physiology such that subsequent activation via its receptor for antigen would trigger apoptosis rather than activation. The implications for therapy are obvious: a drug that inhibited apoptosis might be useful in preventing cell loss, allowing time for other forms of therapy to clear the infection. In cancer, conversely, a way of stimulating apoptosis in the malignant cells would be therapeutically valuable. How to achieve this effect with adequate specificity will be a challenge[88], as it is with current chemotherapeutic agents (most of which are thought to kill cells by mechanisms other than apoptosis, although this is by no means a rule). Of course, the cytotoxic T cell already knows how to do it; these cells are certainly one of nature's most brilliant inventions. Apoptosis-triggering antibodies, of which APO-1 is the first of what will surely be a series, are another exciting potential therapy, as are immunotoxins bearing apoptosis-inducing ligands. At the same time that the therapeutic potential of apoptosis is being explored, it is necessary to seek the mechanisms and molecules involved. Evolution has spent millions of years developing a superb system for the elimination of unwanted cells. It will not take us quite as long to understand how it works, although we should bear in mind that we are relatively ignorant about the other two fundamental cellular processes, division and differentiation. A cell equals any other structure on earth in complexity, and does not give up its secrets willingly.

REFERENCES

1. J.F.R. Kerr, A.H. Wyllie, and A.R. Currie. Apoptosis: a basic biological phenomenon with wide-ranging implications in tissue kinetics, *Br. J. Cancer* 26:239 (1972).
2. H.M. Ellis and H.R. Horvitz. Genetic control of programmed cell death in the nematode C. elegans, *Cell* 44:817 (1986).
3. R.A. Lockshin and J. Beaulaton. Programmed cell death, *Life Sci.* 15:1549 (1974).
4. G.J. Deenen, I. van Balen, and D. Opstelten. In rat B lymphocyte genesis sixty percent is lost from the bone marrow at the transition of nondividing pre-B cell to sIgM+ B lymphocyte, the stage of Ig light chain gene expression, *Eur. J. Immunol.* 20:557 (1990).
5. J.S. Savill, A.H. Wyllie, J.E. Henson, M.J. Walport, P.M. Henson, and C. Haslett. Macrophage phagocytosis of aging neutrophils in inflammation. Programmed cell death in the neutrophil leads to its recognition by macrophages, *J. Clin. Invest.* 83:865 (1989).
6. E.J. Jenkinson, R. Kingston, C.A. Smith, G.T. Williams, and J.J.T. Owen. Antigen-induced apoptosis in developing T cells: a mechanism for negative selection of the T cell receptor repertoire, *Eur. J. Immunol.* 19:2175 (1989).
7. R.C. Duke and J.J. Cohen. IL-2 addiction: withdrawl of growth factor activates a suicide program in dependent T cells, *Lymphokine Res.* 5:289 (1986).
8. A. Glucksmann. Cell deaths in normal vertebrate ontogeny, *Biol. Revs.* 26:59 (1951).
9. R.C. Duke, R. Chervenak, and J.J. Cohen. Endogenous endonuclease-induced DNA fragmentation: An early event in cell-mediated cytolysis, *Proc. Natl. Acad. Sci. U. S. A.* 80:6361 (1983).
10. E. Durkheim, "Suicide, A Study in Sociology", Free Press, Glencoe (1951).
11. J. Yuan and H.R. Horvitz. The Caenorhabditis elegans genes ced-3 and ced-4 act cell autonomously to cause programmed cell death, *Dev. Biol.* 138:33 (1990).
12. A.H. Wyllie. Glucocorticoid-induced thymocyte apoptosis is associated with endogenous nuclease activation, *Nature* 284:555 (1980).
13. C.A. Smith, G.T. Williams, R. Kingston, E.J. Jenkinson, and J.J.T. Owen. Antibodies to CD3/T-cell receptor complex induce death by apoptosis in immature T cells in thymic cultures, *Nature* 337:181 (1989).
14. K.M. Murphy, A.B. Heimberger, and D.Y. Loh. Induction by antigen of intrathymic apoptosis of CD4+ CD8+ TCR^{lo} thymocytes in vivo, *Science* 250:1720 (1990).

15. K.S. Sellins and J.J. Cohen. Gene induction by gamma-irradiation leads to DNA fragmentation in lymphocytes, *J. Immunol.* 139:3199 (1987).
16. D.J. McConkey, P. Hartzell, S.K. Duddy, H. Hakansson, and S. Orrenius. 2,3,7,8-Tetrachlorodibenzo-p-dioxin kills immature thymocytes by Ca++-mediated endonuclease activation, *Science* 242:256 (1988).
17. K.S. Sellins and J.J. Cohen. Hyperthermia induces apoptosis in thymocytes, *Radiat. Res.* 126:88 (1991).
18. C.J. Sanderson, Morphological aspects of lymphocyte mediated cytotoxicity, *in:* "Mechanisms of Cell-Mediated Cytotoxicity", W.R. Clark, and P. Golstein, eds. Plenum Press, New York (1982).
19. G.C. Godman, A.F. Miranda, A.D. Deitch, and S.W. Tanenbaum. Action of cytochalasin D on cells of established lines. III Zeiosis and movements at the cell surface, *J. Cell Biol.* 64:644 (1975).
20. L. Fesus, P.J. Davies, and M. Piacentini. Apoptosis: molecular mechanisms in programmed cell death, *Eur. J. Cell Biol.* 56:170 (1991).
21. M.A. Kolber, K.O. Broschat, and B. Landa-Gonzalez. Cytochalasin B induces cellular DNA fragmentation, *FASEB J.* 4:3021 (1990).
22. N.P. Matylevich, B.A. Korol, P.A. Nelipovich, V.N. Afanasev, and S.R. Umansky. [D2O inhibition of interphase thymocyte death], *Radiobiologiia* 31:27 (1991).
23. Blumenthal, D.K. and Krebs, E.G. Calmodulin-binding domains. In: *Calmodulin*, edited by Cohen, P. and Klee, C.B. Amsterdam: Elsevier, 1988, p. 341–355.
24. S.J. Martin, G.A. O'Brien, W.K. Nishioka, A.J. McGahon, A. Mahboubi, T.C. Saido, and D.R. Green. Proteolysis of fodrin (non-erythroid spectrin) during apoptosis, *J. Biol. Chem.* 270:6425 (1995).
25. V.A. Fadok, D.R. Voelker, P.A. Campbell, J.J. Cohen, D.L. Bratton, and P.M. Henson. Exposure of phosphatidylserine on the surface of apoptotic lymphocytes triggers specific recognition and removal by macrophages, *J. Immunol.* 148:2207 (1992).
26. V.A. Fadok, J.S. Savill, C. Haslett, D.L. Bratton, D.E. Doherty, P.A. Campbell, and P.M. Henson. Different populations of macrophages use either the vitronectin receptor or the phosphatidylserine receptor to recognize and remove apoptotic cells, *J. Immunol.* 149:4029 (1992).
27. J.S. Savill, V. Fadok, P. Henson, and C. Haslett. Phagocyte recognition of cells undergoing apoptosis, *Immunol. Today* 14:131 (1993).
28. L.C. Meagher, J.S. Savill, A. Baker, R.W. Fuller, and C. Haslett. Phagocytosis of apoptotic neutrophils does not induce macrophage release of thromboxane B2, *J. Leukoc. Biol.* 52:269 (1992).
29. A.H. Wyllie, Cell death: a new classification separating apoptosis from necrosis, *in:* "Cell Death in Biology and Pathology", I.D. Bowen and R.A. Lockshin, eds., Chapman & Hall, London (1981).
30. J.A. Cidlowski. Glucocorticoids stimulate ribonucleic acid degradation in isolated rat thymic lymphocytes in vitro, *Endocrinology* 111:184 (1982).
31. R.G. MacDonald and J.A. Cidlowski. Glucocorticoid-stimulated protein degradation in lymphocytes: quantitation by sodium dodecyl sulfate-polyacrylamide gel electrophoresis, *Arch. Biochem. Biophys.* 212:399 (1981).
32. T. Crompton, M.C. Peitsch, H.R. MacDonald, and J. Tschopp. Propidium iodide staining correlates with the extent of DNA degradation in isolated nuclei, *Biochem. Biophys. Res. Commun.* 183:532 (1992).
33. C. Dive, C.D. Gregory, D.J. Phipps, D.L. Evans, A.E. Milner, and A.H. Wyllie. Analysis and discrimination of necrosis and apoptosis (programmed cell death) by multiparameter flow cytometry, *Biochim. Biophys. Acta* 1133:275 (1992).
34. M.G. Ormerod, M.K. Collins, G. Rodriguez-Tarduchy, and D. Robertson. Apoptosis in interleukin-3-dependent haemopoietic cells. Quantification by two flow cytometric methods, *J. Immunol. Methods* 153:57 (1992).
35. X.-M. Sun, R.T. Snowden, D.N. Skilleter, D. Dinsdale, M.G. Ormerod, and G.M. Cohen. A flow-cytometric method for the separation and quantitation of normal and apoptotic thymocytes, *Anal. Biochem.* 204:351 (1992).
36. W.G. Telford, L.E. King, and P.J. Fraker. Comparative evaluation of several DNA binding dyes in the detection of apoptosis-associated chromatin degradation by flow cytometry, *Cytometry* 13:137 (1992).
37. Z. Darzynkiewicz, S. Bruno, G. Del Bino, W. Gorczyca, M.A. Hotz, P. Lassota, and F. Traganos. Features of apoptotic cells measured by flow cytometry, *Cytometry* 13:795 (1992).
38. G.M. Cohen, X.-M. Sun, R.T. Snowden, D. Dinsdale, and D.N. Skilleter. Key morphological features of apoptosis may occur in the absence of internucleosomal DNA fragmentation, *Biochem. J.* 286:331 (1992).
39. K.S. Sellins and J.J. Cohen. Cytotoxic T lymphocytes induce different types of DNA damage in target cells of different origins, *J. Immunol.* 147:795 (1991).
40. G.M. Cohen, X.M. Sun, H. Fearnhead, M. MacFarlane, D.G. Brown, R.T. Snowden, and D. Dinsdale. Formation of large molecular weight fragments of DNA is a key committed step of apoptosis in thymocytes, *J. Immunol.* 153:507 (1994).

41. D.J. McConkey, P. Hartzell, P. Nicotera, and S. Orrenius. Calcium-activated DNA fragmentation kills immature thymocytes, *FASEB J.* 3:1843 (1989).
42. Y.F. Shi, M.G. Szalay, L. Paskar, M. Boyer, B. Singh, and D.R. Green. Activation-induced cell death in T cell hybridomas is due to apoptosis. Morphologic aspects and DNA fragmentation, *J. Immunol.* 144:3326 (1990).
43. T. Crompton. IL3-dependent cells die by apoptosis on removal of their growth factor, *Growth Factors* 4:109 (1991).
44. M.D. Jacobson, J.F. Burne, and M.C. Raff. Programmed cell death and Bcl-2 protection in the absence of a nucleus, *EMBO J.* 13:1899 (1994).
45. J.J. Cohen. Programmed cell death in the immune system, *Adv. Immunol.* 50:55 (1991).
46. S.V. Lennon, S.J. Martin, and T.G. Cotter. Induction of apoptosis (programmed cell death) in tumour cell lines by widely diverging stimuli, *Biochem. Soc. Trans.* 18:343 (1990).
47. J.J. Cohen and R.C. Duke. Glucocorticoid activation of a calcium-dependent endonuclease in thymocyte nuclei leads to cell death, *J. Immunol.* 132:38 (1984).
48. G.I. Evan, A.H. Wyllie, C.S. Gilbert, T.D. Littlewood, H. Land, M. Brooks, C.M. Waters, L.Z. Penn, and D.C. Hancock. Induction of apoptosis in fibroblasts by c-myc protein, *Cell* 69:119 (1992).
49. Y.F. Shi, J.M. Glynn, L.J. Guilbert, T.G. Cotter, R.P. Bissonnette, and D.R. Green. Role for c-myc in activation-induced apoptotic cell death in T cell hybridomas, *Science* 257:212 (1992).
50. L. Rao, M. Debbas, P. Sabbatini, D.M. Hockenbery, S.J. Korsmeyer, and E. White. The adenovirus E1A proteins induce apoptosis, which is inhibited by the E1B 19-kDa and Bcl-2 proteins, *Proc. Natl. Acad. Sci. U. S. A.* 89:7742 (1992).
51. J. Yuan and H.R. Horvitz. The Caenorhabditis elegans cell death gene ced-4 encodes a novel protein and is expressed during the period of extensive programmed cell death, *Development* 116:309 (1992).
52. J. Yuan, S. Shaham, S. Ledoux, H.M. Ellis, and H.R. Horvitz. The C. elegans cell death gene ced-3 encodes a protein similar to mammalian interleukin-1 beta-converting enzyme, *Cell* 75:641 (1993).
53. P. Li, H. Allen, S. Banerjee, S. Franklin, L. Herzog, C. Johnston, J. McDowell, M. Paskind, L. Rodman, and J. Salfeld. Mice deficient in IL-1 beta-converting enzyme are defective in production of mature IL-1 beta and resistant to endotoxic shock, *Cell* 80:401 (1995).
54. M.K. Squier, A.C. Miller, A.M. Malkinson, and J.J. Cohen. Calpain activation in apoptosis, *J. Cell Physiol.* 159:229 (1994).
55. D. Ginsberg, D. Michael-Michalovitz, and M. Oren. Induction of growth arrest by a temperature-sensitive p53 mutant is correlated with increased nuclear localization and decreased stability of the protein, *Mol. Cell. Biol.* 11:582 (1991).
56. E. Yonish-Rouach, D. Resnitzky, J. Lotem, L. Sachs, A. Kimchi, and M. Oren. Wild-type p53 induces apoptosis of myeloid leukaemic cells that is inhibited by interleukin-6, *Nature* 352:345 (1991).
57. P. Shaw, R. Bovey, S. Tardy, R. Sahli, B. Sordat, and J. Costa. Induction of apoptosis by wild-type p53 in a human colon tumor-derived cell line, *Proc. Natl. Acad. Sci. U. S. A.* 89:4495 (1992).
58. D.P. Lane. p53, guardian of the genome, *Nature* 358:15 (1992).
59. T.G. Graeber, C. Osmanian, T. Jacks, D.E. Housman, C.J. Koch, S.W. Lowe, and A.J. Giaccia. Hypoxia-mediated selection of cells with diminished apoptotic potential in solid tumors, *Nature* 379:88 (1996).
60. N. Itoh, S. Yonehara, A. Ishii, M. Yonehara, S. Mizushima, M. Sameshima, A. Hase, Y. Seto, and S. Nagata. The polypeptide encoded by the cDNA for human cell surface antigen Fas can mediate apoptosis, *Cell* 66:233 (1991).
61. A. Oehm, I. Behrmann, W. Falk, M. Pawlita, G. Maier, C. Klas, M. Li-Weber, S. Richards, J. Dhein, B.C. Trauth, and et al.. Purification and molecular cloning of the APO-1 cell surface antigen, a member of the tumor necrosis factor/nerve growth factor receptor superfamily. Sequence identity with the Fas antigen, *J. Biol. Chem.* 267:10709 (1992).
62. R. Watanabe-Fukunaga, C.I. Brannan, N.G. Copeland, N.A. Jenkins, and S. Nagata. Lymphoproliferation disorder in mice explained by defects in Fas antigen that mediates apoptosis, *Nature* 356:314 (1992).
63. K.M. Debatin, C.K. Goldmann, R. Bamford, T.A. Waldmann, and P.H. Krammer. Monoclonal-antibody-mediated apoptosis in adult T-cell leukaemia, *Lancet* 335:497 (1990).
64. E. Rouvier, M.F. Luciani, and P. Golstein. Fas involvement in Ca(2+)-independent T cell-mediated cytotoxicity, *J. Exp. Med.* 177:195 (1993).
65. T. Suda and S. Nagata. Purification and characterization of the Fas-ligand that induces apoptosis, *J. Exp. Med.* 179:873 (1994).
66. F. Vignaux and P. Golstein. Fas-based lymphocyte-mediated cytotoxicity against syngeneic activated lymphocytes: a regulatory pathway? *Eur. J. Immunol.* 24:923 (1994).

67. G.H. Fisher, F.J. Rosenberg, S.E. Straus, J.K. Dale, L.A. Middleton, A.Y. Lin, W. Strober, M.J. Lenardo, and J.M. Puck. Dominant interfering Fas gene mutations impair apoptosis in a human autoimmune lymphoproliferative syndrome, *Cell* 81:935 (1995).
68. K.B. Elkon. Apoptosis in SLE--too little or too much? *Clin. Exp. Rheumatol.* 12:553 (1994).
69. J.D. Mountz, J. Wu, J. Cheng, and T. Zhou. Autoimmune disease. A problem of defective apoptosis, *Arthritis Rheum.* 37:1415 (1994).
70. D. Bellgrau, D. Gold, H. Selawry, J. Moore, A. Franzusoff, and R.C. Duke. A role for CD95 ligand in preventing graft rejection, *Nature* 377:630 (1995).
71. D.L. Vaux, S. Cory, and J.M. Adams. Bcl-2 gene promotes haemopoietic cell survival and cooperates with c-myc to immortalize pre-B cells, *Nature* 335:440 (1988).
72. T.J. McDonnell, G. Nunez, F.M. Platt, D.M. Hockenbery, L. London, J.P. McKearn, and S.J. Korsmeyer. Deregulated Bcl-2-immunoglobulin transgene expands a resting but responsive immunoglobulin M and D-expressing B-cell population, *Mol. Cell Biol.* 10:1901 (1990).
73. D.L. Vaux, H.L. Aguila, and I.L. Weissman. Bcl-2 prevents death of factor-deprived cells but fails to prevent apoptosis in targets of cell mediated killing, *Int. Immunol.* 4:821 (1992).
74. L.T. Zhong, T. Sarafian, D.J. Kane, A.C. Charles, S.P. Mah, R.H. Edwards, and D.E. Bredesen. bcl-2 inhibits death of central neural cells induced by multiple agents, *Proc. Natl. Acad. Sci. U. S. A.* 90:4533 (1993).
75. J.C. Reed. Bcl-2 and the regulation of programmed cell death, *J. Cell Biol.* 124:1 (1994).
76. L.H. Boise, M. Gonzalez-Garcia, C.E. Postema, L. Ding, T. Lindsten, L.A. Turka, X. Mao, G. Nunez, and C.B. Thompson. bcl-x, a bcl-2-related gene that functions as a dominant regulator of apoptotic cell death, *Cell* 74:597 (1993).
77. S.J. Korsmeyer, J.R. Shutter, D.J. Veis, D.E. Merry, and Z.N. Oltvai. Bcl-2/Bax: a rheostat that regulates an anti-oxidant pathway and cell death, *Semin. Cancer Biol.* 4:327 (1993).
78. R. Buttyan, C.A. Olsson, J. Pintar, C. Chang, M.G. Bandyk, P.Y. Ng, and I.S. Sawczuk. Induction of the TRPM-2 gene in cells undergoing programmed death, *Mol. Cell. Biol.* 9:3473 (1989).
79. G.A. Garden, M. Bothwell, and E.W. Rubel. Lack of correspondence between mRNA expression for a putative cell death molecule (SGP-2) and neuronal cell death in the central nervous system, *J. Neurobiol.* 22:590 (1991).
80. G.P. Owens, W.E. Hahn, and J.J. Cohen. Identification of mRNAs associated with programmed cell death in immature thymocytes, *Mol. Cell. Biol.* 11:4177 (1991).
81. T.J. Mahalik, W.E. Hahn, G.H. Clayton, and G.P. Owens. Programmed cell death in developing grafts of fetal substantia nigra, *Exp. Neurol.* 129:27 (1994).
82. S. Valera, N. Hussy, R.J. Evans, N. Adami, R.A. North, A. Suprenant, and G. Buell. A new class of ligand-gated ion channel defined by P2x receptor for extracellular ATP, *Nature* 371:516 (1994).
83. G.P. Owens and J.J. Cohen. Identification of genes involved in programmed cell death, *Cancer Metastasis Revs.* 11:149 (1992).
84. L. Meyaard, S.A. Otto, R.R. Jonker, M.J. Mijnster, R.P.M. Keet, and F. Miedema. Programmed death of T cells in HIV-1 infection, *Science* 257:217 (1992).
85. J.C. Ameisen and A. Capron. Cell dysfunction and depletion in AIDS: the programmed cell death hypothesis, *Immunol. Today* 12:102 (1991).
86. A.G. Laurent-Crawford, B. Krust, S. Muller, Y. Riviere, M.A. Rey-Cuille, J.M. Bechet, L. Montagnier, and A.G. Hovanessian. The cytopathic effect of HIV is associated with apoptosis, *Virology* 185:829 (1991).
87. N.K. Banda, J. Bernier, D.K. Kurahara, R. Kurrle, N. Haigwood, R.P. Sekaly, and T.H. Finkel. Crosslinking CD4 by human immunodeficiency virus gp120 primes T cells for activation-induced apoptosis, *J. Exp. Med.* 176:1099 (1992).
88. J.J. Cohen and M. al-Rubeai. Apoptosis-targeted therapies: the 'next big thing' in biotechnology? *Trends. Biotechnol.* 13:281 (1995).

3

MECHANISMS FOR RECOGNITION AND PHAGOCYTOSIS OF APOPTOTIC LYMPHOCYTES BY MACROPHAGES

Robert A. Schlegel,[1] Melissa Callahan,[1] Stephen Krahling,[1]
Deepti Pradhan,[1] and Patrick Williamson[2]

[1] Department of Biochemistry and Molecular Biology
Penn State University
University Park, Pennsylvania 16802
[2] Department of Biology
Amherst College
Amherst, Massachusetts 01002

SUMMARY

Erythrocytes have an asymmetric distribution of phospholipids across the bilayer of their plasma membranes, maintained by an ATP-dependent aminophospholipid translocase, and dissipated by activation of a non-specific lipid flipsite. Loss of asymmetry provokes recognition by the reticuloendothelial system. In vitro, enhanced phagocytosis of erythrocytes with a symmetric bilayer can be inhibited by artificial lipid vesicles made of phosphatidylserine (PS), indicating that macrophages recognize the PS that appears on the erythrocyte surface upon loss of asymmetry. It is becoming increasingly clear that these same fundamental membrane structure/function relationships established in the erythrocyte paradigm also apply to lymphocytes. All evidence suggests that lymphocytes maintain an asymmetric transbilayer distribution of phospholipids in their plasma membranes, maintained by an aminophospholipid translocase. Asymmetry is lost as part of the program of cell death, by down-regulation of the translocase and activation of the non-specific lipid flipsite, exposing PS on the cell surface. That PS exposure has functional consequences is demonstrated by the ability of artificial lipid vesicles containing PS to inhibit enhanced phagocytosis of apoptotic lymphocytes by macrophages. However, other signals besides PS are also involved in recognition of apoptotic lymphocytes. Studies with other inhibitors indicate that macrophages also utilize integrin-mediated and lectin-like recognition systems, although each is restricted to either unactivated or activated macrophages. These results indicate that although many fundamental features of recognition by the reticuloendothelial system may be analogous in erythrocytes and lymphocytes, the signals for recognition of apoptotic lymphocytes are more complex and involve multiple recognition systems.

Mechanisms of Lymphocyte Activation and Immune Regulation VI, Edited by Gupta and Cohen
Plenum Press, New York, 1996

1. THE ERYTHROCYTE PARADIGM

The erythrocyte has long been a paradigm for the study of the structure and function of plasma membranes. The mature erythrocyte has no other membrane, unlike other cells where the plasma membrane may represent as little as 10% of total cellular membrane. Our strategy has been to investigate the fundamental structural elements of the erythrocyte plasma membrane and how they relate to function, and then to ask whether the conclusions which are developed apply to other more complex blood cells. Until recently, the lipid bilayer has often been considered as a two-dimensional solvent, existing solely to provide a matrix in which exciting proteins, such as receptors, are suspended. However, lipids are increasingly being recognized as active participants in many important cellular processes such as membrane fusion and trafficking, and as signaling molecules. Although some phospholipids and their metabolic products are now generally recognized as intracellular second messengers, it is less well appreciated that phospholipids can also serve as recognition signals when presented on the surface of cells.

It has been known for 20 years that the phospholipids of the erythrocyte plasma membrane are not randomly distributed across the bilayer. The aminophospholipids, phosphatidylserine (PS) and phosphatidylethanolamine (PE), are concentrated exclusively or primarily, respectively, in the inner leaflet of the membrane bilayer, whereas the choline phospholipids, phosphatidylcholine (PC) and sphingomyelin, are concentrated primarily in the outer leaflet.[1] As the lifespan of erythrocytes in the human circulation is 120 days, and the rate of transbilayer diffusion of phospholipids occurs on the time scale of hours, it is clear that this asymmetric distribution must be actively maintained by the cell. In large part, this asymmetry is maintained by an ATP-dependent aminophospholipid translocating activity in the plasma membrane, which transports PS, and at a lower rate PE, but not PC, from the outer to the inner leaflet of the bilayer.[2]

This energy-requiring activity for maintaining normal lipid asymmetry is complemented in erythrocytes by a separate pathway which mediates randomization of transbilayer lipid distribution. Elevating cytosolic Ca^{2+} concentrations activates this pathway, abolishing asymmetry.[3] This redistribution results from the rapid, bidirectional diffusion of all major classes of phospholipids, aminophospholipids and choline phospholipids alike, across the erythrocyte bilayer.[4] At the moment, the nature of the non-specific flipsite activated by Ca^{2+} is a matter of some controversy, with a complex of Ca^{2+} and phosphatidylinositol 4,5 bisphosphate[5] and a protein[6] as candidates. Regardless of the mechanism, loss of asymmetry by erythrocytes has readily detectable biological consequences, including increased adherence to endothelial cells[7] and the development of a procoagulant surface as a result of external exposure of PS.[8] The likelihood that these effects might manifest themselves in vivo is reduced by a third consequence of loss of asymmetry: the reticuloendothelial system can recognize erythrocytes which present PS on their surface and eliminate them from the circulation.[9] In vitro, macrophages phagocytose erythrocytes with PS exposed on their surface[10] by a mechanism inhibitable by artificial PS vesicles,[11] suggesting that the exposure of PS itself provokes this recognition.

2. EXTENSION OF THE PARADIGM TO LYMPHOCYTES

Are these principles of plasma membrane structure and biological function applicable to other blood cells, and particularly lymphocytes?

2.1. Phospholipid Asymmetry

The first principle - an asymmetric distribution of plasma membrane phospholipids - is difficult to generalize directly, because the biochemical assays for determining the distribution of endogenous phospholipids in erythrocytes are confounded by the lipid in the internal membranes of complex cells such as lymphocytes, where the plasma membrane represents only a small portion of total cellular membrane. The first evidence that lymphocytes maintain an asymmetric transbilayer distribution of phospholipids was derived from an indirect assay. Merocyanine 540 (MC540) is fluorescent lipophilic molecule which intercalates into membranes in which the phospholipids are loosely packed.[12] Because of its net charge, the probe does not penetrate the bilayer, and therefore reports on the packing of phospholipids in the outer leaflet of cells when applied externally. When applied to erythrocytes, it binds sparingly because of the tight packing of the saturated fatty acid sidechains of the phospholipids found in the outer leaflet.[13] However, upon loss of asymmetry, the less saturated fatty acid sidechains of the phospholipids normally residing in the inner leaflet are brought to the outer leaflet, and binding of MC540 is increased.[14] When MC540 was applied to blood leukocytes, it bound to lymphocytes, monocytes, and neutrophils in the same amounts per unit surface area as to erythrocytes,[15] suggesting that the surface of all these blood cells were equivalent at least with respect to the tightness of packing of phospholipids. This early evidence for the similarity in lipid bilayer organization of erythrocytes and other blood cells has since been supplemented by other experiments indicating, for example, that the lymphocyte surface is not procoagulant, implying that PS is confined to the internal leaflet of the lymphocyte plasma membrane bilayer. These and other experiments, discussed below, suggest that the phospholipids of the lymphocyte membrane have an asymmetric distribution.

2.2. Loss of Asymmetry

Does the second principle of the erythrocyte paradigm - loss of lipid asymmetry as a physiologically relevant process - apply to lymphocytes as well? In fact, there is now much evidence that the asymmetric distribution of phospholipids is abolished when lymphocytes undergo apoptosis, or programmed cell death. Lymphocytes induced in vitro to undergo apoptosis bind increased amounts of MC540.[16] Importantly, the same is true for apoptotic lymphocytes generated in vivo. Thymocytes from mice injected with hydrocortisone contain a significant subpopulation of cells with increased MC540 staining per unit surface area, which were shown to overlap substantially with the subpopulation of cells judged apoptotic by reduction in DNA staining.[17] The same holds true for B lymphocytes undergoing apoptosis.[18] In addition, procoagulation assays[16] and measurements of the binding of annexin,[19,20] a protein whose binding in the presence of Ca^{2+} is specific for PS, imply that apoptotic, but not non-apoptotic, lymphocytes display PS on their surface.

2.3. Mechanisms for Maintenance and Loss of Asymmetry

If the normal lipid distribution in the lymphocyte plasma membrane is asymmetric, with PS sequestered in the inner leaflet, and if loss of asymmetry and exposure of PS occur during apoptosis, are the mechanisms controlling the transbilayer distribution of phospholipids in lymphocytes the same as in erythrocytes? The mechanism for establishing and maintaining the normal asymmetric distribution appears to be the same: the presence of the aminophospholipid translocase in the lymphocyte plasma membrane was first suggested by the demonstration that pig lymphocytes specifically translocate spin-labeled analogs of aminophospholipids inward from the outer leaflet of the plasma membrane.[21] Similarly, DO11.10 T cell hybridoma cells clear a fluorescent analog of PS (NBD-PS) from their surface in a matter of minutes, whereas a PC analog is not transported at all.[22] Moreover, induction of apoptosis in these cells by anti-CD3 monoclonal antibody down-regulates this activity, detected as a subpopulation of cells no longer able to transport NBD-PS. At about the same

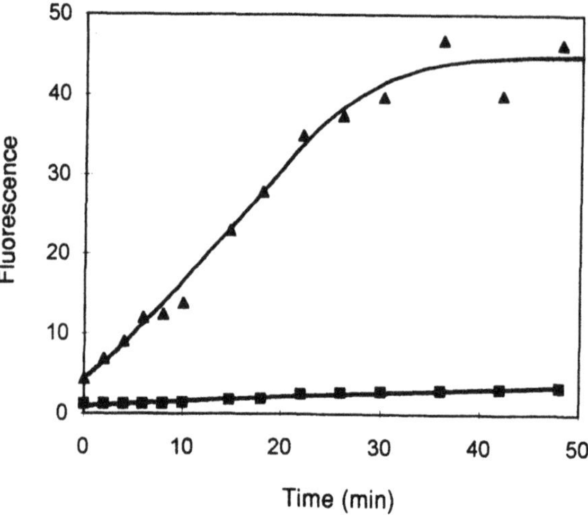

Figure 1. Internalization of fluorescent phospholipid analogs by untreated mouse thymocytes. Thymocytes at 2.5 x 10^6/ml were labeled with 1 uM NBD-PS or NBD-PC for 1 min at 0°, centrifuged, the supernatant removed and the cells resuspended in buffer at room temperature. After the specified times at room temperature, aliquots were removed into a solution of BSA to remove probe remaining in the outer leaflet, and the fluorescence of cells resulting from internalized probe measured by flow cytometry.[22] ▲--▲, NBD-PS; ■--■, NBD-PC.

time in the apoptotic program, a non-specific flipsite like that observed in erythrocytes is activated, detected as a subpopulation of cells in which NBD-PC is rapidly internalized.[22] Together these results indicate that down-regulation of the translocase and up-regulation of the flipsite are coordinated elements of the apoptotic program of these cells. Importantly (see below), the timing of these events is such that PS appears on the surface of apoptotic cells several hours before the onset of cell lysis.

These events also take place in primary thymocytes. As shown in Figure 1, NBD-PS is translocated inward and cleared from the surface of mouse thymocytes in about 30 min, whereas NBD-PC is not translocated at all over this same time frame. After 4 hr of treatment in vitro with dexamethasone to induce apoptosis, a subpopulation of thymocytes, unable to translocate NBD-PS, detected by their reduced fluorescence intensity, has appeared (Figure 2); by 6 hr of treatment additional cells have been recruited into this subpopulation. Also at

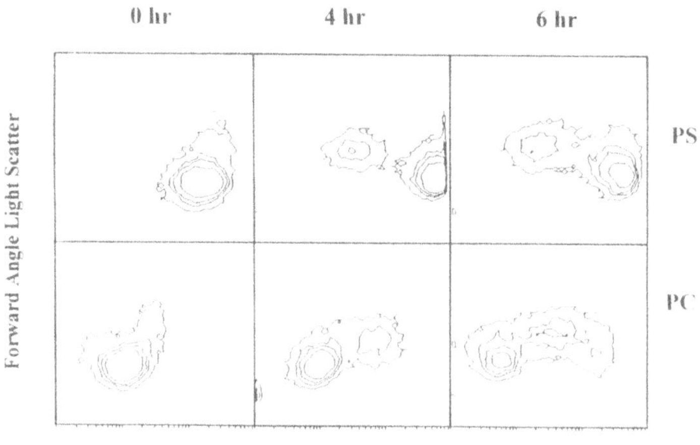

Figure 2. Internalization of fluorescent phospholipid analogs by mouse thymocytes induced to undergo apoptosis. Thymocytes at 5 x 10^7/ml were incubated at 37° with or without 10^{-6} M dexamethasone.[17] At 0, 4 and 6 hr, aliquots of both cultures were removed and internalization of NBD-PS or NBD-PC measured at 30 min as described in the legend to Figure 1.

Table 1. Percent of mouse thymocytes with reduced NBD-PS transport following induction of apoptosis approximates percent of cells with increased NBD-PC transport

Time of induction (hr)	% reduced NBD-PS fluorescence	% increased NBD-PC fluorescence
0	0	0
4	27	31
6	38	50

4 hr a subpopulation of cells which internalize NBD-PC, detected by their increased fluorescence intensity, has appeared; by 6 hrs additional cells have been recruited into this subpopulation. As shown in Table 1, at each timepoint the fraction of cells unable to transport NBD-PS roughly approximates the fraction of cells which have internalized NBD-PC, suggesting that the cells in which the translocase is down-regulated largely overlap the cells in which the flipsite has been activated.

2.4. PS as a Recognition Signal

Finally, the analogy with erythrocytes raises the question of whether the appearance of PS on the surface of apoptotic lymphocytes is responsible for their recognition and phagocytosis. This question is particularly pointed because an important functional aspect of apoptosis is the generation of a signal for recognition of apoptotic cells prior to the onset of cell lysis, necessary to obviate an inflammatory response. In fact, the preferential phagocytosis of apoptotic lymphocytes by macrophages in vitro is inhibited by artificial lipid vesicles containing PS.[16] Interestingly, however, this is not the case for all macrophages. Whereas uptake by elicited mouse peritoneal macrophages is inhibitable, uptake by unactivated mouse bone marrow macrophages, by human monocyte-derived macrophages and by cells of the J774 mouse macrophage line are all unaffected by the presence of PS vesicles.[23,24] What is remarkable about these findings is that all these unactivated macrophages preferentially phagocytose PS vesicles themselves,[11,25,26] indicating they have a receptor which recognizes PS. What's more, PS is able to completely inhibit the enhanced uptake, by unactivated macrophages, of erythrocytes with PS exposed on their surface,[11] indicating that PS is the signal, and the only signal, by which these cells are recognized.

Why then can't unactivated macrophages recognize PS on the surface of apoptotic lymphocytes? The answer to this question appears to be that they can: PS presented on the surface of erythrocytes can block the uptake of apoptotic thymocytes by J774 macrophages, even though PS vesicles cannot.[24] However, surprisingly, such erythrocytes cannot inhibit the uptake of apoptotic lymphocytes by activated macrophages, even though PS vesicles are effective in this case. Since both activated and unactivated macrophages can preferentially phagocytose vesicles containing PS, erythrocytes with PS on their surface, and apoptotic lymphocytes, whereas the two classes of macrophages show reciprocal sensitivity to vesicles and erythrocytes as inhibitors, these results suggest that the PS receptors on the two classes of macrophages differ from each other.

However, these agents are not the only ones capable of inhibiting phagocytosis of apoptotic lymphocytes. Uptake by unactivated macrophages can be blocked by antibody to the vitronectin receptor and by the tetrapeptide RGDS,[23] indicating that recognition of apoptotic lymphocytes by these macrophages involves integrin-mediated interactions. This sensitivity is, however, restricted to unactivated macrophages, as RGDS has no effect on uptake by activated macrophages.[23,24] In contrast, N-acetylglucosamine is able to inhibit

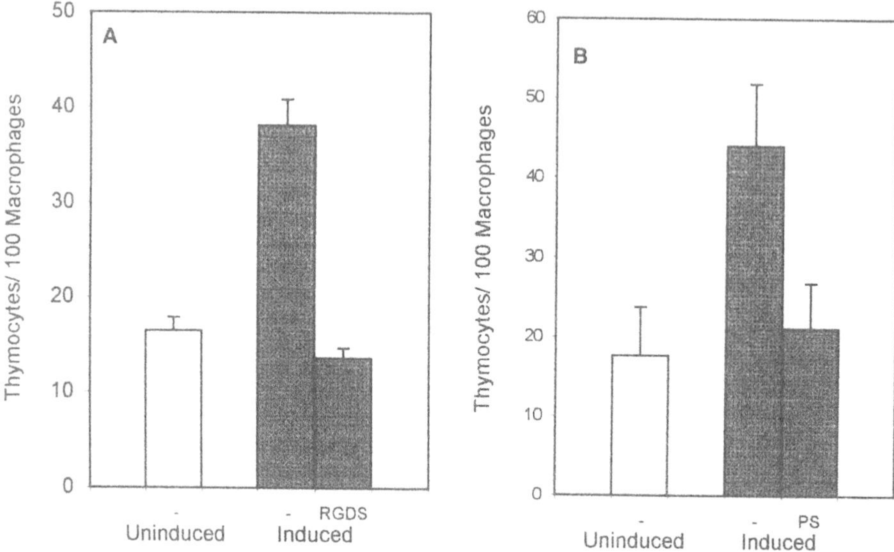

Figure 3. Phagocytosis of DO11.10 T hybridoma cells by macrophages. DO11.10 cells were incubated in the presence or absence of anti-CD3 antibody for 6 hr at 37°C,[22] 10^6 cells added to monolayer cultures of 3×10^5 J774 macrophages (a) or elicited mouse peritoneal macrophages (b) in the presence or absence of RGDS or PS vesicles, and the number of DO11.10 cells phagocytosed in 30 min enumerated.[24]

phagocytosis by activated, but not by unactivated, macrophages.[24] These results hold for T cell hybridoma cells as well as primary lymphocytes. As shown in Figure 3, DO11.10 cells, induced to undergo apoptosis by anti-CD3 antibody, are phagocytosed by activated mouse peritoneal macrophages and by unactivated J774 macrophages in several-fold greater numbers than uninduced cells. RGDS inhibits uptake by unactivated macrophages, whereas PS vesicles inhibit uptake by activated macrophages.

In summary, both classes of macrophages are able to recognize PS, though apparently by different receptors, and each class has an additional mechanism for recognition involving either integrins or a lectin-like receptor. However, the effects of these inhibitors are not additive when used in combination,[24] suggesting that recognition is of a single complex signal on the apoptotic cell surface, of which PS is but one component, and is by a receptor complex. These possibilities are avenues for future investigations.

ACKNOWLEDGMENT

This work was supported by a grant from the American Heart Association, Pennsylvania Affiliate.

REFERENCES

1. P. Williamson and R.A. Schlegel, Back and forth: the regulation and function of transbilayer phospholipid movement in eukaryotic cells (Review). *Mol. Memb. Biol.* 11:199 (1994).
2. M. Seigneuret and P.F. Devaux, ATP-dependent asymmetric distribution of spin-labelled phospholipids in the erythrocyte membrane: relation to shape change. *Proc. Natl. Acad. Sci. USA*, 81:3751 (1984).

3. P. Williamson, L. Algarin, J. Bateman, H.R. Choe, and R.A. Schlegel, Phospholipid asymmetry in human erythrocyte ghosts. *J. Cell. Physiol.* 123:209 (1985).
4. P. Williamson, A. Kulick, A. Zachowski, R.A. Schlegel, and P.F. Devaux, Ca^{++} induces transbilayer redistribution of all major phospholipids in human erythrocytes. *Biochemistry* 31:6355 (1992).
5. J.-C. Sulpice, A. Zachowski, P.F. Devaux, and F. Giraud, Requirement for phosphatidylinositol 4,5-bisphosphate in the Ca^{2+}-induced phospholipid redistribution in the human erythrocyte membrane. *J. Biol. Chem.* 269:6347 (1994).
6. E.M. Bevers, T. Wiedmer, P. Confurius, J. Zhao, E.F. Smeets, R.A. Schlegel, A.J. Schroit, H.J. Weiss, P. Williamson, R.F.A. Zwaal, and P.J. Sims, The complex of phosphatidylinositol 4,5-bisphosphate and calcium ions is not responsible for Ca^{2+}-induced loss of phospholipid asymmetry in the human erythrocyte: A study in Scott syndrome, a disorder of calcium-induced phospholipid scrambling. *Blood* 86:1983 (1995).
7. R.A. Schlegel, T.W. Prendergast, and P. Williamson, Membrane phospholipid asymmetry as a factor in erythrocyte-endothelial cell interaction. *J. Cell. Physiol.* 123:215 (1985).
8. J. Connor, C. Bucana, I. Fidler, and A.J. Schroit, Differentiation-dependent expression of phosphatidylserine in mammalian plasma membranes: quantitative assessment of outer-leaflet lipid by prothrombinase complex formation. *Proc. Natl. Acad. Sci. USA* 86:3184 (1989).
9. T.M. Allen, P. Williamson, and R. A. Schlegel, Phosphatidylserine as a determinant of reticuloendothelial recognition of liposome models of the erythrocyte surface. *Proc. Natl. Acad. Sci. USA* 85:8067 (1988).
10. L. McEvoy, P. Williamson, and R.A. Schlegel, Membrane phospholipid asymmetry as a determinant of erythrocyte recognition by macrophages. *Proc. Natl. Acad. Sci. USA* 83:3311 (1986).
11. D. Pradhan, P. Williamson, and R.A. Schlegel, Phosphatidylserine vesicles inhibit phagocytosis of erythrocytes with a symmetric distribution of phospholipids. *Mol. Memb. Biol.* 11:181 (1994).
12. P. Williamson, K. Mattocks, and R.A. Schlegel, Merocyanine 540: A fluorescent probe sensitive to lipid packing. *Biochim. Biophys. Acta* 732:387 (1983).
13. R.A. Schlegel, B.M. Phelps, A. Waggoner, L. Terada, and P. Williamson, Binding of merocyanine 540 to normal and leukemic erythroid cells. *Cell* 20:321 (1980).
14. P. Williamson, J. Bateman, K. Kozarsky, K. Mattocks, N. Hermanowicz, H.R. Choe, and R.A. Schlegel, Involvement of spectrin in the maintenance of phase-state asymmetry in the erythrocyte membrane. *Cell* 30:725 (1982).
15. L. McEvoy, R.A. Schlegel, P. Williamson, and B. Del Buono, Merocyanine 540 as a flow cytometric probe of membrane lipid organization. *J. Leuk. Biol.* 44:374 (1988).
16. V.A. Fadok, D.R. Voelker, P.A. Campbell, J.J. Cohen, D.L. Bratton, and P.M. Henson, Exposure of phosphatidylserine on the surface of apoptotic lymphocytes triggers specific recognition and removal by macrophages. *J. Immunol.* 148:2207 (1992).
17. R.A. Schlegel, M. Stevens, K. Lumley-Sapanski, and P. Williamson, Altered lipid packing identifies apoptotic thymocytes. *Immunol. Lett.* 36:283 (1993).
18. D.A. Mower, Jr., D.W. Peckham, V.A. Illera, J.K. Fishbaugh, L.L. Stunz, and R.F. Ashman, Decreased membrane phospholipid packing and decreased cell size precede DNA cleavage in mature mouse B cell apoptosis. *J. Immunol.* 152:4832 (1994).
19. G. Koopman, C.P.M. Reutelingsperger, G.A.M. Kuijten, R.M.J. Keehnen, S.T. Pals, and M.H.J. van Oers, Annexin V for flow cytometric detection of phosphatidylserine expression on B cells undergoing apoptosis. *Blood* 84:1415 (1994).
20. S.J. Martin, C.P.M. Reutelingsperger, A.J. McGahon, J.A. Rader, R.C.A.A. van Schie, D.M. LaFace, and D.R. Green, Early redistribution of plasma membrane phosphatidylserine is a general feature of apoptosis regardless of the initiating stimulus: inhibition by overexpression of Bcl-2 and Abl. *J. Exp. Med.* 182:1545 (1995).
21. A. Zachowski, A. Herrmann, A. Paraf, and P.F. Devaux, Phospholipid outside-inside translocation in lymphocyte plasma membranes is a protein-mediated phenomenon. *Biochim. Biophys. Acta* 897:197 (1987).
22. B. Verhoven, R.A. Schlegel, and P. Williamson, Mechanisms of phosphatidylserine exposure, a phagocytic recognition signal, on apoptotic T lymphocytes. *J. Exp. Med.* 182:1597 (1995).
23. V.A. Fadok, J.S. Savill, C. Haslett, D.L. Bratton, D.E. Doherty, P. Campbell, and P.M. Henson, Different populations of macrophages use either the vitronectin receptor or the phosphatidylserine receptor to recognize and remove apoptotic cells. *J. Immunol.* 149:4029 (1992).
24. D. Pradhan, S. Krahling, P. Williamson, and R.A. Schlegel, Recognition of apoptotic lymphocytes by macrophages: multiple receptors for a complex signal. (Manuscript submitted) (1996).

25. T.M. Allen, G.A. Austin, A. Chonn, L. Lin, and K.C. Lee, Uptake of liposomes by cultured mouse bone marrow macrophages: influence of liposome composition and size. *Biochim. Biophys. Acta* 1061: 56 (1991).
26. K.-D. Lee, S. Nir, and D. Papahadjopoulos, Quantitative analysis of liposome-cell interactions in vitro: rate constants of binding and endocytosis with suspension and adherent J774 cells and human monocytes. *Biochemistry* 32:889 (1993).

4

CYTOTOXIC LYMPHOCYTE KILLING ENTERS THE ICE AGE

Seamus J. Martin, Gustavo P. Amarante-Mendes, and Douglas R. Green

Division of Cellular Immunology
La Jolla Institute for Allergy and Immunology
11149 North Torrey Pines Road, La Jolla, California 92037

INTRODUCTION

Apoptosis, a mode of cell death that culminates in early recognition (i.e. before membrane rupture) of dying cells by phagocytes, appears to have been highly conserved throughout evolution. Apoptosis can be triggered by a diverse array of both physiological and pathological stimuli all of which seem to engage the same cellular machinery that is responsible for the destruction of the cell from within. Thus, although the proximal signalling events that can result in apoptosis can vary from one stimulus to another, it is likely that these signals all converge at some point on a common set of effector molecules which we will call '*the executioner*'.

THE KISS OF DEATH

In the context of the immune system, apoptosis plays a critical role in many situations where cells are required to die; during negative selection, peripheral deletion, B cell affinity maturation, and cytotoxic lymphocyte (T and NK-mediated) killing. Although the existence of a machinery for death within all cells has only recently gained widespread acceptance, the fact that such a machinery exists is now well established.

It has been known for some years that the major subset of cytotoxic lymphocytes (CL's) induce apoptosis in their targets by discharging cytoplasmic granules loaded with perforin and a number of proteases and other enzymes—the so called granzymes—into the intercellular space between the CL and the target cell (1–5). Perforin, a pore-forming protein which on its own can cause necrotic cell death at high (most likely non-physiological) concentrations (6), probably serves as a conduit for the entry of granzymes into the target cell, although this has not yet been proven. The essential role of perforin in CL-mediated killing was recently demonstrated in perforin-deficient mice (7). These animals exhibit a reduction in the ability to reject tumors and are severely impaired in viral clearance even though they express normal numbers, ratios and distribution of lymphocyte subsets. In addition, CTLs generated *in vivo* or *in vitro* from perforin-deficient mice are able to kill only against Fas (CD95)-positive targets.

Although CL granules contain several granzymes, only one of these—granzyme B—has been shown to be sufficient, in combination with perforin, to induce apoptosis in target cells with rapid kinetics (8, 9). Furthermore, mice carrying a targeted disruption of granzyme B exhibit severe defects in CTL and NK-mediated killing suggesting that this protein plays a critical role in this process (10, 11). Once again, granzyme B-deficient mice retained the ability to induce death in a Fas-dependent manner. In striking contrast to these data, CL killing was preserved in granzyme A-deficient mice (12).

GRANZYME B AND ICE ARE ASPASES

Granzyme B is a serine protease that preferentially cleaves after aspartate (Asp) residues and was the first protease with this substrate specificity to be described (8). Until recently, the means whereby granzyme B triggers apoptosis upon delivery into the cytoplasm of a target cell was completely unknown, since no substrate for this protease had been described.

In 1993 a major piece in the cell death puzzle fell into place with the report from Horvitz's lab that the *C. elegans* cell death-associated gene *ced-3* was homologous to a mammalian cysteine protease, interleukin-1β converting enzyme (ICE), that, like granzyme B, exhibited a preference for cleaving after Asp residues (13, 14). This discovery was quickly followed by the discovery of a veritable trove of related proteases (see Table I), all of which are also thought to cleave after Asp. It has also recently been confirmed that CED-3 is indeed an Asp-cleaving protease (15, 16).

DEATH'S STING MAY BE A PROTEASE

While the discovery that proteases may be central components of the cell death machinery came initially as a surprise, the evidence for protease involvement in apoptosis

Table 1. Members of the ICE/CED-3 family of proteases described to date

Protease	Reference
CED-3	13, 15
ICE	14
Tx/ICH-2/ICE$_{rel}$II	48–50
ICE$_{rel}$III	50
Mch2	51
CPP32/Yama/Apopain	52, 29, 34
NEDD-2/ICH-1	53, 24
MCH3/ICE-LAP3/CMH-1	54–56

ICE/CED-3 family proteases are constitutively present within the cell as inactive precursor proteins that require cleavage at specific Asp residues to achieve their active configuration (14; Figure 1). This presented the intriguing possibility that granzyme B may trigger apoptosis by engaging endogenous ICE/CED-3 proteases within the target. However, before we deal with the outcome of experiments that were conducted to test this possibility we will first consider the evidence for a role for ICE/CED-3 family proteases in apoptosis.

Table 2. Proteins that undergo proteolysis during apoptosis

Substrate	Reference
Actin	57
Adenomatous polyposis coli protein (APC)	58
Cytosolic phospholipase A2 (cPLA2)	59
DNA-dependent protein kinase	60
Fodrin	40
Gas2	61
GDP dissociation inhibitor (GDI)	45
Lamins A, B, C	17, 38
Nuclear-mitotic apparatus protein (NuMA)	62, 63
Plasminogen activator inhibitor 2 (PAI2)	64
Poly (ADP-Ribose) polymerase (PARP)	18, 33
Protein kinase C δ	65
Sterol regulatory element-binding proteins	66
U1 small nuclear ribonucleoprotein (U1snRNP)	67

was already present in the literature and has dramatically increased since. Perhaps the earliest indications that proteases may participate in apoptosis came from studies that observed proteolytic cleavage of several proteins during this process (17, 18). Additional evidence was provided by the observations that protease inhibitors could block or delay apoptosis (18–21). Strong support for a role of ICE-related proteases in apoptosis has subsequently been provided by studies that have examined the effects on apoptosis of CrmA, a cowpox-derived serpin (serine protease inhibitor) that was known for its ability to act as a pseudo-substrate for, and thereby neutralize, ICE (22). Several groups have independently reported that CrmA potently blocks several forms of apoptosis, including that induced by ligation of Fas as well as due to CTL attack (23–31). Recently, another pseudosubstrate for ICE-family proteases was discovered in the form of the baculovirus p35 protein, which is known for its ability to act as a potent repressor of apoptosis (15, 32). Furthermore, numerous reports are now appearing in the literature that peptide inhibitors of ICE/CED-3 family proteases (YVAD, DEVD, VAD) inhibit many forms of apoptosis (25, 26, 33–36). All of this evidence for protease involvement in apoptosis is supported by the almost weekly discovery of new proteins that are cleaved during this process (**Table II**). Together these data heavily implicate proteases as central components of the cell death machinery and suggest that ICE/CED-3 family proteases may well comprise the *executioner* itself (37).

DEATH BY A THOUSAND CUTS

So how do proteases produce the apoptotic phenotype in a cell? The short answer is that we do not know. However, we can make some informed guesses by looking at the nature of the substrates that become cleaved during apoptosis. As can be seen from Table II most of the substrates that have so far been documented to undergo proteolytic cleavage during apoptosis are localized in the nucleus. It is likely that many of these cleavage events are not contributory to the apoptotic phenotype but are merely bystander effects due to these proteins possessing the appropriate cleavage site. However, it is relatively easy to conceive how cleavage of some of these proteins may directly lead to changes in the cell that frequently occur during apoptosis. For example, cleavage of the nuclear lamins, proteins that are largely responsible for the maintenance of the integrity of the nuclear envelope, could be directly responsible for collapse (condensation) of the nucleus (17, 38, 39). Proteolysis of fodrin, a cytoskeletal protein that plays a major role in the cortical cytoskeleton, could lead to the plasma membrane blebbing that is typical of apoptosis (40). Proteases could also serve to degrade an inhibitor of the endonuclease that cleaves DNA at internucleosomal sites during apoptosis. Indeed, a protease with nuclease-activating properties has been described (41). Clearly, further studies are required in order to fully investigate the impact of each of the cleavage events upon the cell and how they participate in the cell death process.

CELL-FREE APOPTOSIS

To return to the question of whether the role of granzyme B in CL killing is to activate endogenous ICE/CED-3 family proteases within the target cell, we decided to explore this possibility using a mammalian cell-free model of apoptosis that we and others have recently established (42–44).

The advantage of using a cell-free system in this context was that it enabled us to ask whether the simple addition of granzyme B to cell-free extracts of Jurkat cells was sufficient to trigger apoptosis-associated changes in naive nuclei added to these extracts. Addition of recombinant proteins, inhibitory peptides, or membrane impermeable drugs to cell-free extracts is a relatively simple matter when compared with the difficulty of microinjecting these agents.

In our earlier studies using the cell-free model, we and others had demonstrated that post-nuclear extracts from cells that had been briefly exposed (45–60 min) to pro-apoptotic stimuli such as anti-Fas (CD95) antibodies or UV irradiation could reproduce changes typical of apoptosis in nuclei added to these extracts and were thus ‘primed’ for apoptosis (42–44). These changes included chromatin condensation, internucleosomal DNA cleavage, as well as cleavage of fodrin and poly (ADP-ribose) polymerase (PARP). The nature of the priming factor within these extracts was suspected to be a protease of the ICE/CED-3 family (42, 44), and recent studies indicate that this idea appears to be correct (35).

In our experiments with granzyme B, we used post-nuclear extracts from healthy (unprimed) Jurkat cells to ask whether granzyme B could trigger apoptotic events in these extracts. Strikingly, this was found to be the case (45). Granzyme B-mediated apoptosis was characterized by nuclear changes typical of apoptosis, as well as degradation of PARP, fodrin, U1 small nuclear ribonucleoprotein (U1snRNP), and GDP dissociation inhibitor (GDI), all of which are known to undergo proteolysis during many forms of apoptosis. We then explored whether granzyme B-induced apoptosis could be inhibited by tetrapeptide aldehyde inhibitors of ICE/CED-3 family proteases (YVAD or DEVD) by adding these inhibitors to the cell-free extracts prior to the addition of granzyme B. These experiments revealed that DEVD, but not YVAD, significantly inhibited granzyme B-induced nuclear as well as proteolytic events in the cell free system, strongly suggesting that the ability of granzyme B to trigger apoptosis was dependent upon its ability to activate a DEVD-inhibitable protease within these extracts (45). Since DEVD was initially described as an inhibitor of CPP32 (34), this suggested that CPP32 or a closely related ICE/CED-3 protease was activated by granzyme B in these extracts.

GRANZYME B: A TRIPSWITCH FOR THE EXECUTIONER

To directly test the possibility that granzyme B could process and thereby activate CPP32, we transcribed and translated CPP32 and ICE *in vitro* and asked whether granzyme B could process either of these proteins. As mentioned previously, ICE/CED-3 proteases are typically present within cells as inactive precursor proteins that require proteolysis at specific

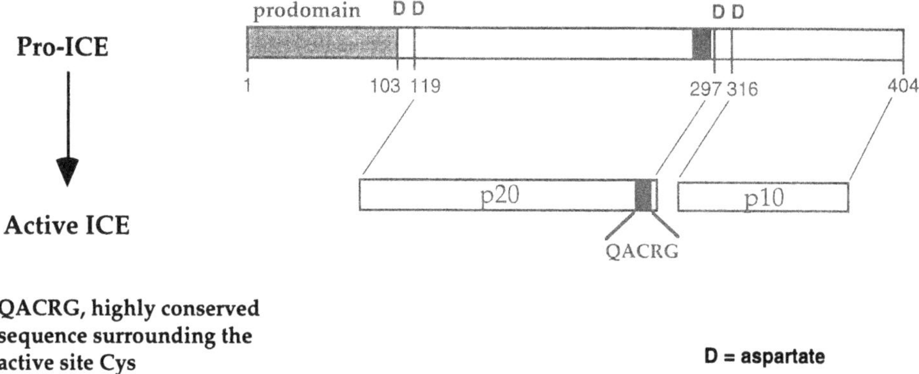

Figure 1. ICE/CED-3 proteases require processing at aspartate residues to achieve their active configuration.

Asp residues to achieve their active heterodimeric forms, which further associate to form tetramers (Figure 1). *In vitro* transcribed and translated ICE was not cleaved by granzyme B, even at high concentrations of the latter, as also reported by others (45, 46). In striking contrast, CPP32 proved to be very sensitive to degradation by granzyme B, yielding degradation products of 17 kDa and 12 kDa which are typical of the active CPP32 protease (29, 34, 35, 45). Titration and inhibitor experiments revealed that granzyme B-induced maturation of CPP32 was achieved via a two step processing mechanism whereby the initial cleavage occured between the large and small subunits of the CPP32 proenzyme, with removal of the prodomain being achieved via an autoproteolytic step (Figure 2). These observations were also confirmed in the cell-free system (45). Similar observations, with respect to grazyme B-mediated cleavage of CPP32 were also recently reported by Bleakley's group (47).

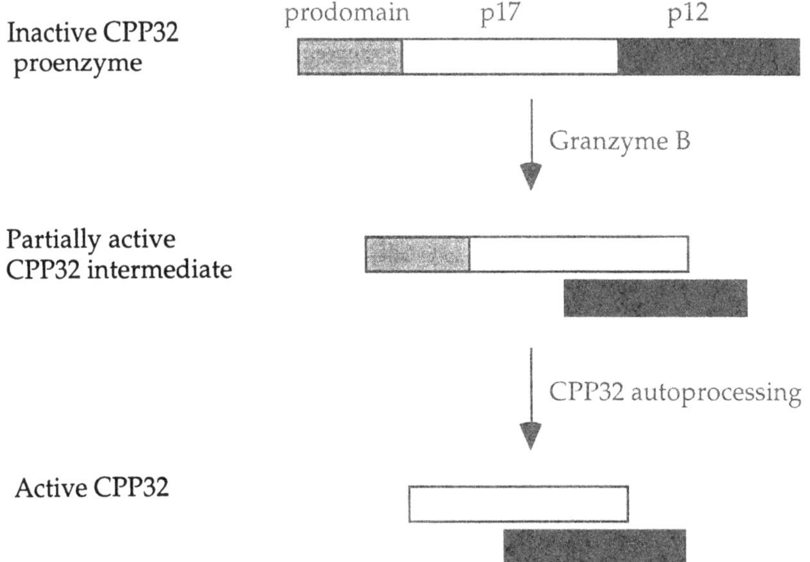

Figure 2. Granzyme B activates CPP32 via a two-step process.

CONCLUSIONS

The experiments described in this overview indicate that upon delivery into the cytoplasm, granzyme B engages the target cell's own death machinery, effectively bypassing the normal control mechanisms that prevent spontaneous activation of the *executioner* within a healthy cell. These experiments do not rule out a role for other granzymes in the CL killing process, but imply that they are not absolutely essential for the death of the target. In addition, whether granzyme B can also activate other ICE/CED-3 family proteases is unknown but is the subject of current investigations.

ACKNOWLEDGMENTS

Some of the research described in this overview was supported by a Wellcome Trust International Travelling Fellowship (041080) to S.J.M., grants from the National Institutes of Health (GM52735) and American Cancer Society (CB-82) to D.R.G., and a a Brazilian Research Council (CNPq) Fellowship to G.P.A-M.

REFERENCES

1. Russell, J. H., Masakowski, V., Rucinsky, T., and Phllips, G., Mechanisms of immune lysis III. Characterization of the nature and kinetics of the cytotoxic T lymphocyte-induced nuclear lesion in the target, *J. Immunol.*, **128:**2087 (1982).
2. Duke, R. C., Chervenak, R., and Cohen, J. J., Endogenous endonuclease-induced DNA fragmentation; an early event in cell-mediated cytolysis, *Proc. Natl. Acad. Sci. USA* **80:**6361 (1983).
3. Henkart, P. A., Mechanism of lymphocyte-mediated cytotoxicity, *Annu. Rev. Immunol.* **3:**31 (1985).
4. Smyth, M. J., and Trapani, J. A., Granzymes: exogenous proteinases that induce target cell apoptosis, *Immunol. Today* **16:**202 (1995).
5. Liu, C.- C., Walsh, C. M., and Young, J. D. -E., Perforin: structure and function. *Immunol. Today* **16:**194 (1995).
6. Duke, R. C., Sellins, K. S., and Cohen, J. J., Cytotoxic lymphocyte-derived lytic granules do not induce DNA fragmentation in target cells, *J. Immunol.* **141:**2191 (1988).
7. Kagi, D., Vignaux, F., Ledermann, B., Burki, K., Depraetere, V., Nagata, S., Hengartner, H., and Goldstein, P., Fas and perforin as major mechanisms of T-cell mediated cytotoxixity, *Science* **265:**528 (1994).
8. Shi, L., Kraut, R. P., Aebersold, R., and Greenberg, A. H., A natural killer cell granule protein that induces DNA fragmentation and apoptosis, *J. Exp. Med.* **175:**553- (1992).
9. Shi, L., Kam, C. M., Powers, J. C., Aebersold, R., and Greenberg, A. H., Purification of three cytotoxic lymphocyte granule serine proteases that induce apoptosis through distinct substrate and target cell interactions, *J. Exp. Med.* **176:**1521 (1992).
10. Heusel, J. W., Wesselschmidt, R. L., Shresta, S., Russell, J. H., and Ley, T. J., Cytotoxic lymphocytes require granzyme B for the rapid induction of DNA fragmentation and apoptosis in allogenic target cells, *Cell* **76:**977 (1994).
11. Shresta, S., MacIvor, D. M., Heusel, J. W., Russell, J. H., and Ley, T. J., Natural killer and lymphokine-activated killer cells require granzyme B for the rapid induction of apoptosis in susceptible target cells, *Proc. Natl. Acad. Sci. USA* **92:**5679 (1995).
12. Ebnet, K., Hausmann, M., Lehmann-Grube, F., Mullbacher, A., Kopf, M., Lamers, M., and Simon, M. M., Granzyme A-deficient mice retain potent cell-mediated cytotoxicity, *EMBO J.* **14:**4230 (1995).
13. Yuan, J., Shaham, S., Ledoux, S., Ellis, H. M., and Horvitz, H. R. (1993). The *C. elegans* cell death gene *ced-3* encodes a protein similar to mammalian interleukin-1β-converting enzyme. *Cell* **75,** 641–652.
14. Thornberry, N. A., Bull, H. G., Calaycay, J. R., Chapman, K. T., Howard, A. D., Kostura, M. J., Miller, D. K., Molineaux, S. M., Weidner, J. R., Aunins, J., Elliston, K. O., Ayala, J. M., Casano, F. J., Chin, J., Ding, G. J.-F., Egger, L. A., Gaffney, E. P., Limjuco, G., Palyha, O. C., Raju, S. M., Ralando, A. M., Salley, J. P., Yamin, T. T., Lee, T. D., Shivley, J. E., MacCross, M., Mumford, R. A., Schmidt, J. A., and

Tocci, M. J., A novel heterodimeric cysteine protease is required for interleukin-1β processing in monocytes, *Nature* **356:**768 (1992).
15. Xue, D., and Horvitz, H. R., Inhibition of the *Caenorhabditis elegans* cell-death protease CED-3 by a CED-3 cleavage site in baculovirus p35 protein, *Nature* **377:**248 (1995).
16. Hugunin, M., Quintal, L. J., Mankovich, J. A., and Ghayur, T., Protease activity of *in vitro* transcribed and translated *C. elegans* cell death gene (*ced-3*) product, *J. Biol. Chem.* **271**:3517 (1996).
17. Kaufmann, S. H., Induction of endonucleolytic DNA cleavage in human acute myelogenous leukemia cells by etoposide, camptothecin, and other cytotoxic anticancer drugs; a cautionary note, *Cancer Res.* **49:**5870 (1989).
18. Kaufmann, S. H., Desnoyers, S., Ottaviano, Y., Davidson, N. E., and Poirier, G. G., Specific proteolytic cleavage of poly(ADP-ribose) polymerase: an early marker of chemotherapy-induced apoptosis, *Cancer Res.* **53:**3976 (1993).
19. Sarin, A., Adams, D. H., and Henkart, P. A., Protease inhibitors selectively block T cell receptor-triggered programmed cell death in a murine T cell hybridoma and activated peripheral T cells, *J. Exp. Med.* **178:**1693 (1993).
20. Sarin, A., Clerici, M., Blatt, S. P., Hendrix, C. W., Shearer, G. M., and Henkart, P. A., Inhibition of activation-induced programmed cell death and restoration of defective immune responses of HIV+ donors by cysteine proteases, *J. Immunol.* **153:**862 (1994).
21. Fearnhead, H. O., Rivett, A. J., Dinsdale, D., and Cohen, G. M., A pre-existing protease is a common effector of thymocyte apoptosis mediated by diverse stimuli, *FEBS Lett.* **357:**242 (1995).
22. Ray, C. A., Black, R. A., Kronheim, S. R. , Greenstreet, T. A., Sleath, P. R., Salvesen, G. S., and Pickup, D. J., Viral inhibition of inflammation: cowpox virus encodes an inhibitor of the interleukin-1β converting enzyme, *Cell* **69:**597 (1992).
23. Gagliardini, V., Fernandez, P.-A., Lee, R. K. K., Drexler, H. C. A., Rotello, R. J., Fishman, M. C., and Yuan, J., Prevention of vertebrate neuronal death by the CrmA gene, *Science* **263:**826 (1994).
24. Wang, L., Miura, M., Bergeron, L., Zhu, H., and Yuan, J., Ich-1, an Ice/ced-3-related gene, encodes both positive and negative regulators of programmed cell death, *Cell* **78:**739 (1994).
25. Enari, M., Hug, H., and Nagata, S., Involvement of an ICE-like protease in Fas-mediated apoptosis., *Nature* **375:**78 (1995).
26. Los, M., Van de Craen, M., Penning, L. C., Schenk, H., Westendorp, M., Baeuerle, P. A., Droge, W., Krammer, P. H., Fiers, W., and Schulze-Osthoff, K., Requirement for an ICE/CED-3 protease for Fas/APO-1-mediated apoptosis, *Nature* **275:**81 (1995).
27. Miura, M., Friedlander, R. M., and Yuan, J., Tumor necrosis factor-induced apoptosis is mediated by a CrmA-sensitive cell death pathway, *Proc. Natl. Acad. Sci. USA***92**:8318 (1995).
28. Tewari, M., and Dixit, V. M., Fas- and tumor necrosis factor-induced apoptosis is inhibited by the poxvirus CrmA gene product, *J. Biol. Chem.* **270**:3255 (1995).
29. Tewari, M., Quan, L. T., O'Rourke, K., Desnoyers, S., Zeng, Z., Beidler, D. R., Poirier, G. G., Salvesen, G. S., and Dixit, V. M., YAMA/CPP32β, a mammalian homolog of Ced-3, is a CrmA-inhibitable protease that cleaves the death substrate poly(ADP-ribose) polymerase, *Cell* **81:**801 (1995).
30. Tewari, M., Beidler, D. R., and Dixit, V. M., CrmA-inhibitable cleavage of the 70-kDa protein component of the U1 small nuclear ribonucleoprotein during Fas- and Tumor necrosis factor-induced apoptosis, *J. Biol. Chem.* **270:**18738 (1995).
31. Tewari, M., Telford, W. G., Miller, R. A., and Dixit, V. M., CrmA, a poxvirus-encoded serpin, inhibits cytotoxic T-lymphocyte-mediated apoptosis, *J. Biol. Chem.* **270:**3255 (1995).
32. Bump, N. J., Hackett, M., Hugunin, M., Seshagiri, S., Brady, K., Chen, P., Ferenz, C., Franklin, S., Ghayur, T., Li, P., Licari, P., Mankovich, J., Shi, L., Greenberg, A. H., Miller, L. K., and Wong, W. W., Inhibition of ICE family proteases by baculovirus anti-apoptotic protein p35, *Science* **269:**1885 (1995).
33. Lazebnik, Y. A., Kaufmann, S. H., Desnoyers, S., Poirier, G. G., and Earnshaw, W. C., Cleavage of poly(ADP-ribose) polymerase by a proteinase with properties like ICE, *Nature* **371:**346 (1994).
34. Nicholson, D. W., Ali, A., Thornberry, N. A., Vaillancourt, J. P., Ding, C. K., Gallant, M., Gareau, Y., Griffin, P. R., Labelle, M., Lazebnik, Y. A., Munday, N. A., Raju, S. M., Smulson, M. E., Yamin, T.-T., Yu, V. L., and Miller, D. K., Identification and inhibition of the ICE/CED-3 protease necessary for mammalian apoptosis, *Nature* **376:**37 (1995).
35. Schlegel, J., Peters, I., Orrenius, S., Miller, D. K., Thornberry, N. A., Yamin, T.-T., and Nicholson, D. W., CPP32/apopain is a key interleukin-1β converting enzyme-like protease involved in fas-mediated apoptosis, *J. Biol. Chem.* **271:**1841 (1996).
36. Pronk, G. J., Ramer, K., Amiri, P., and Williams, L. T., Requirement of an ICE-like protease for induction of apoptosis and ceramide generation by reaper, *Science* **271:**808 (1996).

37. Martin, S. J., and Green, D. R., Protease activation during apoptosis: death by a thousand cuts?, *Cell* **82**:349 (1995).
38. Neamati, N., Fernandez, A., Wright, S., Kiefer, J., and McConkey, D. J., Degradation of lamin B1 precedes oligonucleosomal DNA fragmentation in apoptotic thymocytes and isolated thymocyte nuclei, *J. Immunol.* **154**:3788 (1995).
39. Lazebnik, Y. A., Takahashi, A., Moir, R. D., Goldman, R. D., Poirier, G. G., Kaufmann, S. H., and Earnshaw, W. C., Studies of the lamin proteinase reveal multiple parallel biochemical pathways during apoptotic execution, *Proc. Natl. Acad. Sci. USA* **92**:9042 (1995).
40. Martin, S. J., O'Brien, G. A., Nishioka, W. K., McGahon, A. J., Saido, T., and Green, D. R., Proteolysis of Fodrin (nonerythroid spectrin) during apoptosis, *J. Biol. Chem.* **270**:6425 (1995).
41. Wright, S. C., Wei, Q. S., Zhong, J., Zheng, H., Kinder, D. H., and Larrick, J. W., Purification of a 24 kD protease from apoptotic tumor cells that activates DNA fragmentation, *J. Exp. Med.* **180**:2113 (1994).
42. Chow, S. C., Weis, M., Kass, G. E. N., Holmstrom, T. H., Eriksson, J. E., and Orrenius, S., Involvement of multiple proteases during Fas-mediated apoptosis in T lymphocytes, *FEBS Lett.* **364**:134 (1995).
43. Enari, M., Hase, A., and Nagata, S., Apoptosis by a cytosolic extract from Fas-activated cells, *EMBO J.* **14**:5201 (1995).
44. Martin, S. J., Newmeyer, D. D., Mathias, S., Farschon, D., Wang, H.-G., Reed, J. C., Kolesnick, R. N., and Green, D. R., Cell-free reconstitution of Fas-, UV radiation- and ceramide-induced apoptosis, *EMBO J.* **14**:5191 (1995).
45. Martin, S. J., Amarante-Mendes, G. P., Shi, L., Chuang, T.-H., Casiano, C. A., O'Brien, G. A., Fitzgerald, P., Tan, E. M., Bokoch, G. M., Greenberg, A. H., and Green, D. R., The cytotoxic cell protease granzyme B initiates apoptosis in a cell-free system by proteolytic processing and activation of the ICE/CED-3 family protease, CPP32, via a novel two-step mechanism, *EMBO J.* In Press.
46. Darmon, A. J., Ehrman, N., Caputo, A., Fujinaga, J., and Bleackley, R. C., The cytotoxic T cell proteinase granzyme B does not activate Interleukin-1β-converting enzyme, *J. Biol. Chem.* **269**:32043 (1994).
47. Darmon, A. J., Nicholson, D. W., and Bleackley, R. C., Activation of the apoptotic protease CPP32 by cytotoxic T-cell-derived granzyme B, *Nature* **377**:446 (1995).
48. Faucheu, C., Diu, A., Chan, A. W. E., Blanchet, A.-M., Miossec, C., Herve, F., Collard-Dutilleul, V., Gu, Y., Aldape, R. A., Lippke, J. A., Rocher, C., Su, M. S.-S., Livingston, D. J., Hercend, T., and Lalanne, J.-L., A novel human protease similar to the interleukin-1β converting enzyme induces apoptosis in transfected cells, *EMBO J.* **14**:1914 (1995).
49. Kamens, J., Paskind, M., Hugunin, M., Talanian, R. V., Allen, H., Banach, D., Bump, N., Hackett, M., Johnston, C. G., Li, P., Mankovich, J. A., Terranova, M., and Ghayur, T., Identification and characterization of Ich-2, a novel member of the interleukin-1β-converting enzyme family of cysteine proteases, *J. Biol. Chem.* **270**:15250 (1995).
50. Munday, N. A., Vaillancourt, J. P., Ali, A., Casano, F. J., Miller, D. K., Molineaux, S. M., Yamin, T. T., Yu, V. L. and Nicholson, D. W., Molecular cloning and pro-apoptotic activity of ICErelII and ICErelIII members of the ICE/CED-3 family of cysteine proteases, *J. Biol. Chem.* **270**:15870 (1995).
51. Fernandes-Alnemri, T., Litwack, G., and Alnemri, E. S., Mch2, a new member of the apoptotic Ced-3/Ice-like protease gene family. Cancer Res. **55**:2737 (1995).
52. Fernandes-Alnemri, T., Litwack, G., and Alnemri, E. S., CPP32, a novel apoptotic protein with homology to *Caenorhabditis elegans* cell death protein Ced-3 and mammalian interleukin-1β-converting enzyme, *J. Biol. Chem.* **269**:30761 (1994).
53. Kumar, S., Kinoshita, M., Noda, M., Copeland, N. G., and Jenkins, N. A., Induction of apoptosis by the mouse Nedd2 gene, which encodes a protein similar to the product of the Caenorhabditis elegans cell death gene ced-3 and the mammalian IL-1β converting enzyme, *Genes and Develop.* **8**:1613 (1994).
54. Fernandes-Alnemri, T., Takahashi, A., Armstrong, R., Krebs, J., Fritz, L., Tomaselli, K. J., Wang, L., Yu, Z., Croce, C. M., Earnshaw, W. C., Litwack, G., and Alnemri, E. S., Mch, 3 a novel human apoptotic cysteine protease highly related to CPP32, Cancer Res. **55**:6045 (1995).
55. Duan, H., Chinnaiyan, A. M., Hudson, P. L., Wing, J. P., He, W.-W., and Dixit, V. M., ICE-LAP3, a novel mammalian homologue of the *Caenorhabditis elegans* cell death protein Ced-3 is activated during Fas- and tumor necrosis factor-induced apoptosis, *J. Biol. Chem.* **271**:1621 (1996).
56. Lippke, J. A., Gu, Y., Sarnecki, C., Caron, P. R., and Su, M. S.-S., Identification and characterization of CPP32/Mch2 homolog 1, a novel cysteine protease similar to CPP32, *J. Biol. Chem.* **271**:1825 (1996).
57. Kayalar, C., Ord, T., Testa, M. P., Zhong, L.-T., and Bredesen, D. E., Cleavage of actin by ICE to reverse DNAse I inhibition, *Proc. Natl. Acad. Sci. USA.* In press.
58. Browne, S. J., Williams, A. C., Hague, A., Butt, A. J., and Paraskeva, C., Loss of APC protein expressed in human colonic epithelial cells and the appearance of a specific low-molecular weight form is associated with apoptosis *in vitro*, *Int. J. Cancer* **59**:56 (1994).

59. Voelkel-Johnson, C., Entingh, A. J., Wold, W. S. M., Gooding, L. R., and Laster, S. M., Activation of intracellular proteases is an early event in TNF-induced apoptosis, *J. Immunol.* **154:**1707 (1995).
60. Casciola-Rosen, L. A., Anhalt, G. J., and Rosen, A., DNA-dependent protein kinase is one of a subset of autoantigens specifically cleaved early during apoptosis, *J. Exp. Med.* **182:**1625 (1995).
61. Brancolini, C., Bendetti, M., and Schneider, C., Microfilament reorganization during apoptosis: the role of Gas2, a possible substrate for ICE-like proteases, *EMBO J.* **14:**5179 (1995).
62. Weaver, V. M., Carson, C. E., Walker, P. R., Chaly, N., Lach, B., Raymond, Y., Brown, D. L., and Sikorska, M., Degradation of nuclear matrix and DNA cleavage in apoptotic thymocytes, *J. Cell Sci.* **109:**45 (1996).
63. Hsu, H.-L., and Yeh, N.-H., Dynamic changes of NuMA during the cell cycle and possible appearance of a truncated form of NuMA during apoptosis, *J. Cell Sci.* **109:**277 (1996).
64. Jensen, P. H., Cressey, L. I., Gjertsen, B. T., Madsen, P., Mellgren, G., Hokland, P., Gliemann, J., Doskeland, S. O., Lanotte, M., and Vintermyr, O. K., Cleaved intracellular plasminogen activator inhibitor 2 in human myeloleukemia cells is a marker of apoptosis, *Br. J. Cancer* **70:**834 (1994).
65. Emoto, Y., Manome, Y., Meinhardt, G., Kisaki, H., Kharbanda, S., Robertson, M., Ghayur, T., Wong, W. W., Kamen, R., Weichselbaum, R., and Kufe, D., Proteolytic activation of protein kinase C δ by an ICE-like protease in apoptotic cells, *EMBO J.* **14:**6148 (1995).
66. Wang, X., Pai, J., Weidenfeld, E. A., Medina, J. C., Slaughter, C. A., Goldstein, J. L., and Brown, M. S., Purification of an interleukin-1β converting enzyme-related cysteine proteases that cleaves sterol regulatory element-binding proteins between the leucine zipper and transmembrane domains, *J. Biol. Chem.* **270:**18044 (1995).
67. Casciola-Rosen, L. A., Miller, D. K., Anhalt, G. J., and Rosen, A., Specific cleavage of the 70-kDa protein component of the U1 small nuclear ribonucleoprotein is a characteristic biochemical feature of apoptotic cell death, *J. Biol. Chem.* **269:**30757(1994).

CROSS-TALK BETWEEN CERAMIDE AND PKC ACTIVITY IN THE CONTROL OF APOPTOSIS IN WEHI-231

Steven J. Chmura,[1] Edwardine Nodzenski,[2] Mary A. Crane,[3]
Subbulakshmi Virudachalam,[2] Dennis E. Hallahan,[2]
Ralph R. Weichselbaum,[2] and Jose Quintans[1]

[1] Department of Pathology
[2] Department of Radiation and Cellular Oncology
[3] Division of Biological Sciences
University of Chicago and the Pritzker School of Medicine
Chicago, Illinois 60637

1. SUMMARY

WEHI-231, a murine B-cell lymphoma, readily undergoes programmed cell death following surface immunoglobulin (Ig) cross-linking [1]. Ceramide has been shown to induce apoptosis in WEHI-231 following its exposure to anti-Ig antibodies, dexamethasone, and irradiation [2]. Recently, Haimovitz-Friedman et al. have demonstrated in endothelial cells that PMA not only prevented ceramide mediated apoptosis, but inhibited the generation of ceramide following irradiation [3]. In this paper we use highly specific PKC inhibitors to explore the connection between PKC activity, ceramide signaling and apoptosis. Both chelerythrine chloride and calphostin C triggered rapid apoptosis in WEHI-231 and acted in synergy with exogenous ceramide to induce apoptosis. Detailed studies of chelerythrine's mechanism of action revealed that 30 minutes following addition of 10μM chelerythrine, sphingomyelin and phosphatidylcholine (PC) mass decreased confirming our previous findings of neutral, but not acidic, sphingomyelinase activation following treatment with PKC inhibitors [4]. The novel observation that inhibition of PKC isoforms present in WEHI-231 leads to a rapid rise in cellular ceramide as a result of sphingomyelin hydrolysis further suggests an antagonistic relationship between PKC activity and ceramide in the signaling events preceding apoptosis.

2. INTRODUCTION

Homeostasis in multicellular organisms requires controlled cellular proliferation, differentiation, and death. The signaling pathways governing a cell's fate integrate both intracellular and extracellular signals through such events as protein phosphorylation and gene transcription. Like cellular proliferation and differentiation, apoptosis is regulated by a complex network of cytokines, hormones, and cell-cell interactions and is defined by characteristic morphological changes including cellular and nuclear shrinkage, cytoplasmic blebbing, and chromatin condensation. Apoptosis occurs in response to viral infections, DNA damaging agents, and cytokines both *in-vivo* and *in-vitro*. Recent investigations provide convincing evidence that irradiation [3], immune signals [2], and various cytokines [5] may trigger apoptosis through the hydrolysis of membrane sphingomyelin generating the lipid second messenger ceramide. A transient intracellular elevation of ceramide and its metabolites has been linked to apoptosis in numerous cell lines [6][7]and appears to be a common signaling event in tumor necrosis factor alpha (TNFα), Fas, immunoglobulin M (IgM) cross-linking, and dexamethasone induced apoptosis [2][8][9]

Recent studies indicate that TNFα, interleukin-1, and other activators of sphingomyelin transduce their signals through the p54 family of kinases that ultimately regulates the activity of c-Jun [10] through a c-abl independent pathway [11]. Ceramide may also interact with other cellular signaling pathways including those involving the p42 isoform of the mitogen-activated protein (MAP) kinases as well as stimulating both a serine/threonine kinase and phosphatase [12]. Publications by Venable et al. and Gomez-Munoz et al. demonstrate that ceramide generation following sphingomyelin hydrolysis directly inhibits phospholipase D resulting in decreased 1,2 diacylglycerol (DAG) and inactivation of the protein kinase C (PKC) mitogenic pathway [13][14]. Furthermore, it appears that PKCα, a calcium, phosphatidylserine (PS), and diacylglycerol (DAG) dependent isoform is specifically inhibited by exogenous ceramide and through the hydrolysis of sphingomyelin [15].

Protein kinase C activity appears to be required for mitogenesis and may be inhibited by sphingomyelin hydrolysis. Kolesnick et al. demonstrated that sphingomyelinase activation and subsequent generation of ceramide prevents PKC activation [16]. A metabolite of ceramide, sphingosine, may induce apoptosis through inhibition of PKC [17]. In support of the idea that ceramide and PKC act in divergent and possibly antagonistic pathways are reports that inhibition of PKC not only radiosensitizes endothelial cells by enhancing apoptosis, but that activation of PKC minimized the intracellular elevation of ceramide and blocked apoptosis following exposure to ionizing radiation [18][19]. Using various sublines of WEHI-231 our laboratory demonstrated that phorbol esters prevented the generation of ceramide and subsequent apoptosis following surface Ig cross-linking, again suggesting that PKC may modulate ceramide generation [20].

Collectively, these studies suggest an antagonistic relationship between protein kinase C and ceramide in triggering apoptosis following the exposure of a cell to a wide variety of stresses. To investigate the relationship between PKC activation and ceramide production, we studied the effects of two specific inhibitors of PKC, chelerythrine chloride and calphostin C, on the viability of WEHI-231 cells. We report here that PKC inhibition with chelerythrine chloride or calphostin C induced rapid apoptosis excluding the often assumed requirement of new protein synthesis for this process. These specific PKC inhibitors acted in synergy with exogenous ceramide inducing apoptosis. Studies detailing the mechanism of the observed synergistic effects revealed that exogenous ceramide inhibits the activity of PKC isoforms present in WEHI-231. Furthermore PKC inhibition led to a rapid decrease in sphingomyelin mass and ceramide accumulation confirming our previous report of sphingomyelinase activation following chelerythrine treatment [4]. To confirm a causal relationship between PKC inhibition and ceramide accumulation we generated WEHI-231 variants under selective pressures. Cells selected for resistance to chelerythrine failed to generate ceramide and undergo apoptosis when exposed to this agent.

3. MATERIALS AND METHODS

3.1 Drugs and Reagents

Chelerythrine [IC_{50} (PKC) = .66μM, IC_{50} (PKA) = 170μM], Calphostin C [IC_{50} (PKC) = .50nM, IC_{50} (PKA) = 50μM], 7-amino-actinomycin D, ATP, and propidium iodide were purchased from Sigma Chemical Corp., St. Louis, MO. Chelerythrine and Calphostin C were dissolved in sterile water and methyl sulfoxide (DMSO), respectively. C2 Ceramide and its inactive isomer C2 dihydroceramide, were purchased from Matreya Chemicals, PA. Thin-layer chromatography plates were purchased from Whatman (10 cm x 10 cm LHP-K TLC plate). Autoradiography film was from Dupont. [^{3}H] palmitic acid (60 Ci/mmol), [^{14}C] sphingomyelin (60 Ci/mmol), and [γ-^{32}P] ATP were purchased from DuPont NEN. All solvents were HPLC grade.

3.2 Cell Culture

WEHI-231 JM were grown at a density of 3 - 6 x 10^5 cells/ml in RPMI 1640 culture medium containing, 1% penicillin-streptomycin, 10% heat inactivated fetal calf serum (Gibco/BRL, Grand Island, NY), and 5 x 10^{-5} M β-mercaptoethanol at 5% CO_2. The JM subline used for the experiments was derived by clonal selection of cells susceptible to anti-IgM induced apoptosis. Cells were periodically checked for susceptibility to anti-Ig induced apoptosis [1].

WEHI 231 SC (selected with chelerythrine) was derived from continuous passage of cells in 10μM chelerythrine. Each week, cells were checked for their continued resistance to the inhibitor and ceramide sensitivity.

3.3 Viability Assay via Propidium Iodide Exclusion

2.5–3.5x10^5 cells were cultured in 24-well tissue culture plates for all experiments. Cells were treated with varying concentrations of C2 ceramide, PMA, chelerythrine, or calphostin C as described below. Control cultures were exposed to ethanol and DMSO, the solvents used for ceramide and calphostin C respectively. At the indicated time points, cells were harvested, washed once in phosphate buffered saline (PBS), and resuspended in PBS containing 50μl of 100μg/ml propidium iodide. Cells were analyzed via flow cytometry (FACS) on a FACScan (Becton-Dickson) using Lysis II software with FL2 and FSH compensation set to 50% and 25% respectively.

3.4 7-Amino-Actinomycin D (7-AAD) Assay for Apoptosis

This procedure was performed similarly to that published by Schmid et al [21]. 5x10^5 cells were collected at the appropriate time points and washed once in PBS/Azide .01%. Cells were resuspended in 500μl of PBS/Azide containing 20μg/ml of 7-amino-actinomycin D and incubated at 4°C in the dark. The red fluorescence was filtered through the 650nm long pass filter (FL-3). Three distinct regions were gated corresponding to live (R1), early apoptosis (R2), late apoptosis/dead (R3).

3.5 DAPI Staining for Nuclear Visualization

5x10^5 cells were centrifuged at 1000 rpm and resuspended in approximately 100 μl cell culture media. The cell suspension was then mixed with 100 μl of DAPI (4',6-iamidino-2-phenylindole, Sigma D9542, 1 ug/ml in PBT [= PBS + 1% Triton X-100]). One drop of this mixture was placed on a microslide with a coverslip. The cells were viewed by fluorescence microscopy using an Olympus BX-40 microscope with a 100 Watt Mercury lamp, a 40 X Fluorite objective, N.A. 0.75, (Leco #1-UB527) and a UV filter cube (ex 330–385 nm, em 420 nm, wide band pass, Leco #U-M536). Images were photographed using an Optronics cooled low-light video camera (Leco #DEI-470TB) to which was attached a 2X coupler (Leco #HR200-CMT). The image was printed on a Sony UP-1800MD printer, at 4 exposures per print.

3.6 Ceramide and DAG Quantification by DAG Kinase Reaction

Quantification of ceramide by diacylglycerol kinase was based on the incorporation of [^{32}P] into ceramide by diacylglycerol kinase to produce ceramide -1-phosphate as described by Preiss, *et al.* [22] and modified by Dressler and Kolesnick [23]. After irradiation, lipids were extracted and resuspended by bath sonnication in 20 µl of 7.5% n-octyl-ß-D-glucopyranoside, 5mM cardiolipin, 1 mM DETAPAC. 70 ul of a reaction mix was added to give a final concentration of 0.05M imidazole/HCl pH 6.6, 0.05M NaCl, 12.5 mM $MgCl_2$, 1 mM EGTA and DAG kinase at a concentration of 0.7 U/ml. The reaction was started by the addition of 10µl of [γ-^{32}P]ATP (1.0 µCi/tube) in 5mM ATP and incubated at room temperature for 30 min. The reaction was stopped by the extraction of lipids with 450µl of $CHCl_3$:CH_3OH (1:2) and 20µl of 1% $HClO_4$. The monophase was mixed and after 10 min., 150 µl $CHCl_3$ and 150 µl of 1% $HCl0_4$ were added and the tubes vortexed and centrifuged at 5000g for 5 min. The lower organic phase was washed twice with 1% $HClO_4$ and then dried in a Speed Vac Apparatus. The phosphorylation products were resolved on a 10 cm x 10 cm LH P-K TLC plate (Whatman), previously run in acetone, using $CHCl_3$:CH_3OH:CH_3COOH (65:15:5 v/v) as a solvent. Ceramide (SIGMA) was phosphorylated concomitantly with the samples for a standard curve and after autoradiography, bands corresponding to these standards were scraped and quantified by liquid scintillation counting. Data was calculated as pmoles of ceramide/10^6 cells. Autoradiographs were quantitated using an Epson 1200c 30 bit scanner and NIH Image software (public domain program developed at the NIH and available from the Internet via FTP from zippy.nimh.nih.gov) for both the ceramide-1-phosphate (c-1-P) and phosphatidic acid (PA) spots which represent ceramide and 1,2-diacylglycerol levels respectively. Images were calibrated to both an external and internal controls corresponding to the known ceramide standards. Both scintillation counting and density analysis provided similar results.

3.7 Labeling, Extraction, and Analysis of [^{3}H] Palmitic Acid Labeled Lipids

Cells are prelabeled with [^{3}H] palmitate (10µCi/ml) for 24 hours to isotopic equilibrium in 15ml glass tubes to minimize changes in cellular ceramide and sphingomyelin due to handling and changing of media ([24]). Following treatment of cells with corresponding agents, cells were pelleted in the 15ml glass tubes and media poured off. Total lipids were extracted via a modified Folch method [25]. Briefly, 1ml of MeOH: 2N HCL (100:6, v/v) was used to resuspend the cell pellet. 2ml of chloroform and .6ml of water were added to each tube, vortexed 3 times, and centrifuged for 20 minutes at 1000 x g producing a bi-layer. At this point, the upper layer and the interphase are removed via gentle aspiration, and an equal amount of the organic phase was taken from each tube, transferred to a microcentrifuge tube, and evaporated to dryness in a Speed Vac apparatus. Samples were are reconstituted with 30 µl of chloroform:methanol (95:5) and a 10 µl aliquot was streaked on a 10 cm x 10 cm LHP-K TLC plate (Whatman), previously run in a wash solvent of $CHCl_3$:CH_3OH: H_20 (60:40:10, v/v). Sphingolipids were resolved in $CHCl_3$:MeOH:CH_3COOH:H_20 (32.5:12.5:4.4:2.25). Plates were run to the top, air dried and sample lanes were covered with a glass plate while marker lanes were iodine stained. Plates were sprayed with En^3Hance (DuPont) and autoradiographed. Bands corresponding to markers (sphingomyelin and phosphatidylcholine) (SIGMA) were scraped and quantified by liquid scintillation counting. Sphingomyelin and phosphatidylcholine measurements were expressed as percent of control. Autoradiographs were quantitated using scanning densitometry as for ceramide.

3.8 Protein Kinase C Assay via Phosphotransferase

Two assays were employed to quantitate PKC phosphotransferase activity. For the ceramide experiments, 30μM C2-ceramide was added to exponentially growing cell WEHI-231 cells in complete media. Phosphotransferase activity of PKC isoforms was quantified as previously described [26]. Briefly, cells were kept on ice and total cellular protein was extracted at corresponding time points. [γ-^{32}P]ATP was added to protein extract from whole cell lysates and a synthetic peptide derived from myelin basic protein, Gln-Lys-Arg-Pro-Ser(8-)Gln-Arg-Ser-Lys-Tyr-Leu. PKC activity was inhibited by peptides derived from the psuedosubstrate regions which served as internal controls. Precipitation of the peptide on phosphocellulose was followed by scintillation counting. Enzyme activity is quantified as pmols [^{32}P] incorporated per minute and normalized to 10^5 cells per assay. This peptide fragment is phosphorylated by both group A and B isoforms of PKC.

We verified that chelerythrine inhibits PKC group A and B isoforms (PS dependent) using the Non-Radioactive Protein Kinase assay kit (Panvera). This ELISA kit utilizes an antibody (YC-10) raised against a peptide fragment derived from Glial fibrillary acidic protein (GFAP) to quantitate the activity of PKC isoforms. Total cellular lysates were prepared as above. Increasing concentrations of chelerythrine were added to the lysates and peptide. At the indicated time points, the reaction was stopped and the ELISA was performed per the protocol. O.D. of each well was read at 492nm with a microplate reader.

4. RESULTS

4.1 Chelerythrine and Calphostin C Trigger Apoptosis

To investigate the effects of PKC inhibitors on the viability of WEHI-231, we added increasing concentrations of chelerythrine chloride and calphostin C to cell cultures grown in complete media. We examined cell viability at multiple time points (12, 24, and 48 hours) using the propidium iodide exclusion assay (PI) and staining with 7-amino-actinomycin D (7-AAD). In our experience the 7-AAD and PI methods deliver comparable end-point estimates of cellular viability [27] while the 7-AAD method detects changes in membrane integrity earlier [21] than PI. At 24 hours both inhibitors caused death of WEHI-231 in a dose dependent manner (Figure 1A). Representative results of PI exclusion and 7-AAD assays at 24 hours are shown in figure 1B and 1C, respectively. Similar results were observed at 48 hours (not shown).

Chelerythrine, a potent active site inhibitor of PKC activity, does not alter the distribution of PKC throughout the cellular compartments [28]. We confirmed that chelerythrine inhibited the phosphotransferase activity of PKC using WEHI-231 whole cell lysates and a peptide derived from glial fibrillary acidic protein [29]. The reaction was carried out in the presence of phosphatidylserine and calcium for the experimental group to ensure PKC group A and B activity. Controls were run with EGTA in exchange for PS to eliminate

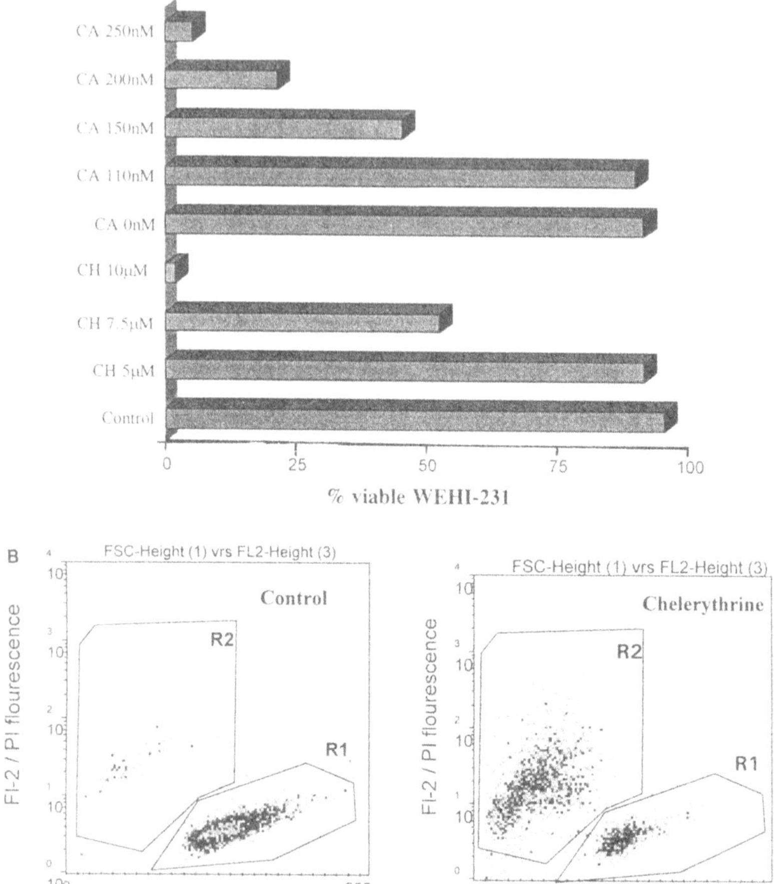

Figure 1. Effects of chelerythrine and Calphostin C on the viability of WEHI-231 and PKC activity. Viability of WEHI measured by propidium iodide fluorescence (A) using FACS analysis 24 hours following treatment with chelerythrine (CH) or calphostin C (CA). FACS analysis represents WEHI-231 viability using propidium iodide (B) or 7-AAD staining (C). 24 hours following the addition of 10μM chelerythrine to $5x10^5$ cells, cells were gated as alive (R1) or dead (R2) as demonstrated in (B). Regions gated R1 through R3 represent live, early apoptotic, and late apoptotic / necrotic, respectively (C). PKC activity was measured using the Panvera NRPK assay 30 minutes following treatment with increasing concentration of chelerythrine (D). The inhibitor was added to cell protein extracts as described in the materials and methods. Control represents no inhibitor with EGTA in place of PS. Bars represent average of duplicate experiments.

activity of the PS and Ca^{++} dependent isoforms. Within 30 minutes following the addition of 10μM chelerythrine, phosphotransferase activity of PKC was abolished (Figure 1D) demonstrating that chelerythrine inhibits the PKC group A and B isoforms (α,β,δ,ε) shown to be expressed in WEHI-231 [30]. In addition to chelerythrine, calphostin C has also been shown to be a highly specific inhibitor of PKC isoforms both *in-vitro* and *in-vivo* at the concentrations used for our experiments [31][32].

In order to establish that apoptotic death occurred as a result of PKC inhibition, we looked for DNA degradation characteristic of apoptosis using gel electrophoresis of low molecular weight DNA. Visible DNA laddering and terminal transferase dUTP end labeling

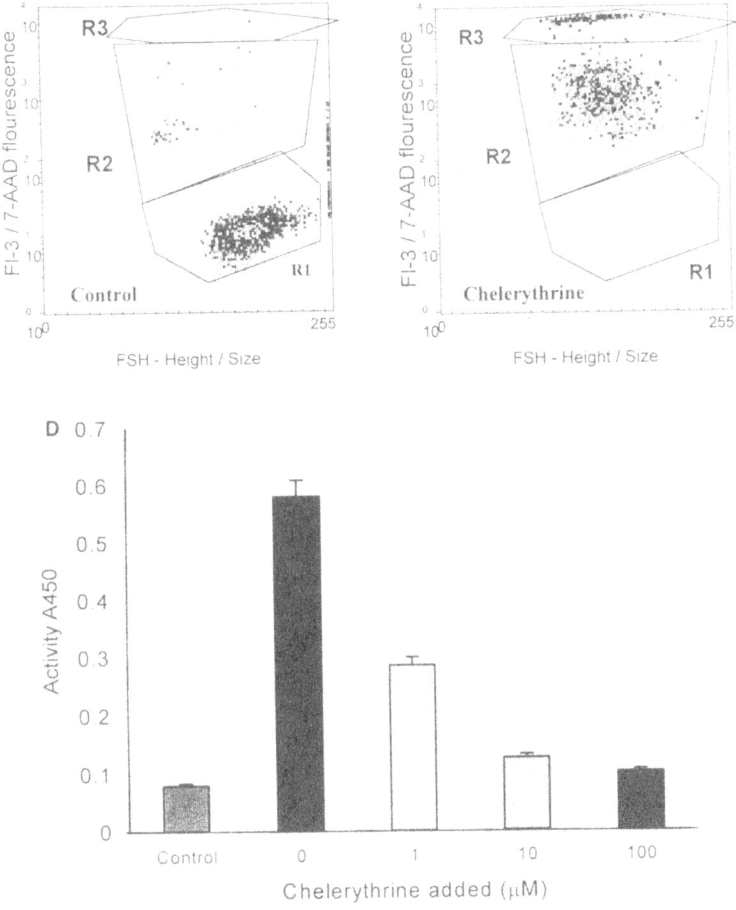

Figure 1. *(Continued)*

(data not shown) characteristic of apoptosis were observed following treatment with 10μM chelerythrine or 250nM calphostin C. Two hours following treatment with either chelerythrine (15μM) or calphostin C (300nM) greater than 90% of WEHI-231 showed nuclear condensation typical of apoptosis using DAPI florescent staining of DNA (figure 2) confirming that death was via apoptosis and suggesting that new protein synthesis is not required to trigger apoptosis in WEHI-231.

4.2 C2 Ceramide Acts in Synergy with Protein Kinase C Inhibitors Inducing Apoptosis

C2-ceramide (30μM) kills greater than 90% of WEHI-231 JM subline in 24 hours under standard culture conditions (16). Since both ceramide and protein kinase C inhibition induce apoptosis in WEHI-231, we investigated whether their effects are interactive. We selected concentrations of C-2 ceramide and protein kinase C inhibitors which by themselves did not effect cell viability. A combination of 20μM ceramide plus either 5μM chelerythrine or 110nM calphostin C killed greater than 80% of WEHI-231 within 24 hours (figure 3) while addition of each component separately caused less than 20% death.

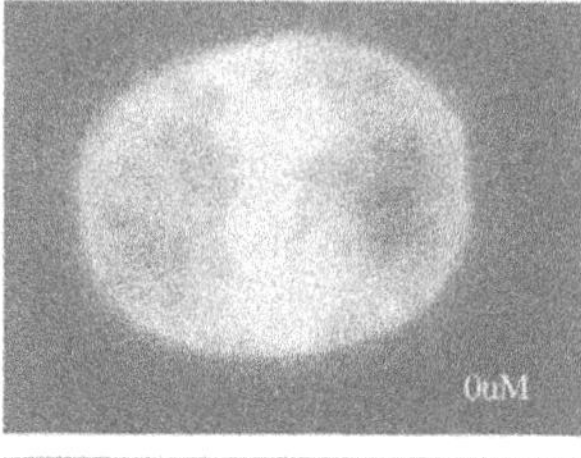

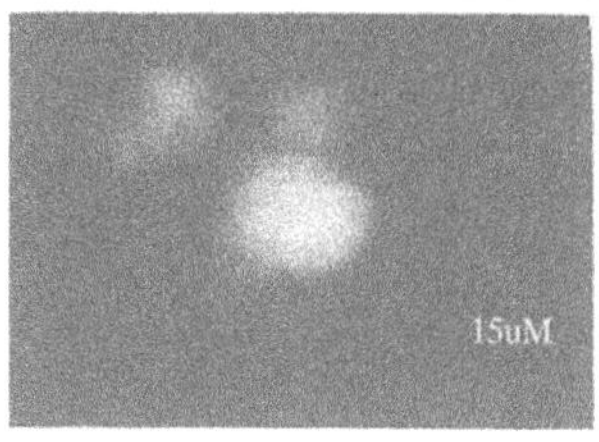

Figure 2. Chelerythrine induces rapid morphological characteristics of apoptosis. DAPI fluorescent staining of WEHI-231 nuclei demonstrating the rapid nuclear effects of treatment with 15mM chelerythrine chloride for 2 hours. Similar changes were seen with Calphostin C (300nM) and lower doses of either agent at 24 hours.

Since PKC inhibitors act in synergy with ceramide to induce apoptosis, we examined whether protein kinase C activation by phorbol esters would increase cell survival following the addition of C2 ceramide. Figure 4 compares the viability of cells following a 24 hour incubation with varying concentrations of C2 ceramide and PMA (2ng/μl). Higher concentration of PMA (5μg/ml) were found to decrease cell viability and enhance ceramide mediated killing presumably through the depletion of PKC over time [33]. All cells were assayed for viability using propidium iodide exclusion assay [34]. Preincubation of cells for 30 minutes with 2ng/ml of PMA increased viability following the addition of 35μM ceramide. Following the addition of ceramide nearly 90% of the control cells were dead whereas over 80% of the PMA treated cells remained viable at 24 hours. This result is consistent with the ability of PMA to limit acute ceramide toxicity in U937 cells as previously reported by Obeid et al [6]. Similar results were seen at 48 hours (not shown).

4.3 Exogenous C2-Ceramide Inhibits PKC Activity in WEHI-231

Since PKC inhibitors increased the intracellular accumulation of ceramide, we examined whether ceramide would cause a reduction in PKC activity as reported by other investigators [16][15][23]. We employed the phosphotransferase assay to quantify the activity of PKC from intact cells following the addition of 30μM C2-ceramide. Within 5 minutes of treatment, PKC activity decreased to 25% of its baseline value and was completely abolished within 30 minutes (figure 5). No effect was observed with C2-dihydroceramide. It appears that the addition of exogenous short-chain ceramide analogs to WEHI-231 results in the inhibition of PKC isoforms which provided an initial mechanism for the synergistic effects of PKC inhibitors and ceramide.

4.4 Protein Kinase C Inhibition Generates Ceramide through the Breakdown of Sphingomyelin and Activation of a Neutral Sphingomyelinase

Previous studies demonstrated that PKC activation inhibits ceramide generation and leads to increased PC and sphingomyelin accumulation with a corresponding decrease in ceramide mass [35]. Our finding that PKC inhibitors act in synergy to cause apoptosis led us to examine whether inhibition of PKC with chelerythrine would generate ceramide. Since calphostin C and chelerythrine both appeared to act similarly in our studies, chelerythrine

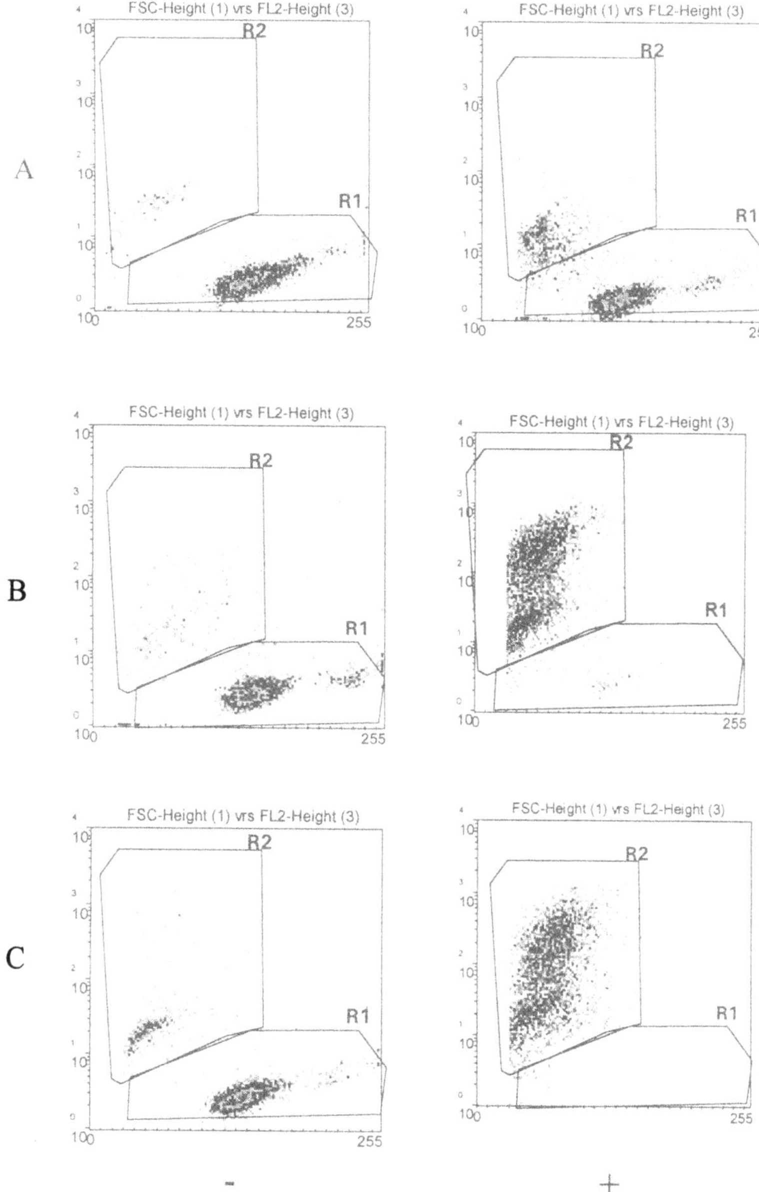

Figure 3. Chelerythrine (CH) and calphostin C (CA) act in synergy with ceramide (C2) causing death. Viability was assayed by propidium iodide fluorescence using FACS analysis 24 hours following treatment with the indicated compounds. The regions gated R1 and R2 represent the population of cells which are viable and not viable respectively. Ceramide (20μM) was added to the (+) lanes. (A)7% and 18% dead. (B) 5μM chelerythrine, 7% dead. 20μM C2 ceramide and 5μM chelerythrine, 85% dead. (C) 20μM C2 ceramide and 110nM calphostin C, 95% dead. Dihydro-ceramide showed no effect on viability (not shown).

was used for the remaining studies due to its greater solubility and its substantially lower cost compared with calphostin C.

We employed the DAG kinase assay to quantify ceramide and DAG production following PKC inhibition [4][22][23]. Approximately 30 minutes following the addition of chelerythrine, ceramide generation peaked to 100% over baseline and fell to nearly baseline

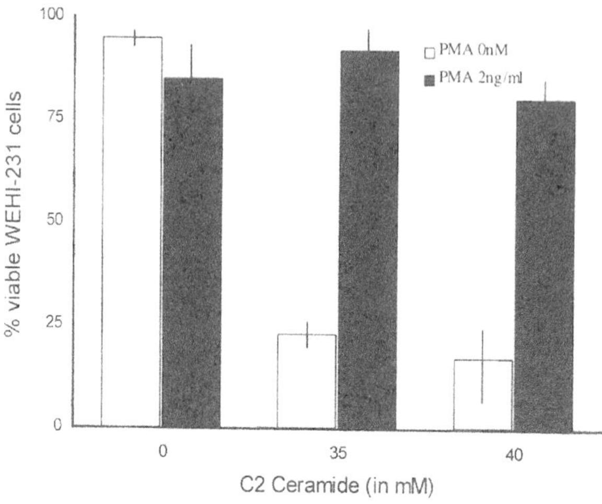

Figure 4. Effects of C2 ceramide and PMA on the viability of WEHI-231. Cellular viability was measured by propidium iodide fluorescence as in figure 1. Each bar represents a single experiment with the dotted lines representing the range of values obtained from 3 individual experiments.

levels within hours. DAG increased to 280% of control (figure 6A) within 30 minutes, in agreement with other reports that various PKC inhibitors or downregulation of PKC with prolonged PMA treatment will potentiate DAG production [16][36]. Thus PKC inhibition induces ceramide and DAG accumulation within the first hour following treatment with the chelerythrine. It should be noted that the kinetics of ceramide accumulation are faster following PKC inhibition than that reported for anti-Ig induced apoptosis (1 hour vs. 24 hours) [2]. In both instances ceramide accumulation preceded apoptosis.

To investigate the source of ceramide and DAG accumulation, two types of experiments were conducted aimed at exploring the metabolism of sphingolipids and their precursors. To examine the sphingomyelin and phosphocholine levels in WEHI-231, $4x10^6$ cells were labeled

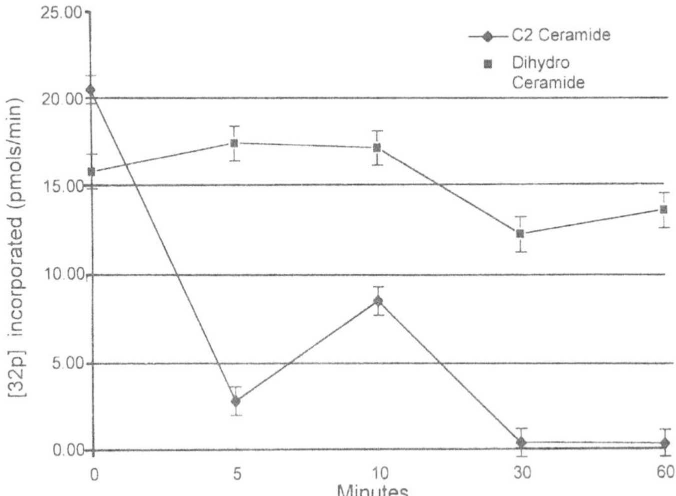

Figure 5. Effects of C2 ceramide versus Dihydro-ceramide on protein kinase C activity. Exponentially growing WEHI-231 were treated with exogenous C2-ceramide. At the appropriate time point, cells were harvested and assayed for phosphotransferase activity as described in the materials and methods. Activity of the group A and B isoforms of protein kinase C plotted against time following the addition of either C2 ceramide (■) or dihydro-ceramide (◆). Activity is measured by the amount of $[\alpha-^{32}p]$ATP incorporated per minute.

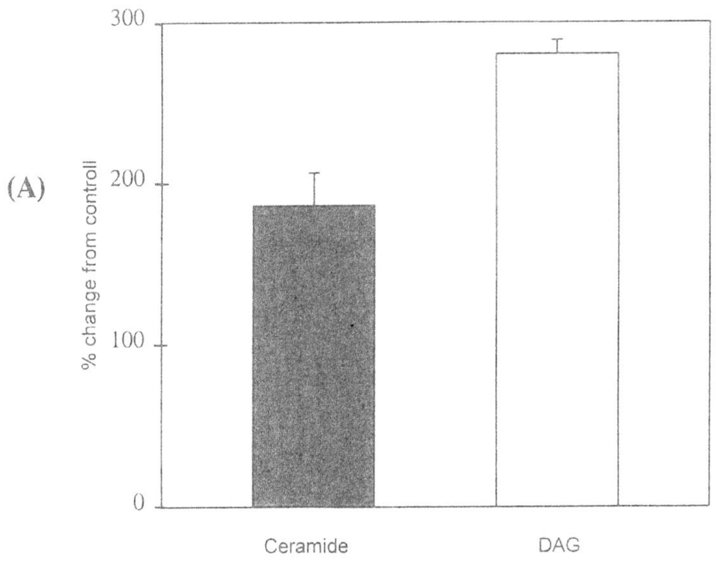

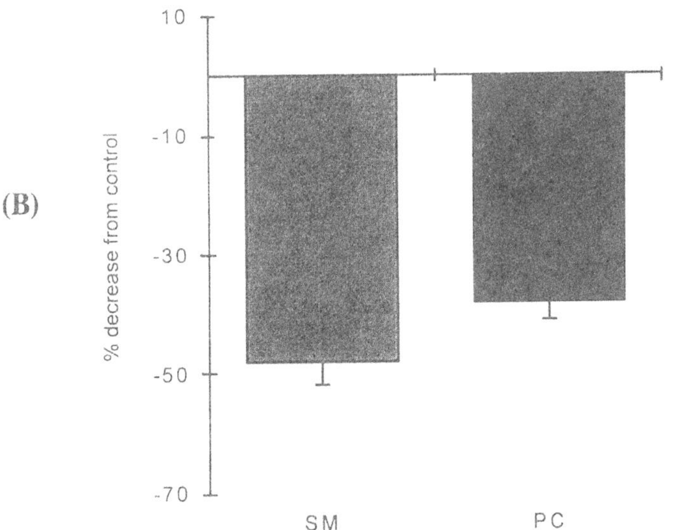

Figure 6. Chelerythrine causes an increase in ceramide production through the activation of a neutral sphingomyelinase and decreases in sphingomyelin and phosphatidylcholine mass. (A) Ceramide production was quantitated using the DAG kinase assay (A) for JM cells and total lipid extraction was used to assay sphingomyelin mass (B). Following addition of 10μM chelerythrine, cells were lysed at the appropriate time points as described above. Each data point represents the average percentage change in ceramide or sphingomyelin generation from baseline derived from at least 4 independent experiments. Standard deviations depicted by error bars. No change in viability or cell size was detectable at the 2 hour time point as assayed by propidium iodide exclusion and FACS.

with [^{3}H]-palmitate (10μCi/ml) for 24 hours, total lipids extracted, and separated using thin-layer chromatography detailed in [2]. As depicted in figures 6b, sphingomyelin mass decreased concurrently with the increase in ceramide reaching its lowest levels 60 minutes (55% of control) following treatment with the inhibitor. Phosphatidylcholine also decreased to 62% of control in agreement with previous reports demonstrating that a prolonged DAG accumulation is the result of PC hydrolysis and not the hydrolysis of phosphatidylinositol 4,5 bisphosphate (PIP_2) [37]. Thus inhibition of PKC causes not only a rapid increase in intracellular ceramide and DAG, but a concurrent decrease in both sphingomyelin and PC mass similar to that observed by other investigators [38][3]. While decreased sphingomyelin synthesis may account for some of the changes observed in the labeled cells, these experiments suggest that ceramide is produced by hydrolysis of sphingomyelin.

4.5 Cells Resistant to Chelerythrine Produce Less Ceramide following Protein Kinase C Inhibition

To provide additional evidence to support our contention that the generation of ceramide is linked to PKC inhibition, we took advantage of the selectability of WEHI-231 cells [1] and cloned a cell line resistant to the PKC inhibitor chelerythrine. We hypothesized that cells

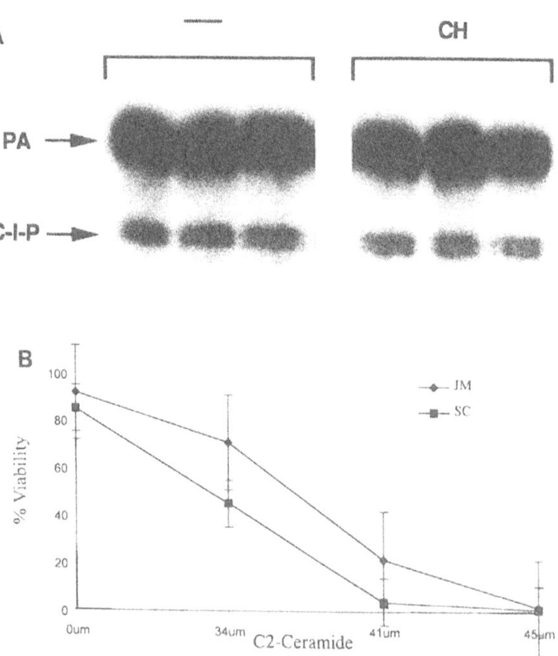

Figure 7. The WEHI-231 SC subline is specifically resistant to chelerythrine and but is not resistant to exogenous ceramide nor sIg cross-linking. (A) Autoradiograph depicting the lack of ceramide (C-1-P) and DAG (PA) accumulation following 10mM chelerythrine at 30 minutes using the DAG kinase assay. (B) Viability of JM and SC cells following the addition of 30mM C2-ceramide. (C) Viability of JM, MG (a subline resistant to sIg cross linking), bcl-x (CMV-neo-bcl-x transfectants), and SC cells following the addition of increasing concentrations of chelerythrine. (D) Viability of cells described above following 24 or 48 hour incubation with anti-IgM (BET supernatant). Sublines were assayed by propidium iodide exclusion 24 hours following treatment with C2-ceramide. Data represents represent the mean of three individual experiments +/- S.D.

selected for resistance to chelerythrine cytotoxicity would not generate ceramide when exposed to this inhibitor. We selected the SC subline through successive passages in 10μM chelerythrine. Each week, cells were tested and re-selected for resistance to the inhibitor.

Ceramide generation was studied in the SC subline following the addition of 10μM chelerythrine - a concentration of the inhibitor which does not affect the viability of these cells but kills the wild-type JM cell line (figure 1A). As demonstrated in figure 7A, ceramide and DAG production were dramatically reduced in the resistant subline with a peak ceramide

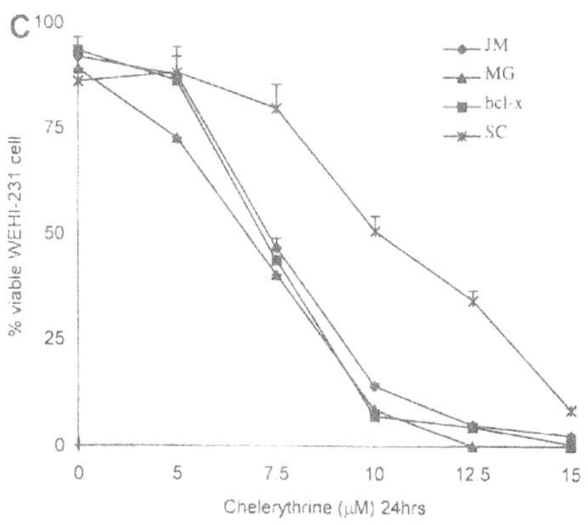

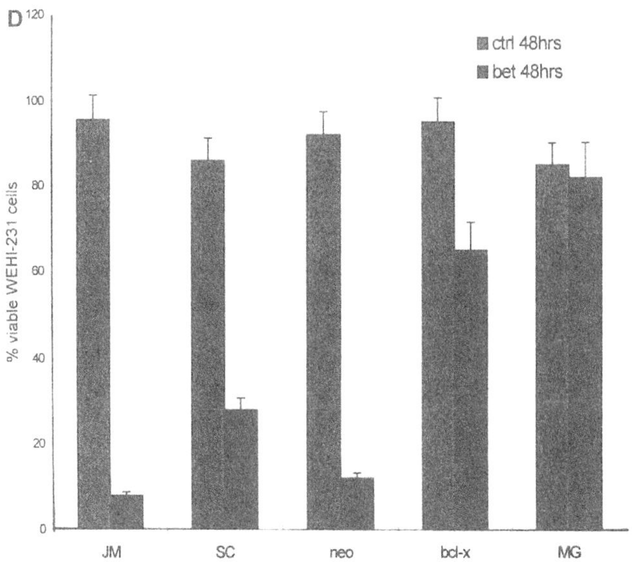

Figure 7. (*Continued*)

accumulation reaching only 20% above its baseline level compared to a near 100% increase in the JM line. The data imply that a direct relationship exists between protein kinase C levels and ceramide production. The SC cells remain viable 24 hours following addition of chelerythrine demonstrating that in the absence of ceramide production, the cells fail to die. Addition of 30μM exogenous ceramide, however, causes apoptosis suggesting that the ceramide signaling pathway was not perturbed in these cells (figure 7B). Finally, chelerythrine can be easily visualized on the TLC plate from lipid preparations of treated cells discounting the possibility that the method of resistance in the SC subline results from decreased inhibitor uptake.

Previous studies have documented differential mechanisms of resistance to apoptosis in WEHI-231 cells. The WEHI-231 MG subline produces only low levels of ceramide when compared to the JM subline following surface Ig cross-linking (sIg) and is resistant to anti-Ig induced apoptosis [1]. We were interested to find if resistance to chelerythrine was specific to the SC subline or merely a result of the selection process. Figure 7C demonstrates that the MG subline previously shown to be resistant to anti-Ig induced death is not resistant to chelerythrine. In addition we studied the susceptibility of WEHI-231 overexpressing the anti-apoptotic gene bcl-x [39]. Figure 7C also demonstrates that bcl-x transfectants lack resistance to chelerythrine although they are resistant to sIg cross-linking and immunosuppressants [39]. Collectively, these findings demonstrate that the resistance of the SC subline is specific. This conclusion was reinforced by the results of the reciprocal experiment that tested the susceptibility of SC cells to anti-Ig induced apoptosis. Figure 7D shows that the SC subline is not resistant to sIg cross-linking induced apoptosis. Thus chelerythrine selection selectively effects a pathway that is uniquely altered in the SC subline but not in the MG and bcl-x sublines.

5. DISCUSSION

PKC, a family of at least 13 serine/threonine protein kinases, plays a key role in mitogenic signal transduction [40][41]. Activation of these enzymes by tumor promoting phorbol esters has previously been used to characterize their mitogenic effects [42]. Inhibition of PKC using both nonspecific and highly selective pharmacological inhibitors triggers apoptosis in a variety of cell lines [17][43]. Little is known, however, about the mechanism by which PKC inhibition mediates cell death.

We selected the kinase inhibitors calphostin C and chelerythrine due to their high specificity for the PKC family of enzymes. These inhibitors are at least 100 fold more selective for inhibiting the kinase activity of PKC than for other protein kinases (i.e. PKA, PKG) as opposed to the relatively non-selective yet widely used agents staurosporin and H7 [44]. Chelerythrine and calphostin C compete for the conserved catalytic sites and regulatory domains of PKC respectively and appear to be potent and specific inhibitors of the group A and group B kinases [31][32][45]. Unlike H7 and staurosporin, these reagents do not inhibit other kinases or activate PLD [44] at concentrations used for our experiments. Possibly due to their neutral lipophilic nature, these compounds associate with phospholipids both in the serum and cell membrane. Chelerythrine in particular can be visualized easily on the TLC plate. These compounds were found to be many times more potent in inducing apoptosis when used in serum free medium.

Our observation that inhibition of PKC isoforms present in WEHI-231 leads to a rapid breakdown of sphingomyelin coupled with a rise in cellular ceramide and 1,2-diacylglycerol (DAG) strengthens the idea that an antagonistic relationship exists between PKC activity and ceramide in the signaling events preceding apoptosis. Furthermore, the rapid and sustained rise in intracellular DAG suggests that PKC may negatively regulate DAG production as proposed by Bishop et al [36].

Inhibition of PKC activity causes a sustained generation of ceramide and DAG with a concurrent decrease in sphingomyelin and PC mass consistent with increased sphingomyelin and PC hydrolysis. PKC activation by phorbol esters has been shown to limit ceramide generation [3][20] and, conversely, its inhibition leads to apparent sphingomyelinase activation and ceramide production. While the rapid rise in intracellular ceramide and DAG are suggestive of increased sphingomyelinase activity and PC hydrolysis [46][47], these studies cannot preclude the possibility that PKC inhibition leads to decreased PC and sphingomyelin synthesis.

The mechanism by which ceramide mediates cell death is unknown. Previous studies have demonstrated multiple signaling targets for ceramide, including a serine/threonine kinase (CAPK) and phosphatase (CAPP), the PKC isoform ζ [48], p54 [11], and phospholipase D [13][14]. Hannun et al. speculate that CAPP and PKC act on a common substrate leading to their seemingly antagonistic effects [5]. Recent studies suggest that CAPP activation may downregulate expression of c-myc, opposing the actions of the PKC signaling cascade [49]. Our results provide yet another mechanism for the apoptotic inducing effects of ceramide or its sphingoid metabolites, PKC inhibition.

Control of apoptosis in WEHI-231 appears to rely on a feed-forward mechanism ultimately leading to activation of the ceramide signaling pathway, PKC inhibition, and apoptosis. Our results further suggest that PKC and ceramide may act in antagonistic yet convergent pathways signaling apoptosis in WEHI-231 and that new protein synthesis in not required for apoptosis. Studies are currently underway using antisense probes for the various isoforms of protein kinase C present in WEHI in order to elucidate which of the isoforms present in this cell line contribute to apoptosis and ceramide generation.

SUPPORTING GRANTS AND ACKNOWLEDGMENTS

We would like to thank Ann Koons for her help and technical support. This was supported by NIH grants: GM07183, 5-R01-CA41068, 5-R01-CA42596, PO1-CA-19266, 5T32-CA09594

6. REFERENCES

1. Gottschalk A. , and Quintans J., 1995. Apoptosis in B Lymphocytes: The WEHI-231 perspective. *Immunol. Cell Biol.* **73**: 41–49.
2. Quintans, J., Kilkus, J., McShan, C., Gottschalk, A., Dawson, G. 1994. Ceramide mediates the apoptotic response of WEHI 231 to anti-IgM, Coticosteroids, and Irradiation. *Biochem. Biophys. Res. Commun.* **202**: 710–714.
3. Haimovitz-Friedman A., Kan, C. C., Fuks Z., Kolesnick, R.N. 1994. Ionizing radiation acts on cellular membranes to generate ceramide and initiate apoptosis. *J. Exp Med.* **180**: 525–535.
4. Chmura, S.J., Nodzenski, E., Crane, M., Weichselbaum, R.R, Quintans, J. 1996. Protein kinase C inhibition by chelerythrine chloride induces activation of a neutral sphingomyelinase and apoptosis in a B-cell lymphoma (WEHI-231). Submitted for publication.
5. Hannun, Y.A., and Obeid, L. M. 1994. Characteristics and partial purification of a novel cytosolic, magnesium-independent, neutral sphingomyelinase activated in the early signal transduction of 1 alpha,25-dihydroxyvitamin D3-induced HL-60 cell differentiation. *TIBS.* **20**, 73–77.
6. Obeid, L. M., Linardic, C. M., Karolak, L. A. and Hannun, Y. A. 1993. Programmed Cell Death Induced by Ceramide. *Science* **259**:1769–1771.
7. Hannun, Y. A., Bell, R. M. 1993. Selectivity of ceramide-mediated biology. Lack of activity of erythro-dihydroceramide. *Adv. Lipid Res.* **25**: 27–41.

8. Nagata, S., Golstein, P. 1995. The Fas death factor. *Science*. **267**: 1449–1452.
9. Dressler, K. A., Mathias, S., Kolesnick, R. N. 1992. Tumor necrosis factor-alpha activates the sphingomyelin signal transduction pathway in a cell-free system. *Science*. **255**: 1715–1718.
10. Kyriakis, J. M., Banerjee, P., Nikolakaki, E., Dai, T., Rubie, E. A., Avruch, J., Woodgett, J.R. 1994. The stress-activated protein kinase subfamily of c-Jun kinases. *Nature* **369**: 156–60.
11. Kharbanda, S., Ren, R., Pandey, P., Shafman, T. D., Kyriakis, Weichselbaum, R.R., Kufe, D. W. 1995. Activation of the c-Abl tyrosine kinase in the stress response to DNA-damaging agents. *Nature* **376**: 375–378
12. Kolesnick, R., Golde, D.W. 1994. The sphingomyelin pathway in tumor necrosis factor and interleukin-1 signaling. *Cell*. **77**: 325–331.
13. Gomez-Munoz A., Martin A., O'Brien L., Brindley D. N. 1994. Cell-permeable ceramides inhibit the stimulation of DNA synthesis and phospholipase D activity by phosphatidate and lysophosphatidate in rat fibroblasts. *J. Biol. Chem.* **269**: 18384–18389.
14. Venable M. E., Blobe G. C., Obeid L. M. 1994. Identification of a defect in the phospholipase D/diacylglycerol pathway in cellular senescence. *J. Biol. Chem.* **269**: 26040–26049.
15. Jones, M. J., Murray, A. W. 1995. Evidence that ceramide selectively inhibits protein kinase C-alpha translocation and modulates bradykinin activation of phospholipase D. *J. Biol. Chem.* **270**, 5007–5013.
16. Kolesnick, R. 1989. Sphingomyelinase action inhibits phorbol ester-induced differentiation of human promyelocytic leukemic (HL-60) cells. *J. Biol. Chem.* **264**: 7617–7623.
17. Ohta, H., Yatomi, Y., Sweeney, E. A., Igarashi, Y. 1994. A possible role of sphingosine in induction of apoptosis by tumor necrosis factor-alpha in human neutrophils. *FEBS Lett.* **355**(3):267–270.
18. Haimovitz-Friedman, A., Balaban, N., McLoughlin, M., Ehleiter, D., Michaeli, J. 1994. Protein kinase C mediates basic fibroblast growth factor protection of endothelial cells against radiation-induced apoptosis. *Cancer Res.* **54**: 2591–2597.
19. Fuks, Z., Persaud, R. S., Alfieri, A., McLoughlin, M., Ehleiter, D., Schwartz, J. L., Seddon, A. P., Cordon-Cardo, C. and Haimovitz-Friedman, A.: 1994. Basic Fibroblast Growth Factor Protects Endothelial Cells Against Radiation-induced Programmed Cell Death *In Vitro* and *In Vivo*. *Cancer Res.* **54**: 2582–2590.
20. Gottschalk A., McShan C., Kilkus J., Dawson G., Quintans J. 1995. Resistance to anti-IgM-induced apoptosis iin a WEHI-231 subline is due to insufficient production of ceramide. *Eur. J. of Immun.* **25**, 1032–8.
21. Schmid, I., Uittenbogaart, C., Keld, B., Giorgi, V. 1994. A rapid method for measuring apoptosis and dual-color immunofluorescence by single laser flow cytometry. *J. Immun. Meth.* **170**: 145–57.
22. Preiss, J., Loomis, C. R., Bishop, W. R., Stein, R., Niedel, J. E. and Bell, R. 1986 Quantitative Measurement of Sn-1,2-Diacylglycerols Present in Platelets. *J. Biol. Chem.* **261**: 8597–8600.
23. Dressler KA, and Kolesnick RN. 1990. Ceramide-1-Phosphate, A Novel Phospholipid in Human Leukemia (HL-60) Cells. *J. Biol. Chem.* **265**:14921–14917.
24. Borchardt, R.A., Lee, W.T., Kalen, A., Bell, R.M. 1994. . Growth-dependent regulation of cellular ceramides in human T-cells. *Biochem. Biophys Acta.* **1212:** 327–36.
25. Folch, J.M.m Lees, M., Stanley, H. 1957 Rapid isolation of total cellular lipids. *J. Biol. Chem.* **226:** 497.
26. Hallahan DE, Virudachalam S, Sherman ML, Huberman E, Kufe DW, Weichselbaum RR, 1991. Tumor necrosis factor gene expression is mediated by protein kinase C following activation by ionizing radiation. Cancer Res. **51**(17): 4565–9.
27. Immunology Methods Manual. 1995. Ed. I. Lefkovits, Academic Press. (In Press).
28. Pavlakovic, G., Eyer, C.l. and G.E. Isom. 1995. Neuroprotecitve effects of PKC inhibition against chemical hypoxia. Brain Res. **676**: 205–211.
29. Yano, T., Taura, C., Shibata, M., Hirono, Y., Ando, S., Kusubata, M. 1991. A monoclonal antibody to the phosphorylated form of glial fibrillary acidic protein. Bioch. Biophys. Res. **175**: 1144–51.
30. Haggerty H.G., Monroe, J.G. 1994. A mutant of the WEHI-231 B lymphocyte line that is resistant to phorbol esters is still sensitive to antigen receptor-mediated growth inhibition. *Cell. Immun.* **154:** 166–80.
31. Kobayashi, E., Nakano, H., Morimito, M., Tamaoki, T. 1989. Calphostin C a novel microbial compound and highly potent protein kinase C inhibitor. *Biochem. Biophys. Res. Commun.* **159:** 548–553.
32. Rotenberg, S., Zhu, J., Su, L., Riedel, H. 1995. Deletion analysis of protein kinase C inactivation by calphostin C. *Mol. Carcin.* **12:** 42–49.
33. Xu, J., Rockow, S., Xiong, W., Li, We. 1994. Interferons block PKC dependent - but not independent - activation of Raf-1 and MAP kinases and mitogenesis in NIH 3T3 cells. *Mol. Cell. Biol.* **14:** 8018–8025.
34. Gottschalk, A ., McShan, C ., Merino, R. and Quintans, J. 1993. Physiological cell death in B Lymphocytes: Differential susceptibility of WEHI-231 sublines to anti-Ig induced PCD and lack of correlation with *bcl-2* expression. *Inter. Immun.* **6:** 121–130.

35. Kolesnick, R. N., 1989. Sphingomyelinase action inhibits phorbol ester induced differentiation of human promyelocytic leukemic (HL-60 cells). *J. Biol. Chem.* **264:** 11688–92
36. Bishop, W.R., August, J., Petrin, J.M., Pai, J.K. 1990. Regulation of sn-1,2-diacylglycerol second messenger formation in thrombin stimulated human platelets. *Biochem J.* **15:** 465–73.
37. Ha, K.S., Exton, J.H. 1993. Differential translocation of protein kinase C isozymes by thrombin and platelet-derived growth factor. *J. Biol Chem.* **268:** 10534–9.
38. Boucher, L.M., Wiegmann, K., Futterer, A., Pfeffer, K., Machleidt, T., Schutze, S., Mak, T.W. and Kronke, M. 1995. CD28 signals through acidic sphingomyelinase. *J. Exp. Med.* **181**(6): 2059–68.
39. Gottschalk, A.R., Boise, L.H., Thompson C.B., Quintans, J. 1994. Identification of immunosuppressant-induced apoptosis in a murine B-cell line and its prevention by bcl-x but not bcl-2. *Proc Natl Acad Sci.* **19:** 7350–4.
40. Nishizuka, Y. 1992. Intracellular Signaling by hydrolysis of phospholipids and activation of protein kinase C. *Science*. **258:** 607–614.
41. Magnuson, N. S., Beck, T., Vahidi, H., Hahn, H., Smola, U. and Rapp, U. R. 1994. Protein Kinase C - an enzyme and its relatives. *Semin. Cancer Biol.* **5:** 277–284.
42. Liyanage, M., Frith, D., Livneh, E., Stabel, S. 1992. Protein kinase C group B members PKC-delta, -epsilon, -zeta and PKC-L(eta). Comparison of properties of recombinant proteins in vitro and in vivo. *Biochem. J.* **283:** 781–787.
43. Jarvis, W., Turner, A., Grant, S., 1994. Induction of apoptosis and cell death by Inhibitors of Protein Kinase C. *Cancer Res.* **54:** 1707–1714.
44. Gupta, S. Gomez-Munoz, A., Matowe, W.C., Brindley, D.N. and Ginsberg, J. 1995. Thyroid-stimulating hormone activates phospholipase D in FRTL-5 thyroid cells via stimulation of protein kinase C. *Endocrinology.* **136**(9):3794–9.
45. Herbert, J., Augereau, J.and Maffrand, J., 1990. Chelerythrine is a potent and specific inhibitor of protein kinase C. *Biochem. Biophys. Res. Commun.* **172**: 993–999.
46. Schutze, S., Machledit, T., Kronke, M. 1994. The role of diacylglycerol and ceramide in TNF and IL-1 signal transduction. *J. Leukocyte Biol.* **56:** 533–538.
47. Kolesnick, R.N. 1987. 1,2-Diacylglycerols but not phorbol esters stimulate sphingomyelin hydrolysis in GH3 pituitary cells. *J. Biol Chem.* **262**: 16759–62.
48. Lozano, J., Berra, E., Municio, M. M., Diaz-Meco, M. T., Dominguez, I., Sanz, L., Moscat, J. 1994. Protein kinase C zeta isoform is critical for kappa B-dependent promoter activation by sphingomyelinase. *J. Biol. Chem.* **269**: 19200–19207.
49. Wolff, R. A., Dobrowsky, R. T. Bielawska, A., Obeid, L. M., Hannun, Y. A. 1994. Role of ceramide-activated protein phosphatase in ceramide-mediated signal transduction. *J. Biol. Chem.* **269**: 19605–19609.

6

CELL CYCLE CONTROL OF T CELL APOPTOSIS INDUCED BY ACTIVATION THROUGH THE T CELL ANTIGEN RECEPTOR

N. Jan Chalupny, Liqing Zhu, Xue-zhong Yu, and Claudio Anasetti

Human Immunogenetics Program
Division of Clinical Research
Fred Hutchinson Cancer Research Center
1124 Columbia Street, Seattle, Washington 98104

1. T CELL RECEPTOR-INDUCED APOPTOSIS

The term apoptosis refers to the morphological process of programmed cell death, a physiological event that occurs at specific stages of normal embryonic development as well as in adult life.[1] During programmed cell death, cells undergo profound structural changes. The plasma membrane becomes ruffled or 'blebbed'. The nucleus shrinks, but the morphology of other cytoplasmic organelles remains relatively unchanged. Within the nucleus, the chromatin condenses, and tends to collapse into patches around the nuclear envelope. Condensation of the chromatin is often accompanied by fragmentation of the DNA, caused by internucleosomal cleavage, resulting in the characteristic 'ladder' pattern of bands seen upon electrophoretic separation. The cell may finally break up into apoptotic bodies, which are rapidly phagocytosed.[2–4]

Programmed cell death plays an important role in the development and maintenance of the immune system. Immature thymocytes are deleted if their T cell receptors (TCRs) display too high an affinity for self antigens presented in the thymus. This process of negative selection rids the organism of potentially autoreactive T cells.[5] Signaling through the TCR/CD3 complex induces apoptosis in immature CD4+/CD8+ thymocytes in organ culture[6] and in living animals.[7] Studies of Wang et al have shown that CD3ε transgenic mice are completely immunodeficient since thymocytes die before developing into mature T cells, indicating that increased signaling through the CD3ε chain can induce programmed cell death.[8] In murine T-cell hybridomas[9] and human T cell leukemic lines[10,11], ligand binding to either TCR or CD3ε induces a rise in the concentration of cytoplasmic free calcium, RNA and protein synthesis, increase in cell size, activation of a calcium dependent endonuclease and programmed cell death.[8] Crosslinking of either TCR or CD3ε has resulted in programmed cell death of mature peripheral T cells.[12] Therefore, it has been unequivocally

Mechanisms of Lymphocyte Activation and Immune Regulation VI, Edited by Gupta and Cohen
Plenum Press, New York, 1996

shown that strong TCR/CD3ε-mediated signals can induce programmed cell death both in the thymus and in the periphery.

Stimulation of mature T cells via their TCR/CD3 complexes with the appropriate antigen/MHC complex, as well as the required co-stimulatory signals, usually results in differentiation and proliferation of these cells, but may result in programmed cell death. The Fas deficient MRL -/- lpr mice do not undergo TCR/CD3-induced apoptosis[13], indicating that the interaction of Fas and Fas-ligand (Fas-L) is involved in TCR/CD3-induced T cell apoptosis, as has been suggested by other studies.[14–17] Signaling through TCR/CD3 induces upregulation of Fas-L mRNA as well as increased surface expression of Fas-L. This induction is sensitive to cycloheximide, indicating that protein synthesis is required.[18] Soluble Fas protein can block TCR/CD3-induced apoptosis, but does not affect IL-2 production. Taken together, these results demonstrate that in this system binding of Fas to Fas-L activates the program for apoptosis.

2. CELL CYCLE CONTROL OF TCR/CD3-INDUCED APOPTOSIS

Stimulation of murine T cell hybridomas with specific antigen results in increased IL-2 production, and dose-dependent decreases in ^{3}[H] thymidine uptake and cell growth.[9] Cell cycle analysis revealed that antigen-activation produced a block in the cell cycle which occurred at the G1/S interface. Treatment of cells with non-specific antigen did not have this effect. Further studies demonstrated that treatment of murine T cell hybridomas and the human T cell leukemia cell line Jurkat with mitogenic anti-CD3ε mAb plus accessory cells induced IL-2 production and inhibited ^{3}[H] thymidine uptake and cell growth.[19] Cell cycle analysis of these cells showed that activation by anti-CD3εplus accessory cells causes a cell cycle block at the G1/S interface and a slowing of cells already in S phase.[19]

Studies by Lenardo have shown that pretreatment of both CD4+ and CD8+ mature murine T cells with IL-2 before TCR/CD3 stimulation causes T cells to undergo apoptosis.[20] In addition, specific deletion of Vβ8+ T cells, induced by injection of mice with the superantigen staphylococcus enterotoxin B (SEB), can be blocked by injection of an anti-IL-2Rα chain mAb.[20] These results suggest that pre-exposure to IL-2 can cause T cells stimulated through TCR/CD3 to undergo cell death in vivo as well as in vitro. Apoptosis was induced by TCR/CD3 stimulation in S phase but not in G1 phase cells.[21] This observation suggested that IL-2 is required to drive cells into S phase where they are sensitive to TCR/CD3-induced apoptosis.

Studies in our laboratory have investigated the role of cell cycle control of TCR/CD3-induced apoptosis in human T cell lines and in TCR transgenic murine T cells.

3. TCR/CD3-INDUCED APOPTOSIS IN HUMAN T CELL LEUKEMIA LINES

Transformed cell lines often carry defects in one or more proteins that regulate the cell cycle, and therefore can be used as models to study how cell cycle controls cell death. We studied TCR/CD3-induced cell death in two T cell leukemic lines, Jurkat and Sup-T13.[11] Treatment of these cell lines with soluble anti-CD3ε did not affect cell replication as measured by [^{3}H]thymidine uptake, although treatment with immobilized anti-CD3ε induced production of IL-2 and a block in [^{3}H]thymidine uptake. To determine whether stimulation via the TCR/CD3 caused cell death in Sup-T13 and Jurkat cells, these cells were cultured on anti-CD3ε mAb BC3[22]-coated plates for 3 days. Jurkat cells remained highly

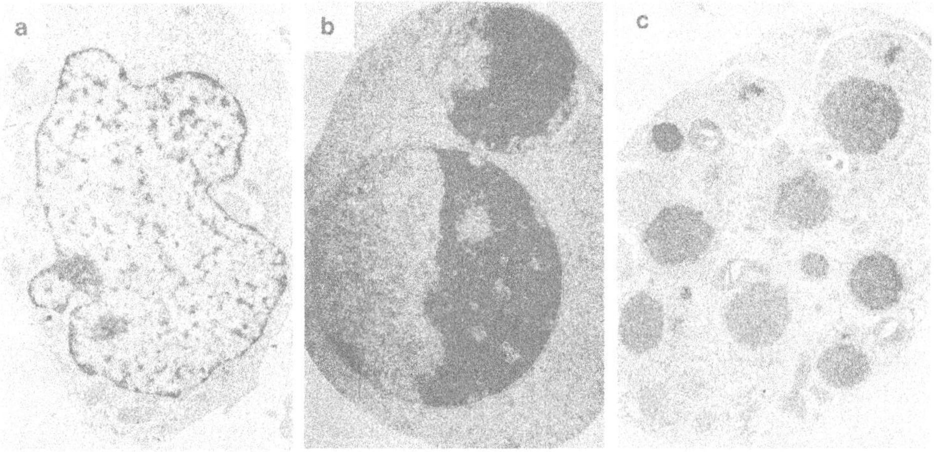

Figure 1. Induction of apoptosis in Sup-T13 cells by stimulation with anti-CD3ε mAb. Panel a: Sup-T13 cell cultured in medium (magnification x 5000). Panel b: Sup-T13 cell stimulated with plastic-bound anti-CD3ε mAb (30 μg/ml in coating solution) for 24 h shows compacted and marginated chromatin (magnification x 8000). Panel c: Sup-T13 cell is fragmented into apoptotic bodies after anti-CD3ε mAb stimulation for 32 h (magnification x 8000).

viable during the period of observation. Other groups have reported death of Jurkat cells after TCR/CD3 stimulation. The p56lck-defective Jurkat cell line JCam1[23] is resistant to TCR/CD3-induced apoptosis and does not upregulate Fas-L mRNA after anti-CD3ε mAb stimulation.[24] Therefore, it is possible that the Jurkat cell line used by us contains this or another defect, although this question has not yet been investigated.

Sup-T13 cells began to die as early as 24 h after anti-CD3ε mAb treatment, and by 72 hours more than 80% of the cells had died. Electron micrographs of Sup-T13 cells cultured on anti-CD3ε mAb for 24 hours revealed that the chromatin had become condensed and marginated (Fig. 1b), while the chromatin in Sup-T13 cells from unstimulated cultures did not become condensed (Fig. 1a). After 32 hours of culture on anti-CD3ε mAb the chromatin of Sup-T13 cells became fragmented into apoptotic bodies (Fig. 1c). The cytoplasmic condensation was accompanied by the appearance of clear vacuoles with relatively normal-appearing organelles. These nuclear changes are characteristic of apoptosis.

The DNA of Sup-T13 cells cultured in the presence of anti-CD3ε mAb also showed the pattern of nuclear DNA fragmentation characteristic of apoptosis after electrophoretic separation on an agarose gel (Fig. 2). The DNA 'ladder' observed contained oligomers whose sizes were multiples of 180–200 base pairs, suggesting that internucleosomal cleavage had occurred. Fragmentation was observed after 5–8 hours in culture with anti-CD3ε mAb. By 18 hours, nuclear DNA was totally degraded and ran as a smear on an agarose gel.

Further experiments were carried out to determine whether TCR/CD3-induced apoptosis of human leukemia T cells was cell cycle dependent. Treatment with immobilized anti-CD3ε mAb blocked Jurkat cells at the G1/S interface without inducing cell death. In contrast, identical treatment of Sup-T13 cells allowed cells to enter S phase, but slowed cell cycle progression and triggered cell death. Treatment of either cell line with soluble anti-CD3ε mAb did not induce cell death. The degraded nuclear DNA of Sup-T13 cells stimulated with immobilized anti-CD3ε mAb appeared as an apoptotic peak displaying hypodiploid DNA by microfluorimetry. DNA fragmentation correlated with entry of cells into S phase. When cells were arrested at the G1/S interface by the DNA synthesis inhibitor

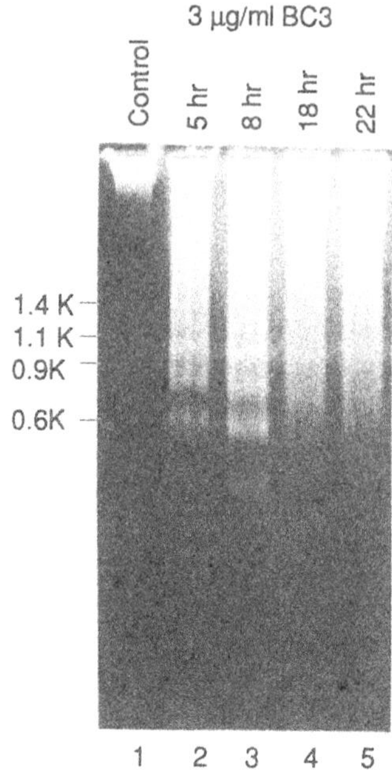

Figure 2. Agarose gel electrophoresis of nuclear DNA of Sup-T13 cells. Genomic DNA of Sup-T13 cells cultured in the absence of anti-CD3ε mAb is shown as control (lane 1). Sup-T13 cells were stimulated in plastic plates coated with 3 mg/ml anti-CD3ε for 5 hours (lane 2), 8 hours (lane 3), 18 hours (lane 4), and 22 hours (lane 5). 10^6 cells were harvested, mixed with RNase, and loaded directly on the gel containing SDS and proteinase K. Molecular weight standards were derived from ΦX174 DNA digested with restriction endonuclease Hae III. The relative mobility of the oligonucleosomal fragments of stimulated Sup T13 DNA reflects integer multiples of a 180-bp to 200-bp unit with respect to the standards.

aphidicolin, apoptosis was prevented (Figure 3). However, when cells were arrested at the G2/M interface by the mitotic inhibitor nocodazole, treatment with anti-CD3ε mAb induced apoptosis and cell death (Table 1). These results demonstrate that DNA synthesis is required for execution of the apoptotic program initiated by TCR/CD3 stimulation of Sup-T13 cells.

To determine the effect of immobilized anti-CD3ε mAb on the ability of Sup-T13 cells to progress through the cell cycle, we synchronized the majority of the cells in S phase and then cultured them with or without immobilized anti-CD3ε mAb (Table 2). Cells were treated with aphidicolin for 18 hours to block them at G1/S, then the aphidicolin was washed away and cells were cultured for 4 hours in medium, allowing the majority of the cells (62%) to progress into S phase. At this point cells were plated on uncoated plastic or on plastic coated with anti-CD3ε mAb. After 6 hours, the percentage of untreated cells in S phase had been reduced to 29%, while the percentage of cells in G2/M had increased from 7.5% to 31%, reflecting progression of these cells through the cell cycle. In contrast, after 6 hours in the presence of immobilized anti-CD3ε mAb, the percentage of cells in S phase had been reduced to only 51%, and the percentage of cells in G2/M had increased from 7.5% to only

Table 1. The mitotic inhibitor nocodazole does not block anti-CD3ε mAb-induced cell death[a]

	% Cell Death	
	Medium	Anti-CD3ε mAb
Medium	0	41
Nocodazole	13	38
Aphidicolin	14	16

[a]Percent cell death was quantitated by trypan blue dye exclusion and the values were normalized to discount baseline cell death. Experiments were performed in triplicate. This table is modified from its previously published form.[11]

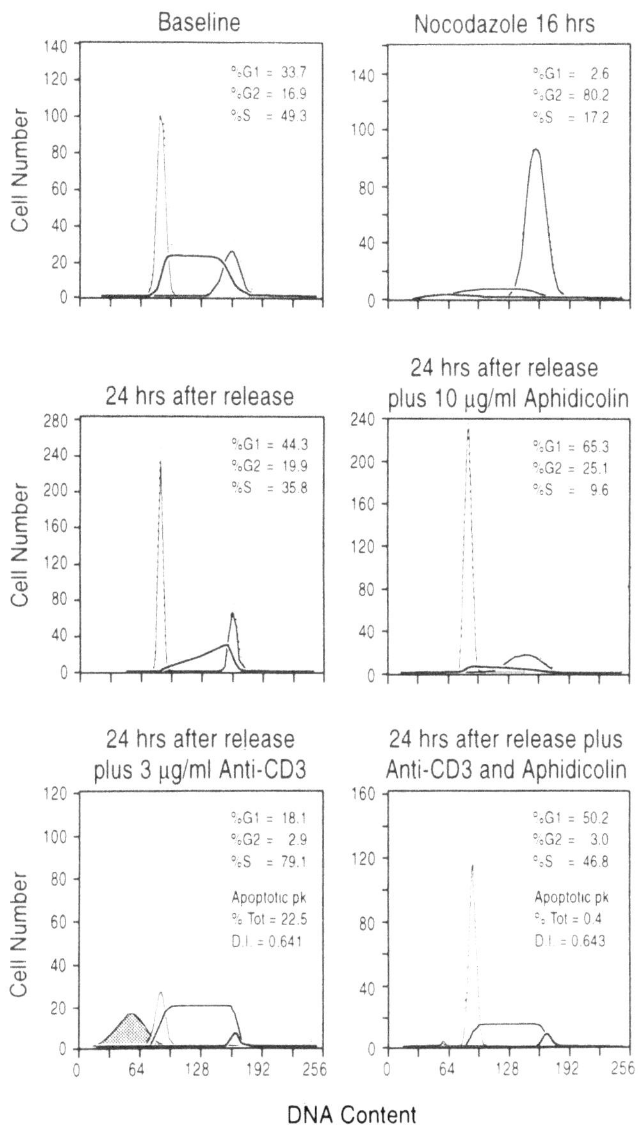

Figure 3. Decreased apoptotic DNA in Sup-T13 cells stimulated with anti-CD3ε mAb in the presence of the DNA synthesis inhibitor aphidicolin. Top left: Normal cycling Sup-T13 cells before any treatment. Top right: Sup-T13 cells were synchronized in G2 with 0.1 μg/ml nocodazole for 16 h. Middle left: Synchronized Sup-T13 cells were allowed to resume cycling for 22 h after washing off nocodazole. Middle right: Synchronized Sup-T13 cells were prevented from entering S phase by treatment with 10 μg/ml aphidicolin for 22 h. Bottom left: Synchronized Sup-T13 cells were stimulated with 3 μg/ml immobilized anti-CD3ε mAb for 22 h and presented an apoptotic peak 22.5% of total nuclear DNA with 0.64 fold the DNA content in G1 phase (as D.I.=0.64 indicates). Bottom right: Synchronized Sup-T13 cells were treated with anti-CD3ε mAb and aphidicolin simultaneously. The apoptotic DNA peak was reduced to 0.4% of total nuclear DNA. No apoptotic peak was observed in Sup-T13 cells cultured in the absence of anti-CD3ε mAb (middle left) or in the presence of aphidicolin alone (middle right). The dotted line in each profile delineates the original data. The continuous line delineates the phases of the cell cycle as calculated by the Multicycle program.

Table 2. Effect of anti-CD3ε mAb stimulation on the progression of Sup-*T*13 through the cell cycle[a]

Hours	% G1	%S	%G2	% Apoptotic DNA
0	21	62	7	9
Unactivated				
2	26	51	14	9
4	27	39	24	9
6	31	29	31	9
18	39	47	9	6
anti-CD3 activated				
2	25	55	8	12
4	23	55	7	15
6	21	51	15	13
18	45	24	10	21

[a]Sup-T13 cells were blocked in G1 phase by culture in 5 μg/ml aphidicolin for 18 hours. Cells were then washed twice in media to remove the aphidicolin and cultured for 4 hours in media plus 10% fetal bovine serum to allow the majority of the cells to enter S phase. Cells were then plated on plastic tissue culture dishes with or without immobilized anti-CD3ε mAb (3 μg/ml). At the indicated time points cells were removed from the plate, lysed, their nuclear DNA stained with propidium iodide and analyzed by flow cytometry.

15%. An increase in the percentage of apoptotic DNA was seen after 18 hours in culture with immobilized anti-CD3ε mAb, but was not seen in cells cultured for 18 hours in the absence of anti-CD3ε mAb. These results demonstrate that in addition to induction of apoptosis, treatment with immobilized anti-CD3ε mAb significantly slows down the progression of Sup-T13 cells through the cell cycle and apoptotic DNA does not appear until cells have progressed through G2.

Current studies in our laboratory have focused on the generation of Sup-T13 mutant cell lines, which are resistant to anti-CD3ε mAb-stimulated cell death. These lines were generated by treatment of the Sup-T13 parent cell line with mutagenizing radiation followed by culture in the presence of anti-CD3ε mAb. When cultured on immobilized anti-CD3ε mAb, these mutant cell lines do not die as rapidly or to the same degree as the Sup-T13 parental cell line. The mutant cell lines produce less IL-2 than Sup-T13 cells in response to anti-CD3ε mAb stimulation. They also have a slower and smaller rise in intracellular calcium in response to anti-CD3ε mAb, although their response to ionomycin is similar in strength to that observed in Sup-T13. Studies by Vignaux et al have shown that calcium is required for TCR/CD3-stimulated expression of Fas-L but not for execution of Fas based cytotoxicity.[18] T cell hybridomas stimulated with anti-CD3ε mAb were able to kill Fas+ target cells in the presence of EGTA-Mg^{2+}, which blocks perforin based lysis. However, if EGTA-Mg^{2+} was present during stimulation with the anti-CD3ε mAb, cytotoxicity was not observed. These results suggest that calcium is required downstream from TCR/CD3 stimulation, but upstream of Fas-based cytotoxicity. The weaker ability of the mutant cell lines to increase intracellular calcium in response to TCR/CD3 stimulation may enable them to resist TCR/CD3-induced apoptosis. Defining the defect of the mutant cell lines may help us to define the mechanism of TCR/CD3-induced apoptosis in Sup-T13 cells.

4. CELL CYCLE CONTROL OF TCR/CD3-INDUCED APOPTOSIS IN TCR TRANSGENIC MURINE T CELLS

To identify the mechanism for TCR/CD3-induced apoptosis in an antigen-specific system, we used CD4+ T cells from the DO10 mouse expressing a transgenic αβ TCR specific for ovalbumin peptide 323–339 (OVA) presented by I-A^d.[25] The transgenic TCR expressed in this system is detected by the clonotypic mAb KJ1–26 and by the anti-Vβ8 mAb, F23.1. This murine line was kindly provided to us by Dr. Dennis Loh. OVA alone induced clonal activation and expansion of peripheral CD4+/TCR-transgene+ cells in vitro. We investigated the effect of a non-mitogenic anti-CD3ε mAb (anti-CD3-Fos) on antigen-specific responses of naive T cells from the TCR transgenic mouse DO10. Anti-CD3-Fos is a genetically engineered F(ab')$_2$-like mAb, derived from the V regions of the hamster anti-mouse CD3εϖ Φ255 μΑβ 145 2X11.[26] Φορ σιμπλιχιτψ, αντι–ΧΔ3–Φοσ ωιλλ βε ρεφερρεδ το ασ αντι–ΧΔ3 Φ(αβ')$_2$ from here on. Treatment of naive T cells with soluble anti-CD3 F(ab')$_2$ did not result in either T cell activation or cell death, whereas activation with antigen followed by treatment with soluble anti-CD3 F(ab')$_2$ resulted in apoptosis of T cells in the S/G2 phases of the cell cycle.

Treatment of CD4+ T cells isolated from DO10 TCR transgenic mice with OVA followed by anti-CD3 F(ab')$_2$ resulted in DNA fragmentation in these cells, demonstrating that binding of anti-CD3 F(ab')2 to antigen-activated T cells induces cell death by apoptosis. Cell death induced by anti-CD3 F(ab')$_2$ plus OVA was selective for CD4+ T cells expressing the TCR transgene specific for OVA and recognized by the anti-TCR mAb KJ1–26.

To investigate whether the Fas/Fas-L interaction is involved in TCR/CD3-induced apoptosis induced by anti-CD3 F(ab')$_2$ in antigen-activated T cells, we generated OVA-specific T cell lines from lymph node cells of the Fas deficient MRL -/- lpr mouse and the wild type MRL +/+ mouse. Treatment with anti-CD3 F(ab')$_2$ alone or in combination with IL-2 induced cell death in the wild type cell lines, but did not induce cell death in the Fas-deficient cell lines, demonstrating that the expression of the Fas protein is required for apoptosis induced by anti-CD3 F(ab')$_2$ in activated murine T cells.

We investigated the correlation between cell cycle entry and induction of apoptosis in the DO10 model. T cells from DO10 mice were stimulated with OVA for 3 days, after which the antigen was washed away and cells were re-cultured with anti-CD3 F(ab')$_2$ at various concentrations. Cell cycle analysis demonstrated that concentrations of anti-CD3 F(ab')$_2$ up to 200 ng/ml increased the percentage of cells in S phase, while the percentage of cells in G1 and G2 showed a parallel decrease. This is consistent with the S phase slow down induced by CD3 signaling in the human leukemic T cell line Sup-T13.[11] At concentrations of anti-CD3 F(ab')$_2$ up to 200 ng/ml, we observed a direct correlation between the concentration of anti-CD3 F(ab')$_2$ and the generation of apoptotic DNA. At concentrations of anti-CD3 F(ab')$_2$ higher than 200 ng/ml we observed less apoptosis. In each experiment, however, there was a correlation between the fraction of cells in S phase and percent cell death or apoptotic DNA (Table 3).

To investigate the hypothesis that susceptibility to cell death correlated with entry into S phase, cell cycle inhibition experiments were performed. T cells were activated by antigen for 72 hours and then rested in medium for 48 hours. Cells were split into four groups and each was recultured under distinct conditions: medium alone, medium plus IL-2, IL-2 plus mimosine, or IL-2 plus nocodazole. Mimosine is an iron-chelator that blocks the cell cycle in G1. Nocodazole is a mitotic inhibitor which blocks the cell cycle in G2. After 48 hours a fraction of viable cells from each of the four cultures was recultured with or without anti-CD3 F(ab')$_2$. These cultures were maintained under the conditions established after the

Table 3. Anti-CD3 $F(ab')_2$ treatment causes an increase in % cells in S phase and an increase in % apoptotic DNA

Anti-CD3 $F(ab')_2$ (ng/ml)	% S	% Apoptotic DNA
0	10	10
8	13	15
40	33	20
200	51	37

cells had been rested for 48 hours (ie. medium alone, medium plus IL-2, IL-2 plus mimosine, or IL-2 plus nocodazole).

Treatment of cells cultured in medium alone with anti-CD3 $F(ab')_2$ did not result in an increase in cell death over that observed in cells cultured in medium alone without anti-CD3 $F(ab')_2$ (Table 4). Treatment of cells cultured in medium plus IL-2 with anti-CD3 $F(ab')_2$ did result in a significant increase in cell death over that observed in cells grown in medium plus IL-2 without anti-CD3 $F(ab')_2$. Cell cycle analysis demonstrated that cells grown in medium alone were mostly in G1 and had only 2.0% in S phase, while cells cultured in medium plus IL-2 had 44.5% in S phase. These results indicated that cell death correlated with cell cycle entry. Mimosine treatment blocked cells stimulated with IL-2 in G1, and restimulation with anti-CD3 $F(ab')_2$ could not induce cell death, indicating that progression from G1 to S is required for programmed cell death in this system. Nocodazole blocked progression of cells from G2 to M, but allowed cells to move from G1 to S to G2. Treatment of cells grown in IL-2 plus nocodazole with anti-CD3 F(ab')2 also resulted in increased cell death, suggesting that death was induced in S or G2 phases of the cell cycle.

5. CONCLUSIONS

Activation of normal resting T cells requires two signals, the association of specific antigen with the TCR/CD3 and the MHC complex and the binding of costimulatory

Table 4. Effect of cell cycle inhibitors on T cell death induced by anti-CD3 F(ab')2[a]

				% cell death	
Stimulus/inhibitor	%G1	%S	%G2	±anti-CD3	medium
Medium	83	2	15	32	32
IL-2	49	44	6	10	65
IL-2 + Mimosine	86	3	11	38	37
IL-2 + Nocodazole	38	8	53	20	60

[a]T cells were stimulated with OVA at 0.3-1 mM. After three days, cells were placed in secondary cultures containing medium for 48 hours to rest. Cells were then split into four parts and recultured for an additional 48 hours under one of the following conditions: medium, 10 U/ml IL-2 alone, IL-2 plus 600 mM mimosine or IL-2 plus 0.1 μg/ml nocodazole. After this incubation step, viable cells were isolated by Ficoll-Hypaque and stained for DNA content with propidium iodide (PI). A fraction of viable cells from each of the four cultures was seeded for the fourth time in a 24-well plate for 24 hours with or without 1 μg/ml anti-CD3 F(ab')2, and were maintained under the conditions established after the 48 hour rest in medium alone. Cell death was determined by PI staining. The results represent one of three similar experiments.

molecules. The coordinated delivery of both signals leads to expression of IL-2 and other cytokine genes, DNA synthesis and cell proliferation.[27] Re-crosslinking of the TCR/CD3 complex in the presence of IL-2, however, induces apoptosis in murine T cells by delivering a cellular activation signal during a critical phase of the cell cycle.

Studies from our laboratory and others have demonstrated cell cycle regulation of TCR/CD3-induced T cell death both in vitro and in vivo. Our studies have shown that T cells from DO10 mice stimulated with specific antigen followed by the non-mitogenic anti-CD3 $F(ab')_2$ results in both an increased percentage of cells in S phase and an increase in apoptosis. This solulble anti-CD3 $F(ab')_2$ can also induce apoptosis in some activated human T cell clones and in normal murine T cells (Yu, X-Z., unpublished data). Cell cycle inhibition experiments have confirmed that blocking activated DO10 T cells at the G1/S interface rendered them insensitive to anti-CD3 $F(ab')_2$-induced apoptosis, whereas cells blocked in G2 remained sensitive to anti-CD3 $F(ab')_2$-induced apoptosis.

In addition to cell cycling, upregulation of Fas and expression of Fas-L are required for activation of T cell apoptosis by anti-CD3ε mAb. Our studies have demonstrated that stimulation with anti-CD3 $F(ab')_2$ alone or in combination with IL-2 did not induce cell death in antigen-activated OVA-specific T cell lines generated from the Fas deficient MRL -/- lpr mouse. However, antigen-activated OVA-specific cell lines generated from wildtype MRL +/+ mice were susceptible to anti-CD3 $F(ab')_2$-induced apoptosis. It remains to be shown whether Fas-mediated T cell apoptosis is also controlled by cell cycle.

T cell activation through the TCR causes production of IL-2, allowing entry into cell cycle and progression of cells into S phase, where they become susceptible to apoptosis induced by restimulation of the TCR/CD3 complex. Studies of this activation pathway are necessary to define the role of cell cycle on susceptibility to TCR/CD3-induced apoptosis via the Fas/Fas-L interaction. A number of recent reports have identified a potential role of cell cycle and signaling proteins. Fotedar et al have reported that synchronized murine T cell hybridomas activated by anti-CD3 mAb stop cycling shortly after they reach G2/M and undergo apoptosis.[28] Cyclin B- and p34cdc2-associated histone H1 kinase activity is persistently elevated after anti-CD3ε mAb stimulation, although p34cdc2 protein levels do not change. Cyclin-B-specific antisense oligonucleotides were able to suppress anti-CD3ε mAb-induced T cell death, but were not able to suppress dexamethasone-induced T cell death.[28] These results suggest that deregulation of p34cdc2 kinase activity is involved in susceptibility to TCR/CD3-induced T cell apoptosis. Cyclin A has been implicated in c-myc-induced apoptosis and also in apoptosis induced by staurosporine, caffeine, 6-dimethylaminopurine and okadaic acid.[29,30] Miyatake et al have reported that anti-TCR induced G1 arrest in T cell hybridomas, which results in apoptosis, is accompanied by down regulation of cyclin D3 expression and cyclin D3-dependent kinase activity.[31] A more complete understanding of the effects of TCR/CD3 stimulation on cell cycle regulation and apoptosis will result in a clearer understanding of the processes of negative selection and tolerance.

ACKNOWLEDGMENTS

Supported by NIH grants CA 18029 and CA 18221, and AI 33484

8. REFERENCES

1. L. Sachs, and J. Lotem, Control of programmed cell death in normal and leukemic cells: new implications for therapy, *Blood*. 82:15 (1993).

2. J.F.R. Kerr, A.H. Wyllie, and A.R. Currie, Apoptosis: a basic biological phenomenon with wide-ranging implications in tissue kinetics, *Br. J. Cancer.* 26:239 (1972).
3. J.J. Cohen, R.C. Duke, V.A. Fadok, and K.S. Sellins, Apoptosis and programmed cell death in immunity, *Annu. Rev. Immunol.* 10:267 (1992).
4. A.H. Wyllie, J.F.R. Kerr, and A.R. Currie, Cell death: the significance of apoptosis, *Int. Rev. Cytol.* 68:251 (1980).
5. J.J. Cohen, Programmed cell death in the immune system, *Adv. Immunol.* 50:55 (1991).
6. C.A. Smith, G.T. Williams, R. Kingston, E.J. Jenkins, and J.J.T. Owen, Antibodies to CD3/T-cell receptor complex induce death by apoptosis in immature T cells in thymic cultures, *Nature* 337:181 (1989).
7. Y.F. Shi, R.P. Bissonnette, N. Parfray, M. Szalay, R.T. Kuboto, and D.R. Green, In vivo administration of monoclonal antibodies to the CD3 T cell receptor complex induces cell death (apoptosis) in immature thymocytes, *J. Immunol.* 146: 3340 (1991).
8. B. Wang, C. Biron, J. She, K. Higgins, M.J. Sunshine, E. Lacy, N. Lonberg, and C. Terhorst, A block in both early T lymphocyte and natural killer cell development in transgenic mice with high copy numbers of the human CD3E gene, *Proc. Natl. Acad. Sci. USA* 91:9402 (1994).
9. J.D. Ashwell, R.E. Cunningham, P.D. Noguchi, and D. Hernandez, Cell growth cycle block of T cell hybridomas upon activation with antigen, *J. Exp. Med.* 165:173 (1987).
10. S. Takahashi, H.T. Maecker, and R. Levy, DNA fragmentation and cell death mediated by T cell antigen receptor/CD3 complex on a leukemia T cell line, *Eur. J. Immunol.* 19:1911 (1989).
11. L. Zhu, and C. Anasetti, Cell cycle control of apoptosis in human leukemic T cells, *J. Immunol.* 154:192 (1995).
12. D. Kaebelitz, T. Pohl, and K. Pechhold, Activation-induced cell death (apoptosis) of mature peripheral T lymphocytes, *Immunol. Today* 147:338 (1993).
13. J.H. Russell, B. Rush, C. Weaver, and R. Wang, Mature T cells of autoimmune lpr/lpr mice have a defect in antigen-stimulated suicide, *Proc. Natl. Acad. Sci. USA* 90:4409 (1993).
14. J. Dhein, H. Waclczak, C. Baumier, K-M. Debatin, and P.H. Krammer, Autocrine T-cell suicide mediated by APO-1/(Fas/CD95), *Nature* 373:438 (1995).
15. T. Brunner, R.J. Mogil, D. LaFace, N.J. Yoo, A. Mahboubl, F. Echeverri, S.J. Martin W.R. Force, D.H. Lynch, C.F. Ware, and D.R. Green, Cell-autonomous Fas (CD95)/Fas-ligand interaction mediates activation-induced apoptosis in T-cell hybridomas, *Nature* 373:441 (1995).
16. S-T. Ju, D.J. Panka, H. Cui, R. Ettinger, M. El-Khatib, D.H. Sherr, B.Z. Stanger, and A. Marshak-Rothstein, Fas (CD95)/Fas L interactions required for programmed cell death after T-cell activation, *Nature* 373:444 (1995).
17. Y.L. Yang, M. Mercep, C.F. Ware, and J.D. Ashwell, Fas and activation-induced Fas ligand mediate apoptosis of T cell hybridomas-inhibition of Fas ligand expression by retinoic acid and glucocorticoids, *J. Exp. Med.* 181:1673 (1995).
18. F. Vignaux, E. Vivier, B. Malissen, V. Depraetere, S. Nagata, and P. Golstein, TCR/CD3 coupling to Fas-based cytotoxicity, *J. Exp. Med.* 181:781 (1995).
19. M. Mercep, J.A. Bluestone, P.D. Noguchi, and J.D. Ashwell, Inhibition of transformed T cell growth in vitro by monoclonal antibodies directed against distinct activating molecules, *J. Immunol.* 140:324 (1987).
20. M.J. Lenardo, Interleukin-2 programs mouse $\alpha\beta$ T lymphocytes for apoptosis, *Nature* 353:858 (1991).
21. S.A. Boehme, and M.J. Lenardo, Propriocidal apoptosis of mature T lymphocytes occurs at S phase of the cell cycle, *Eur. J. Immunol.* 23:1552 (1993).
22. C. Anasetti, P. Tan, J.A. Hansen, and P.J. Martin, Induction of specific nonresponsiveness in unprimed human T cells by anti-CD3 antibody and alloantigen, *J. Exp. Med.* 172:1691 (1990).
23. D.B. Straus, and A. Weiss, Genetic evidence for the involvement of the lck tyrosine kinase in signal transduction through the T cell antigen receptor, *Cell* 70: 585 (1992).
24. N. Oyaizu, S. Than, T.W. McCloskey, and S. Pahwa, Requirement of p56lck in T-cell receptor/CD3-mediated apoptosis and Fas-ligand induction in Jurkat cells, *Biochem. Biophys. Res. Commun.* 213:994 (1995).
25. K.M. Murphy, A.B. Heimberger, and D.Y. Loh, Induction by antigen of intrathymic apoptosis of $CD4^{+}CD8^{+}TCR^{lo}$ thymocytes in vivo, *Science* 250:1720 (1990).
26. S.A. Kostelny, M.S. Cole, and J.Y. Tso, Formation of a bispecific mAb by the use of leucine zippers, *J. Immunol.* 148:1547 (1992).
27. A.K. Abbas, A.H. Lichtman, and J.S. Pober, Molecular basis of T cell antigen recognition and activation, in: "Cellular and Molecular Immunology," W.B. Saunders Co., Philadelphia (1991).
28. R. Fotedar, J. Flatt, S. Gupta, R.L. Margolis, P. Fitzgerald, H. Messier, and A. Fotedar, Activation-induced T-cell death is cell cycle dependent and regulated by cyclin B, *Mol. Cell. Biol.* 15:932 (1995).

29. A.T. Hoang, K.J. Cohen, J.F. Barrett, D.A. Bergstrom, and C.V. Dang, Participation of cyclin A in myc-induced apoptosis, *Proc. Natl. Acad. Sci. USA* 91:6875 (1994).
30. W. Meikrantz, S. Gisselbrecht, S.W. Tam, and R. Schlegel, Activation of cyclin A- dependent protein kinases during apoptosis, *Proc. Natl. Acad. Sci. USA* 91: 3754 (1994).
31. S. Miyatake, H. Nakano, S.Y. Pak, T. Yamazaki, K. Takase, H. Matsushime, A. Kato, and T. Saito, Induction of G1 arrest by down-regulation of cyclin D3 in T cell hybridomas, *J. Exp. Med.* 182:401 (1995).

ROLE OF ANTIBODY SIGNALING IN INDUCING TUMOR DORMANCY

Jonathan W. Uhr,[1] Radu Marches,[1] Emil Racila,[1] Thomas F. Tucker,[1]
Robert Hsueh,[2] Nancy E. Street,[1] Ellen S. Vitetta,[1] and
Richard H. Scheuermann[2]

[1] Department of Microbiology and Cancer Immunobiology Center
[2] Laboratory of Molecular Pathology and Department of Pathology
University of Texas Southwestern Medical Center at Dallas
5323 Harry Hines Boulevard, Dallas, Texas 75235–8576

Cancer dormancy is a well-recognized clinical phenomenon in which tumor cells are present, but the tumor burden does not increase for long periods of time[1–3]. However, tumor cells can regrow many years later. In breast cancer, there is a steady rate of recurrence 10 to 20 years after removal of the primary tumor[3,4] and the recurrent tumor frequently grows at a rapid rate[5]. A particularly pertinent example is the low grade (follicular) form of non-Hodgkin's lymphoma (NHL) in which long-term remissions are common but, eventually, virtually all die of a recurrence. Levy and Miller[5] have treated such patients with monoclonal anti-idiotype (Id) and have achieved remissions in a high proportion of patients. Relapses, many caused by Id-negative variants, are frequent indicating that the antibody (Ab) was particularly effective in inducing dormancy in cells bearing the corresponding idiotope but that hypermutation of V_H and V_L genes eventually allow some tumor cells from the original clone to escape[5–7].

We have characterized a dormancy model of a murine lymphoma in which BALB/c mice hyperimmunized to the BCL_1 idiotype (Id-immune mice) can harbor 10^6 BCL_1 tumor cells for a lifetime; a single BCL_1 cell injected into a non-immune syngeneic mouse results in progressive tumor growth and death. The antitumor immunity is due mainly to the presence of anti-Id Ab which induces cell cycle arrest (CCA) and apoptosis in the tumor cells[8–10].

We have also characterized another model of dormancy by adoptively transferring IgM-specific Ab and/or Id-immune T cells to SCID mice before tumor cell challenge[10]. The results indicate that Ab to IgM is essential but that simultaneous addition of Id-immune T cells enhances the induction and duration of dormancy. Id-immune T cells by themselves gave no detectable antitumor effect. It was also observed that polyclonal anti-IgM antibodies were more effective than monoclonal antibodies (MAbs) in inducing dormancy in SCID mice bearing a murine B lymphoma (BCL_1). For *in vitro* studies, we have used both an *in vitro* adapted line of BCL_1 (3B3) and a human Burkitt's lymphoma cell line (Daudi). Both display very similar responsiveness to anti-μ Ab[10–12].

Mechanisms of Lymphocyte Activation and Immune Regulation VI, Edited by Gupta and Cohen
Plenum Press, New York, 1996

In this paper, we will discuss three issues that are important to an elucidation of the cellular and molecular mechanisms underlying the induction and maintenance of dormancy: 1) The role of hypercrosslinking IgM on dormancy induction *in vivo* and on the induction of apoptosis and CCA *in vitro.* 2) The role of the APO-1 ligand in IgM-mediated apoptosis. 3) The cellular and molecular mechanisms underlying regrowth of tumor (*i.e.*, tumor cell escape).

1. THE EFFECT OF HYPERCROSSLINKING RECEPTORS ON SIGNALING IN B LYMPHOMA CELLS

The first question to be addressed was whether *hypercrosslinking* of mIgM on B lymphoma cells was important for inducing dormancy by increasing either CCA and/or apoptosis. Early in our studies in which SCID mice received passively administered Ab before challenge with BCL_1 cells, it was observed that polyclonal Abs were more effective than MAbs in inducing and maintaining dormancy[10]. For example, when rabbit anti-BCL_1Id was used, 79% of the mice became dormant and the average duration of dormancy was $153^{\pm 60}$ days. In contrast, when rat monoclonal anti-μ was used, 67% of the animals became dormant and the average duration of dormancy was $76^{\pm 22}$ days. This same pattern was observed using other polyclonal and MAbs directed against various epitopes on IgM.

We also found that: 1) Using the 3B3 cell line, the polyclonal Abs were more effective in inducing apoptosis than the monoclonal Abs[10]. Similar results were obtained using a Burkitt's lymphoma cell line (Daudi) and three different monoclonal anti-μ Abs directed against different epitopes on three different domains of human μ chain[12]. The human lymphoma gave responses similar to those of anti-μ in our mouse lymphoma model. 2) A mixture of three of these monoclonal anti-tumor μ Abs mimicked the effect observed using polyclonal Ab[12]. 3) The hypercrosslinking of mIgM using a single monoclonal Ab and a secondary anti-MAb markedly increased apoptosis[10,12]. 4) Hypercrosslinking of CD40 did not induce apoptosis. 5) Simultaneous crosslinking of CD40 *and* IgM increased the proportion of cells undergoing apoptosis[12].

The present results provide further evidence that the anti-growth activity of Abs is a critical factor in the induction of dormancy in the BCL_1 tumor cells. This conclusion is deduced from: 1) the strong correlation between the capacity of polyclonal *vs* monoclonal Abs to induce long-lived dormancy and their capacity to negatively signal *in vitro*. 2) The capacity of $F(ab')_2$ polyclonal antibodies to induce dormancy after challenge with BCL_1. This is a critical point because without this evidence we would not know to what extent insights from *in vitro* analyses were applicable to induction and maintenance of dormancy. Effector mechanisms may also play a role in Ab-mediated dormancy but they are not sufficient. Thus, prior studies[10] have shown that Id-immune T cells can positively modulate Ab-induced dormancy in SCID mice; a similar adjuvant role for ADCC has not been investigated. Hence, optimal maintenance of dormancy may require negative signaling, cellular immunity and NK cell function.

The present results also indicate that the hypercrosslinking of IgM can change the response of individual lymphoma cells by inducing apoptosis. There is considerable evidence from past studies[Reviewed in 13,14] that a highly multivalent antigen can stimulate normal B cells in a T-independent manner and can induce tolerance in immature B cells. Increased crosslinking of IgM by anti-IgM Abs on normal or neoplastic B cells can also lead to increased cell death[12,13,15–21]. Our results confirm and extend these findings by demonstrating that cells that would have been induced to enter CCA following crosslinking of sIgM can be rendered apoptotic after hypercrosslinking[10,12].

2. ROLE OF APO-1 LIGAND IN IGM-MEDIATED APOPTOSIS

Signal transduction initiated by crosslinking of antigen-specific receptors on T as well as on B lymphoma cells induces apoptosis. In T lymphoma cells, such crosslinking results in upregulation of the APO-1 ligand which then interacts with induced or constitutively expressed APO-1 molecules, thereby triggering apoptosis[22–27]. In similar studies with Daudi cells[28], we have shown that crosslinking the membrane immunoglobulin on these lymphoma cells (that constitutively express APO-1) does not upregulate expression of the APO-1 ligand. Further, a noncytotoxic fragment of anti-APO-1 Ab which blocks T cell receptor-mediated apoptosis in T lymphoma cells[25] does not block anti-μ induced apoptosis in B lymphoma cells. Hence, apoptosis induced by signaling *via* membrane IgM is not mediated by the APO-1 ligand.

Although the APO-1 ligand is not involved, the question remains as to whether APO-1 expression plays a role in the induction of apoptosis by anti-Ig in Daudi cells. Perhaps, crosslinking of the mIg complex results in its association with APO-1 resulting in productive oligomerization of APO-1 and the subsequent delivery of a death signal without the need for its ligand. It is also possible that another member of the APO-1 family along with its ligand is responsible for IgM-mediated apoptosis. These putative mechanisms may also be available to T lymphocytes since the suppression by $F(ab')_2$ anti-APO-1 was not complete in the published studies[25] or in our studies[28], even at high concentrations of blocking reagent. Thus, there could be a number of mechanisms for cell death in antigen receptor-activated T and B lymphomas: 1) APO-1 - APO-1 ligand interaction in T cells; 2) APO-1 on B cells interacting with APO-1 ligand on or secreted by T cells[29, Reviewed in 30]; 3) APO-1 activation after antigen-receptor crosslinking in B cells, and, possibly T cells through a direct association with the antigen-receptor itself; and, 4) An APO-independent death pathway(s) in B and perhaps T cells.

Why should B and T cells differ in such a fundamental mechanism? One speculation that would place our results in the framework of the normal immune response would be as follows: The B cell antigen-specific receptors can be readily saturated and extensively crosslinked by circulating self antigens. If this were to result in co-ligation of APO-1 or another molecule in the APO-1 family, the co-ligated molecules could thereby be activated and directly signal apoptosis, resulting in deletion of the cells (tolerance). In the case of T cells, the antigen is presented only in the form of an MHC-peptide fragment complex which may be much less effective in allowing self-antigens to *extensively* crosslink the T cell receptor. Thus, even in anti-CD3-treated T cells, it was necessary to immobilize the Ab in order to obtain significant apoptosis[25–27]. Hence, for deletion of self-reactive T cells, it might be desirable to have an amplifying mechanism, *i.e.*, secretion of the APO-1 ligand, which would increase sensitivity to the induction of tolerance.

3. REGROWTH OF DORMANT TUMOR (ESCAPE)

We have observed the natural history of dormant BCL_1 tumor in more than 100 BALB/c mice that have been followed for up to 600 days after challenge with BCL_1 tumor cells. The results show a steady rate of loss of dormancy suggesting a stochastic process. One likely interpretation is that genetic alterations have appeared in the replicating tumor cells. To test this possibility we have injected escapee clones back into Id-immune mice to determine whether they are still susceptible to the induction of dormancy. The majority of such escapee clones are no longer susceptible to the induction of dormancy in Id-immune mice. A small percentage of such escapees has lost surface IgM and/or the idiotype. (Since the animals have made a polyclonal anti-idiotype response one would have predicted that loss of many idiotopes simultaneously is an unlikely event.)

Cell lines have been developed from some of these mutants to test their responsiveness *in vitro* to anti-idiotype Ab. We presume that in these Id^+ mutant cells, a signaling component other than the immunoglobulin molecule itself has mutated in a way that interrupts the signaling cascade resulting in cell cycle arrest or apoptosis. Analysis of such signaling components such as lyn, syk, and HS1 is underway to elucidate the mechanisms of escape. In those animals which still show susceptibility to the induction of dormancy, we hypothesize that the anti-Id Ab response may have waned. Thus, when anti-idiotypic Ab titers were tested before tumor challenge there was a significant difference in the proportion of mice that became dormant depending on the initial level of serum anti-idiotypic Ab. In those mice which had anti-idiotypic Ab levels >100 μg/ml, a significantly higher proportion of animals become dormant (90%) compared to those that had a lower titer (50% become dormant). Hence, escape in this tumor model appears to be predominantly due to mutational mechanisms in the tumor cell but, in some instances, may involve a decrease in the host immunologic response.

4. CONCLUDING REMARKS

Clinically, anti-IgM is not useful for signaling lymphoma cells. However, Abs have been developed against the extracellular portion of the IgHβ chain[31,32]. We would predict that these Abs should behave similarly to anti-IgM Abs in terms of crosslinking IgM. Thus, a combination of MAbs against different noncrossreacting epitopes on Igβ (and/or Igα) should induce apoptosis and, thereby, imitate the effectiveness of polyclonal Ab. An additional advantage might be gained if Abs with high valency against these extracellular portions of IgHβ could be designed. Finally, as demonstrated here, anti-CD40, which induces CCA in Daudi cells, can increase the proportion of cells undergoing apoptosis when administered together with anti-μ. These results suggest that combinations of the appropriate MAbs against several antigenic epitopes together with a cytotoxic agent might improve therapy of non-Hodgkin's lymphoma.

5. ACKNOWLEDGMENTS

We thank Ms. C. Patterson and Ms. S. Chadwick for their assistance in the preparation of the manuscript and Ms. N. Lane for her expert technical assistance.

This work was supported by National Institutes of Health grants CA28149, CA58321 and CA64679 and a grant from The Meadows Foundation.

6. REFERENCES

1. Stewart, T.H.M., A.C. Hollinshead, and S. Raman. Tumor dormancy initiation maintenance and termination in animals and humans. Can. J. Surg. **34**: 321–325, 1991.
2. Meltzer, A. Dormancy and breast cancer. J. Surg. Oncol. **43**: 181–188, 1990.
3. Berkowitz, H., F. Rosata, and C.P. Neiby. Late recurrence of carcinoma of breast: Case report and literature survey. Amer. Surg. **32**: 287–289, 1966.
4. Henderson, I.C., J.R. Harris, D.W. Kinne, and S. Hellman. Cancer of the breast. **IN:** Cancer: Principles and Practice of Oncology (V.T. DeVita, Jr., S. Hellman, and S.A. Rosenberg, Eds.), Philadelphia: J.B. Lippincott, pp. 1197–1268, 1989.

5. Levy, R. and A.R. Miller. Therapy of lymphoma directed at idiotypes. Monographs. J. Natl. Cancer Inst. **10**: 61–68, 1990.
6. Meeker, T., J. Lowder, M.L. Cleary, S. Stewart, R. Warnke, J. Sklar, and R. Levy. Emergence of idiotype variants during treatment of B-cell lymphoma with anti-idiotype antibodies. N. Engl. J. Med. **312**: 1658–1665, 1985.
7. Brown, S.L., R.A. Miller, S.J. Horning, D. Czerwinski, S.M. Hart, R. McElderry, T. Basham, R.A. Warnke, T.C. Merigan, and R. Levy. Treatment of B-cell lymphomas with anti-idiotype antibodies alone and in combination with alpha interferon. Blood **73**: 651–661, 1989.
8. Uhr, J.W., T. Tucker, R.D. May, H. Siu, and E.S. Vitetta. Cancer dormancy: Studies of the murine BCL_1 lymphoma. Cancer Res. **51**: 5045S-5053S, 1991.
9. Yefenof, E., L.J. Picker, R.H. Scheuermann, T.F. Tucker, E.S. Vitetta, and J.W. Uhr. Cancer dormancy: Isolation and characterization of dormant lymphoma cells. Proc. Natl. Acad. Sci. USA **90**: 1829–1833, 1993.
10. Racila, E., R.H. Scheuermann, L.J. Picker, E. Yefenof, T. Tucker, W. Chang, R. Marches, N.E. Street, E.S. Vitetta, and J.W. Uhr. Tumor dormancy and cell signaling. II. Antibody as an agonist in inducing dormancy of a B cell lymphoma in SCID mice. J. Exp. Med. **181**: 1539–1550, 1995.
11. Scheuermann, R.H., E. Racila, T. Tucker, E. Yefenof, N.E. Street, E.S. Vitetta, L.J. Picker, and J.W. Uhr. Lyn tyrosine kinase signals cell cycle arrest but not apoptosis in B-lineage lymphoma cells. Proc. Natl. Acad. Sci. USA **91**: 4048–4052, 1994.
12. Marches, R., E. Racila, T.F. Tucker, L. Picker, P. Mongini, R. Hsueh, R.H. Scheuermann, and J.W. Uhr. Tumor dormancy and cell signaling. III: Role of hypercrosslinking of IgM and CD40 on the induction of cell cycle arrest and apoptosis in B lymphoma cells. Therap. Immunol., *in press*, 1995.
13. Goodnow, C.C. Transgenic mice and analysis of B-cell tolerance. Annu. Rev. Immunol. **10**: 489–518, 1992.
14. Nossal, G.J. Cellular and molecular mechanics of B lymphocyte tolerance. Adv. Immunol. **52**: 283–331, 1992.
15. Hasbold, J. and G.G.B. Klaus. Anti-immunoglobulin antibodies induce apoptosis in immature B cell lymphomas. Eur. J. Immunol. **20**: 1685–1690, 1990.
16. Parry, S.L., M.J. Holman, J. Hasbold, and G.G.B. Klaus. Plastic-immobilized anti-μ or anti-δ antibodies induce apoptosis in mature murine B lymphocytes. Eur. J. Immunol. **24**: 974–979, 1994.
17. Parry, S.L., J. Hasbold, M. Holman, and G.B. Klaus. Hypercrosslinking surface IgM or IgD receptors on mature B cells induces apoptosis that is reversed by costimulation with IL-4 and anti-CD40. J. Immunol. **152**: 2821–2829, 1994.
18. Scott, D.W., M. Vankataraman, and J.J. Jandinski. Multiple pathways of B lymphocyte tolerance. Immunol. Rev. **43**: 241–280, 1979.
19. Nossal, G.J.V., B.L. Pike, J.M. Teale, J.E. Layton, T.W. Kay, and F.L. Battye. Cell fractionation methods and the target cells for clonal abortion of B lymphocytes. Immunol. Rev. **43**: 185–216, 1979.
20. Cambier, J.C. and J.T. Ransom. Molecular mechanisms of transmembrane signaling in B lymphocytes. Annu. Rev. Immunol. **5**: 175–199, 1987.
21. Gold, M. and A.L. DeFranco. Biochemistry of B lymphocyte activation. Adv. Immunol. **55**: 221–295, 1994.
22. Adachi, M., R. Watanabe-Fukunaga, and S. Nagata. Aberrant transcription caused by the insertion of an early transposable element in an intron of the Fas antigen gene of lpr mice. Natl. Acad. Sci. USA **90**: 1756–1760, 1993.
23. Chu, J.L., J. Drappa, A. Parnassa, and K.B. Elkon. The defect in Fas mRNA expression in MRL/lpr mice is associated with insertion of the retrotransposon, ETn. J. Exp. Med. **178**: 723–730, 1993.
24. Wu, J., T. Zhou, J. He, and J.D. Mountz. Autoimmune disease in mice due to integration of an endogenous retrovirus in an apoptosis gene. J. Exp. Med. **178**: 461–468, 1993.
25. Dhein, J., H. Walczak, C. Bäumler, K.-M. Debatin, and P.H. Krammer. Autocrine T-cell suicide mediated by APO-1/(Fas/CD95). Nature **373**: 438–441, 1995.
26. Brunner, T., R.J. Mogil, D. LaFace, N.J. Yoo, A. Mahboubi, F. Echeverri, S.J. Martin, W.R. Force, D.H. Lynch, C.F. Ware, and D.R. Green. Cell-autonomous Fas (CD95)/Fas-ligand interaction mediates activation-induced apoptosis in T-cell hybridomas. Nature **373**: 441–444, 1995.
27. Ju, S.-T., D.J. Panka, H. Cui, R. Ettinger, M. El-Khatib, D.H. Sherr, B.Z. Stanger, and A. Marshak-Rothstein. Fas (CD95)/FasL interactions required for programmed cell death after T-cell activation. Nature **373**: 444–448, 1995.
28. Racila, E., R. Hsuch, R. Marches, T.F. Tucker, P.H. Krammer, R.H. Scheuermann, and J.W. Uhr. Tumor dormancy and cell signaling. IV. Anti-μ induced apoptosis in human B lymphoma cells is not caused by an APO-1 - APO-1 ligand interaction. Proc. Natl. Acad. Sci. USA, *in press*, 1995.

29. Ju, S.T., H. Cui, D.J. Panka, R. Ettinger, and A. Marshak-Rothstein. Participation of target Fas protein in apoptosis pathway induced by CD4+ Th1 and CD8+ cytotoxic T cells. Proc. Natl. Acad. Sci. USA **91**: 4185–4189, 1994.
30. Berke, G. The CTL's kiss of death. Cell **81**: 9–12, 1995.
31. Vasile, S., J.E. Coligan, M. Yoshida, and B.K. Seon. Isolation and chemical characterization of the human B29 and mb-1 proteins of the B cell antigen receptor complex. Mol. Immunol. **31**: 419–427, 1994.
32. Nakamura, T., M.C. Sekar, H. Kubagawa, and M. Cooper. Signal transduction in human B cells initiated via Igβ ligation. Int. Immunol. **10**: 1309–1315, 1993.

8

REGULATION OF LYMPHOID APOPTOSIS BY Bcl-2 AND Bcl-x_L

Gabriel Nuñez, Ramon Merino, Philip L. Simonian, and
Didier A. M. Grillot

Department of Pathology
The University of Michigan Medical School
Ann Arbor, Michigan 48109

1. Bcl-2 AND Bcl-x_L ARE REGULATORS OF LYMPHOID APOPTOSIS

Naturally occurring cell death is common during lymphoid development and activation. The death of developing lymphoid cells is a highly regulated process that serves to select lymphoid populations that are functionally competent, and to remove cells that are not longer needed or potentially autoreactive[1]. Elimination of self-reactive B and T lymphocytes by apoptosis is thought to play a major role in the establishment of self-tolerance. The latter process is mediated by high avidity interactions between antigen receptors and self-antigens[2]. In contrast, signaling via the Fas receptor appears to play a major role in the elimination of activated lymphocytes during immune responses in peripheral tissues[3]. In addition, to the antigen and Fas receptors, survival of lymphocytes is controlled by certain cytokines and costimulatory signals[4–5]. The intracellular mechanism that regulates and executes the death program is still poorly understood but it is thought that cell death is controlled by a genetic program induced within the dying lymphocyte. Recently, several genes have been identified that appear to play critical roles in lymphoid survival[6]. The *bcl-2* protooncogene was the first member of a growing family of genes that suppresses cell death in lymphoid cells[7]. Constitutive expression of *bcl-2* in lymphoid cells prevents or delays apoptosis induced by multiple stimuli[7]. A role of Bcl-2 in T and B-cell biology was suggested by its highly restricted cellular distribution during development and in mature lymphoid populations[8–9]. Recent evidences suggest that Bcl-2 plays a role in positive selection of thymocytes[10–12]. However, the ability of Bcl-2 to influence negative selection of thymocytes and immature B cells is controversial[13–15]. Analysis of mutant mice deficient in Bcl-2 have revealed that this protein plays a critical role in the maintenance of mature lymphocytes, although it is not required for lymphoid maturation[16–17]. A plausible interpretation is that other proteins similar to Bcl-2 can function as survival signals during lymphoid development. A candidate protein is Bcl-x_L, a product of *bcl-x*, another apoptosis-regulatory gene of the *bcl-2* family that inhibits apoptosis induced by several stimuli[18–19]. In mice and

Mechanisms of Lymphocyte Activation and Immune Regulation VI, Edited by Gupta and Cohen
Plenum Press, New York, 1996

humans, *bcl-x*$_L$ is the dominant mRNA form expressed in lymphoid tissues including the bone marrow and the thymus[18,20–21]. Furthermore, recent analysis of chimeric mice with a disrupted mutation of *bcl-x* revealed a major role for *bcl-x* during lymphoid development[22–23]. This chapter will focus on the role of *bcl-2* and *bcl-x*, as modulators of apoptosis in lymphocytes.

2. Bcl-2 AND Bcl-x EXHIBIT REGULATED EXPRESSION DURING LYMPHOID DEVELOPMENT

Previous studies have established that Bcl-2 is highly regulated during B and T cell development[8–9]. The expression of Bcl-x was studied in developing lymphoid cells by multi-parameter flow cytometric analysis. The Bcl-x protein was expressed at very low levels in B220^{dull}CD43$^+$ pro-B cells, upregulated in pre-B cells, and downregulated again in immature and mature B cells[24]. We confirmed these observations and established that the long form of Bcl-x (Bcl-x$_L$) was the *bcl-x* product expressed in developing B cells by Western blot analysis[24]. Thus, expression of Bcl-x$_L$ is predominantly restricted to pre-B cells during B-cell development. The expression of Bcl-x$_L$ was also highly regulated in developing thymocytes. The expression of Bcl-x$_L$ was low in CD4$^-$CD8$^-$, increased in CD4$^+$CD8$^+$ immature thymocytes, and downregulated in more mature CD4$^+$CD8$^-$ and CD4$^-$CD8$^+$ thymocytes[23,25]. Furthermore, resting mature T cells expressed undetectable levels of Bcl-x$_L$[23,25]. The expression patterns of Bcl-2 and Bcl-x$_L$ in developing lymphoid cells are summarized in Fig. 1. Thus, the pattern of Bcl-x$_L$ expression in developing and mature B and T cells is strikingly different from that of Bcl-2. These results strongly argue that Bcl-2 and Bcl-x$_L$ play different functional roles during lymphoid selection and peripheral homeostasis. Bcl-2 but not Bcl-x$_L$ is highly expressed in very early B and T-cell precursors that undergo extensive rearrangement of Ig and TCR genes. Hence, Bcl-2 appears to play a role in the initial stage of lymphoid development when a diverse pool of pre-B cells and CD4$^+$CD8$^+$ is generated through Ig and TCR gene recombination. Later in development, Bcl-x$_L$ but not Bcl-2 is expressed in pre-B cells and CD4$^+$CD8$^+$ thymocytes. Most of these lymphoid precursors are thought to die during selection. Bcl-x$_L$ could play a role in the maintenance of pre-B cells and CD4$^+$CD8$^+$ thymocytes prior to positive selection. Because Bcl-x$_L$ is downregulated in more mature IgM$^+$IgD$^-$ B cells and single positive thymocytes, these studies suggest that pre-B cells and CD4$^+$CD8$^+$ thymocytes that are not positively selected, downregulate Bcl-x$_L$ and fail to induce Bcl-2. In contrast, the subpopulation of pre-B cells and CD4$^+$CD8$^+$ thymocytes that undergo positive selection, upregulate Bcl-2 during positive selection and migrate to the periphery as mature lymphocytes[12]. Recent studies with Bcl-x-deficient chimeric mice have shown a defect in the survival of pre-B cells and CD4$^+$CD8$^+$ thymocytes arguing that Bcl-x$_L$ plays a critical role in the maintenance of these early population of lymphoid cells[22–23]. Deletion of self-reactive thymocytes and immature B cells occurs at specific stages of B and T-cell development. In developing B cells, downregulation of Bcl-2 and Bcl-x$_L$ at the immature B cell stage coincides with the increased sensitivity of these cells to IgM-induced apoptosis and may facilitate negative selection. Similarly, CD4$^+$CD8$^+$ thymocytes are very sensitive to TCR-mediated apoptosis and their demise during negative selection may be facilitated by the absence of Bcl-2. Because CD4$^+$CD8$^+$ thymocytes express Bcl-x$_L$, it is possible that Bcl-x$_L$ is functionally inactivated in these cells by interacting proteins such as Bad that can form heterodimers with Bcl-x$_L$[26]. Expression of a *bcl-x*$_L$ transgene in CD4$^+$CD8$^+$ thymocytes confers protection against multiple apoptotic stimuli suggesting that the ratio of Bcl-x$_L$ and interacting proteins is critical in determining the susceptibility to apoptosis in CD4$^+$CD8$^+$ thymocytes.

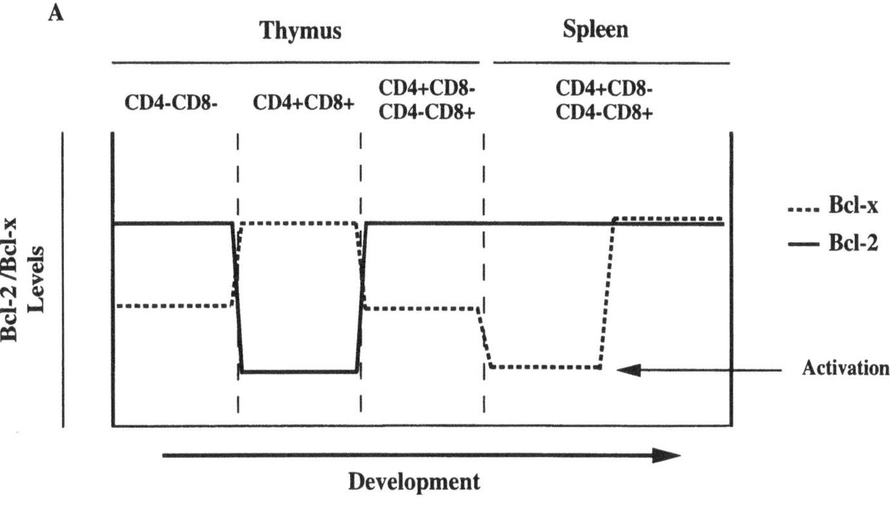

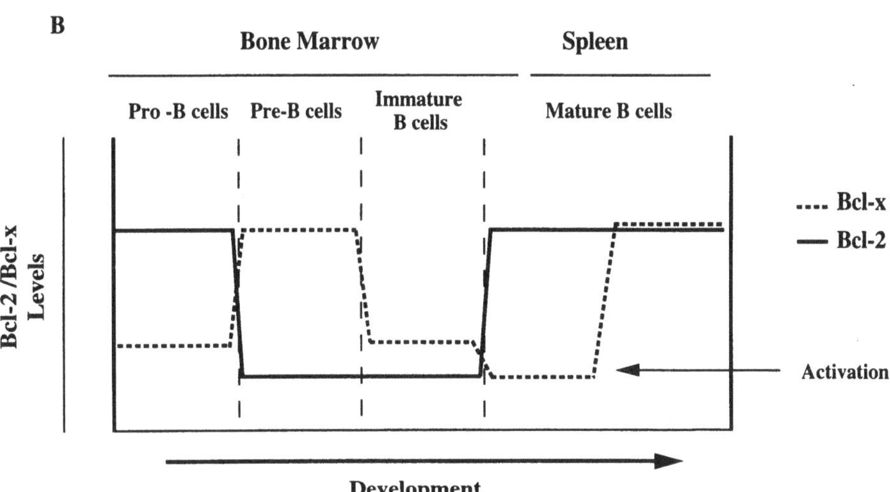

Figure 1. Regulation of Bcl-x and Bcl-2 expression during T and B cell development. Expression analysis was performed by multi-parameter flow cytometry.[(24–25)]

3. DIFFERENTIAL REGULATION OF Bcl-2 AND Bcl-x DURING LYMPHOID ACTIVATION

A major role for Bcl-2 in the maintenance of resting B and T cells in peripheral tissues has been revealed in mutant mice deficient in Bcl-2. Mature lymphocytes in Bcl-2-deficient mice showed accelerated apoptosis in vivo and vitro[16–17]. Significantly, stimulation of Bcl-2-deficient lymphocytes with anti-CD3 inhibited the accelerated apoptosis in vitro, suggesting that activation of mature lymphocytes induces Bcl-2-independent anti-apoptotic signals[27]. We have assessed whether Bcl-x_L is regulated in mature B cells after activation signals. Bcl-x_L was induced by 6 hours and reached maximum expression at 24 hours after surface IgM crosslinking or CD40 stimulation[24]. In contrast, activation of resting B cells failed to increase Bcl-2 expression[24]. These results indicate that activated B cells co-express both Bcl-2 and Bcl-x_L. Similarly, activation of peripheral T cells via CD3 or CD3-CD28 induces Bcl-x_L in human and mice[28–30]. These results argue that a major role of Bcl-x_L in peripheral lymphoid cells is to increase their survival during activation. Under physiological situations, induction of Bcl-x_L may serve as a mechanism to counter more effectively death signals associated with activation and proliferation of mature lymphoid cells. In contrast, Bcl-2 appears to provide a basal level of survival to resting lymphoid cells.

4. TRANSGENIC MICE EXPRESSING CONSTITUTIVELY Bcl-x_L EXHIBIT ACCUMULATION OF LYMPHOID CELLS

Transgenic mice were generated to assess the effect of constitutive Bcl-x_L expression in the animal. Two mouse lines that overexpress Bcl-x_L in the T-cell compartment and one line that expressed Bcl-x_L in developing and mature B cells were studied in detail[24–25]. In transgenic animals in which Bcl-x_L was targeted to T cells, there was an increased in the number of $CD4^+CD8^+$, $CD4^+CD8^-$, and $CD4^-CD8^+$ thymocytes when compared to control littermates[25]. In addition, the size of the peripheral T-cell pool was increased by deregulated Bcl-x_L[25]. This was particularly striking in lymph nodes in which T cells were increased 5.5-fold[25]. In transgenic animals in which Bcl-x_L was targeted to the B-cell compartment, there was an accumulation of pro-B, pre-B, immature and mature B cells in the bone marrow as well as mature B cells in peripheral lymphoid organs[24]. The accumulation pattern of B cell populations was similar in *bcl-x_L* and *bcl-2* transgenic animals[24]. Because activated B cells co-express Bcl-2 and Bcl-x_L, transgenic mice expressing Bcl-2 and Bcl-x_L were generated to assess a functional interaction between these two proteins. We found that both *bcl-x_L* and *bcl-2* transgenes confered a significant increase in the number of absolute mature B cells as compared to single transgenic animals in vivo[24]. Similarly, B cells from double transgenic animals exhibited increased survival in vitro when compared to single *bcl-2* or *bcl-x_L* transgenic animals[24]. In addition, mature B cells co-expressing *bcl-2* and *bcl-x_L* transgenes, but not *bcl-2* or *bcl-x_L* alone, were significantly protected from apoptosis induced by surface IgD crosslinking in vivo in the absence of T cell costimulatory function[24]. Interestingly, B cells from double *bcl-2/bcl-2* transgenic animals exhibited in vitro survival greater than heterozygous *bcl-2* mice and similar to that of *bcl-2/bcl-x_L* transgenic animals (Fig. 2). These results suggest that increased B-cell survival observed in *bcl-2/bcl-x_L*

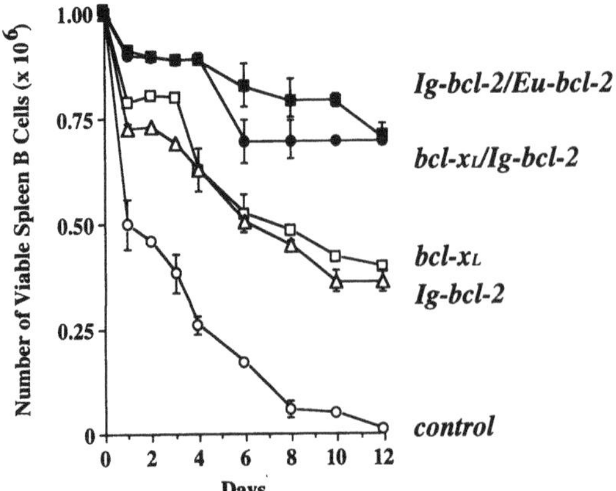

Figure 2. *bcl-x_L* and *bcl-2* promote B cell survival. Analysis was performed using purified splenic B cells from *Ig-bcl-2*[13] and *bcl-x_L*[24] transgenic animals. Double *Ig-bcl-2/Eu-bcl-2* transgenic animals were generated by mating *Eu-bcl-2*[14] and *Ig-bcl-2*[13] mice. Cell survival was assessed by trypan blue exclusion.[24]

transgenic mice is mediated by a gene dose effect and imply that protection against apoptosis is determined in part by the relative amounts of Bcl-2 family members.

5. Bcl-x_L INHIBITS LYMPHOID APOPTOSIS INDUCED BY MULTIPLE STIMULI BUT FAILS TO BLOCK NEGATIVE SELECTION

The survival of thymocytes from *bcl-x_L* transgenic and control littermates was studied after stimuli that induce apoptosis. Thymocytes expressing Bcl-x_L exhibited increased viability in vitro and were resistant to apoptosis induced by glucocorticoids, irradiation, calcium ionophore and CD3 crosslinking in vitro or in vivo[23, 25]. Similarly, mature B cells from *bcl-x_L* transgenic mice exhibited diminished spontaneous apoptosis in vitro and pre-B cells expressing Bcl-x_L were protected from dexamethasone-induced apoptosis in vivo[24]. A critical form of apoptosis is the deletion of thymocytes that express high-affinity TCRs for self-antigens[2]. The ability of Bcl-x_L to inhibit TCR-mediated apoptosis of thymocytes in vivo was assessed in two different models of negative selection. In one model, the effect of Bcl-x_L was determined during clonal deletion induced by the superantigen Mls-1a or endogenous superantigen in the context of the MHC class II molecule I-E. Flow cytometric analysis of thymocytes from appropriate mice revealed that thymocytes reactive to endogenous superantigens were deleted regardless of the expression of the *bcl-x_L* transgene[25]. In the second model, we assessed the ability of *bcl-x_L* to inhibit negative selection of thymocytes specific for the H-Y male antigen in the context of H-2 D^b MHC class I molecules[31]. In the H-Y transgenic system, *bcl-x_L* failed to inhibit clonal deletion induced by H-Y antigen[25]. The inability of *bcl-2* and *bcl-x_L* to inhibit clonal deletion is puzzling in that apoptosis of $CD4^+CD8^+$ thymocytes mediated by CD3 stimulation, a signal that mimics negative selection, is effectively inhibited by both *bcl-2* and *bcl-x_L*[13–14, 25]. It is possible that clonal deletion of thymocytes engages signals such as those generated via CD4-CD8, CD28, CD30 or other surface molecules, in addition to the TCR/CD3 complex. In this scenario, negative selection may involved powerful death signals generated through high avidity self-peptide/TCR interactions or death effectors that are not inhibitable by Bcl-2 or Bcl-x_L. Consistent with this hypothesis, there is experimental evidence that the ability of *bcl-2* and *bcl-x_L* to inhibit apoptosis in B cells is dependent on the avidity of the interaction between the antigen receptor and its ligand (see below).

6. THE AVIDITY OF THE INTERACTION BETWEEN THE SURFACE IGM RECEPTOR AND ITS LIGAND INFLUENCES THE ABILITY OF Bcl-x_L AND Bcl-2 TO INHIBIT ANTI-IGM-INDUCED B CELL APOPTOSIS

The avidity of the antigen receptor with its ligand is critical in determining a biological response in lymphoid cells. For example, high affinity interactions between TCRs and self-antigens mediate negative selection whereas low affinity interactions appear to be involved in positive selection of developing thymocytes[2]. We have assessed the ability of Bcl-2 and Bcl-x_L to inhibit B-cell apoptosis induced by anti-IgM monoclonal antibodies (as ligands) that exhibit different avidity for the IgM receptor. Crosslinking of IgM receptors by anti-IgM induces cell cycle arrest and apoptosis in WEHI-231 lymphoma cells, a model of B-cell tolerance[32]. Incubation of WEHI-231 B cells with varying doses of high (Ak-11, Ak-15 and b7–6) and low avidity (Bet-2 and 331.1.2) anti-IgM monoclonal antibodies

(Mabs) induced apoptosis with similar kinetics[33]. Transfectants that express constitutively Bcl-x_L or Bcl-2 were effectively protected from apoptosis induced by low avidity anti-IgM Mabs. In contrast, Bcl-xL and Bcl-2 were very ineffective in inhibiting apoptosis mediated by high avidity anti-IgM Mabs[33]. The differential response elicited by different anti-IgM Mabs appeared to be due to the avidity of Mabs for surface IgM in that it was independent of the Ig isotype or Cμ-domain specificity of the Mabs[33]. These results have several biological implications. First, they argue that high avidity interactions between antigen receptors and their ligands induce a death signal that is quantitatively (and/or qualitatively) different from that elicited by low avidity interactions. This differential response of high versus low avidity interactions can be revealed and dissociated by two apoptosis inhibitors, Bcl-2 and Bcl-x_L. Under physiological situations, the powerful apoptotic signals triggered by high avidity interactions would ensure the demise of developing cells expressing antigen receptors for self-antigens. Second, several studies have shown that *bcl-2* or *bcl-x_L* are unable to inhibit negative selection of thymocytes in vivo[13–14, 25]. The results obtained in the WEHI-231 system provides a possible explanation for the latter results in that negative selection of thymocytes involves high avidity interactions between self-peptides and TCRs. Third, disregulated expression of Bcl-2, Bcl-x_L or other anti-apoptotic proteins may promote the survival of autoreactive clones, particularly those with antigen receptors that exhibit low or medium avidity for self-antigen.

7. CELL CYCLE ARREST AND ANTI-IGM-INDUCED APOPTOSIS CAN BE DISSOCIATED BY Bcl-x_L

Signaling through IgM in WEHI-231 induces both arrest at the G_0/G_1 phase of the cell cycle which is followed by apoptosis[32–33]. Significantly, WEHI-231 cells over-expressing Bcl-x_L were largely protected from apoptosis induced with either AK-15 or Bet-2, two low avidity anti-IgM Mabs but remained G_0/G_1 arrested[33]. Importantly, incubation of WEHI-231 with recombinant CD40L protected WEHI-231 B cells from apoptosis and overcome cell cycle arrested induced by anti-IgM[33]. Western blot analysis of lysates from WEHI-231 cells incubated with CD40L revealed no induction of Bcl-x_L[33]. However, other studies have shown that Bcl-x_L can be induced by CD40 ligation in

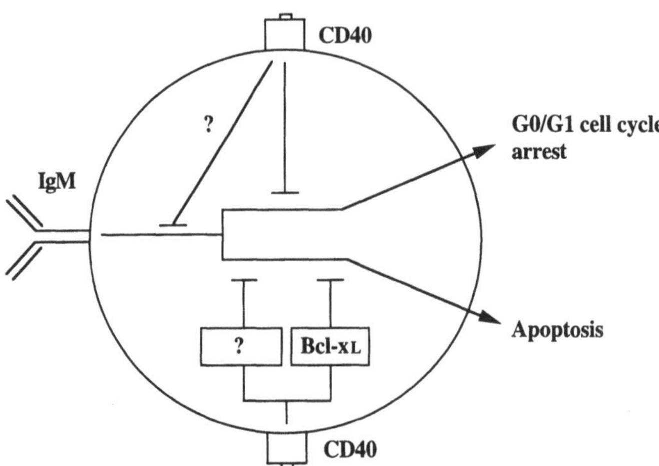

Figure 3. Model for the function of CD40 during anti-IgM-induced cell cycle arrest and apoptosis. See text for further details.

certain sublines of WEHI-231 lymphoma cells[34–36]. Consistent with these results is the observation that activation of resting splenic B cells through CD40 upregulates Bcl-x_L[24]. Together these results indicate that CD40 transduces survival signals to B cells one of which is Bcl-x_L. In addition, our results together with those by other investigations[36] indicate that the ability of CD40 to inhibit apoptosis in WEHI-231 lymphoma B cells is mediated at least in part by a Bcl-x_L-independent mechanism. Furthermore, cell cycle arrest and apoptosis appear to be mediated by independent pathways that can be dissociated by Bcl-2 and Bcl-x_L. The observation that CD40 stimulation inhibits both cell cycle arrest and apoptosis induced by anti-IgM suggests a model (Fig. 3), whereby CD40 transduces different signals that can repress separately apoptosis (such as Bcl-x_L) and cell cycle arrest. An alternative non-exclusive possibility is that CD40 can act at a common site that is upstream in the IgM signaling pathway (Fig. 3).

REFERENCES

1. R.E. Ellis, J.Y. Yuang, and H.R. Horvitz, Mechanisms and functions of cell death. *Annu. Rev. Cell. Biol.* 7:633 (1991).
2. G.J.V. Nossal, Negative selection of lymphocytes, *Cell* 76:229 (1994).
3. D.R. Green and D.W. Scott, Activation-induced apoptosis in lymphocytes. *Curr. Opin. Immunol.* 6:476 (1994).
4. H. Groux, D. Monte, B. Plouvier, A. Capron, and J.C. Ameisen, CD3-mediated apoptosis of human medullary thymocytes and activated peripheral T cells: respective roles of interleukin-1, interleukin-2, interferon-γ and accessory cells, *Eur. J. Immunol.* 23:1623 (1993).
5. J. Hasbold, C. Johnson-Léger, C.J. Atkins, E.A. Clark, and G.G.B. Klaus, Properties of mouse CD40: cellular distribution of CD40 and B cell activation by monoclonal anti-mouse CD40 antibodies, *Eur. J. Immunol.* 24:1835 (1994).
6. G.T. Williams and C.A. Smith, Molecular regulation of apoptosis: genetic controls on cell death, *Cell* 74:777 (1993).
7. G. Nuñez and M.F. Clarke, The Bcl-2 family of proteins: regulators of cell death and survival, *Trends. Cell. Biol.* 4:399 (1994).
8. J. Gratiot-Deans, L. Ding, L. A. Turka, and G. Nuñez, bcl-2 proto-oncogene expression during human T cell developmen. Evidence for biphasic regulation. *J. Immunol.* 151:83 (1993).
9. R. Merino, L. Ding, D.J. Veis, S.J. Korsmeyer, and G. Nuñez, Developmental regulation of the Bcl-2 protein and susceptibility to cell death in B lymphocytes, *EMBO J.* 13:683 (1994).
10. A. Strasser, A.W. Harris, H. von Boehmer, and S. Cory, Positive and negative selection of T cells in T-cell receptor transgenic mice expressing a *bcl-2* transgene, *Proc. Natl. Acad. Sci. USA* 91:1376 (1994).
11. G.P. Linette, M.J. Grusby, S.H. Hedrick, T.H. Hansen, L.H. Glimcher, and S.J. Korsmeyer, Bcl-2 is upregulated at the $CD4^+CD8^+$ stage during positive selection and promotes thymocyte differentiation at several control points, *Immunity* 1:197 (1994).
12. J. Gratiot-Deans, R. Merino, G. Nuñez, and L.A. Turka, Bcl-2 expression during T-cell development: Early loss and late return occur at specific stages of commitment to differentiation and survival, *Proc. Natl. Acad. Sci.*91:10685 (1994).
13. C.L. Sentman, J.R. Shutter, D. Hockenbery, O. Kanagawa, and S.J. Korsmeyer, bcl-2 inhibits multiple forms of apoptosis but not negative selection in thymocytes, *Cell* 67:87 (1991).
14. A. Strasser, A., A.W. Harris, and S. Cory, bcl-2 transgene inhibits T cell death and perturbs thymic self-censorship, *Cell* 67:889 (1991).
15. R.M. Siegel, M. Katsumata, T. Miyashita, D.C. Louie, M.I. Greene, and J.C. Reed, Inhibition of thymocyte apoptosis and negative selection in transgenic mice, *Proc. Natl. Acad. Sci. USA* 89:7003 (1992).
16. D.J. Veis, C.M. Sorenson, J.R. Shutter, and S.J. Korsmeyer, Bcl-2-deficient mice demonstrate fulminant lymphoid apoptosis, polycystic kidneys and hypopigmented hair, *Cell* 75:229 (1993).
17. K. Nakayama, K. Nakayama, I. Negishi, K. Kuida, Y. Shinkai, M.C. Louie, L.E. Fields, P.J. Lucas, V. Stewart, F.W. Alt, and D.Y. Loh, Disappearance of the lymphoid system in Bcl-2 homozygous mutant chimeric mice, *Science* 261:1584 (1993).

18. L.H. Boise, M. Gonzalez-Garcia, C.E. Postema, L. Ding, T. Linsten, L.A. Turka, X. Mao, G. Nuñez, and C.B. Thompson, bcl-x, a bcl-2-related gene that functions as a dominant regulator of apoptotic cell death, *Cell* 74:597 (1993).
19. A.R. Gottschalk, L.H. Boise, C.B. Thompson, and J. Quintans, Identification of immunosuppressant-induced apoptosis in a murine B-cell line and its prevention by bcl-x but not bcl-2. *Proc. Natl. Acad. Sci. USA* 92:7350 (1994).
20. M. Gonzalez-Garcia, R. Perez-Ballestero, L. Ding, L. Duan., L.H. Boise, C.B. Thompson, and G. Nuñez, Bcl-x_L is the major Bcl-x mRNA form expressed during murine development and its product localizes to the outer mitochondria, *Development* 120:3033 (1994).
21. W. Fang, J.J. Rivard, D.L. Mueller, and T.W. Behrens, Cloning and molecular characterization of mouse bcl-x in B and T lymphocytes, *J. Immunol.* 153:4388 (1994).
22. N. Motoyama, F. Wang, K.A. Roth, H. Sawa, K.-i. Nakayama, K. Nakayama, I. Negishi, S. Senju, Q. Zhang, S. Fujii, and D.Y. Loh, Massive cell death of immature hematopoietic cells and neurons in Bcl-x-deficient mice, *Science* 267:1506 (1995).
23. A. Ma, J.C. Pena, B. Chang, E. Margosian, L. Davidson, F.W. Alt, and C.B. Thompson., Bclx regulates the survival of double-positive thymocytes, *Proc. Natl. Acad. Sci. USA* 92:4763 (1995).
24. D.A.M. Grillot, R. Merino, J.C. Pena, W.C. Fanslow, F.D. Finkelman, C.B. Thompson, and G. Núñez, *bcl-x* Exhibits regulated expression during B-cell development and activation, and modulates lymphocyte survival in transgenic mice, *J. Exp. Med.* 183:1852 (1996).
25. D.A.M. Grillot, R. Merino and G. Nuñez, *bcl*-x_L displays a restricted expression during thymic development and inhibits multiple forms of apoptosis but not clonal deletion, *J. Exp. Med.* 182:1973 (1995).
26. E. Yang, J. Zha, J. Jockel, L.H. Boise, C.B. Thompson, and S.J. Korsmeyer, Bad, a heterodimeric partner for Bcl-x_L and Bcl-2, displaces Bax and promotes cell death, *Cell* 80:285 (1995).
27. K. Nakayama, L.B. Dustin, and D.Y. Loh, T-B cell interaction inhibits spontaneous apoptosis of mature lymphocytes in Bcl-2-deficient mice, *J. Exp. Med.* 182:1101 (1995).
28. L.H. Boise, A.J. Minn, P.J. Noel, C.H. June, M.A. Accavitti, T. Lindsten, and C.B. Thompson, CD28 costimulation can promote T cell survival by enhancing the expression of Bcl-X_L, *Immunity,* 3:87 (1995).
29. L.G. Radvanyi, Y. Shi, H. Vaziri, A. Sharma, R. Dhala, G.B. Mills, and R.G. Miller, CD28 costimulation inhibits TCR-induced apoptosis during a primary T cell response, *J. Immunol.* 156:1788 (1996).
30. D.L. Mueller, S. Seiffert, W. Fang and T.W. Behrens, Differential regulation of bcl-2 and bcl-x by CD3, CD28, and the IL-2 receptorin cloned $CD4^+$ helper T cells: a model for the long term survival of memory cells. *J. Immunol.* 156:1764 (1996).
31. P. Kisielow, H. Bluthmann, U.D. Staerz, M. Steinmetz, and H. von Boehmer, Tolerance in T-cell receptor transgenic mice involves deletion of non mature $CD4^+CD8^+$ thymocytes, *Nature* 333:742 (1988).
32. J. Hasbold and G.G.B. Klaus, Anti-immunoglobulin antibodies induce apoptosis in immature B cell lymphomas. *Eur. J. Immunol.* 20:1685 (1990).
33. R. Merino, D.A. M. Grillot, P.L. Simonian, S. Muthukkumar, W.C. Fanslow, S. Bondada, and G. Nuñez, Modulation of anti-gM-induced apoptosis by Bcl-x_L and CD40 in WEHI-231 cells. Dissociation from cell cycle arrest and dependence on the avidity of the antibody-IgM receptor interaction. J. Immunol. 155:3830 (1995).
34. M.S.K. Choi, L.H. Boise, A.R. Gottschalk, J. Quintans, C.B. Thompson, and G.G.B. Klaus, The role of bcl-X_L in CD40-mediated rescue from anti-μ-induced apoptosis in WEHI-231 B lymphoma cells, *Eur. J. Immunol.* 25:1352 (1995).
35. T. Ishida, N. Kobayashi, T. Tojo, S. Shida , T. Yamamoto, and J.I. Inoue, CD40 signaling-mediated induction of Bcl-X_L, Cdk4, and Cdk6. Implication of their cooperation in selective B cell growth. *J Immunol.* 155:5527 (1995).
36. Z. Wang, J.G. Karras, R.G. Howard, and T.L. Rothstein, Induction of *bcl-x* by CD40 engagement rescues sIg-induced apoptosis in murine B cells, *J. Immunol.* 155: 3722 (1995).

9

THE EPSTEIN–BARR VIRUS GENE BHRF1, A HOMOLOGUE OF THE CELLULAR ONCOGENE Bcl-2, INHIBITS APOPTOSIS INDUCED BY GAMMA RADIATION AND CHEMOTHERAPEUTIC DRUGS

N. J. McCarthy,[1,2*] S. A. Hazlewood,[3,1] D. S. Huen,[3] A. B. Rickinson,[3] and G. T. Williams[1,2†]

[1] Department of Biological Sciences
Keele University
Keele, Staffordshire, ST5 5BG, United Kingdom
[2] Department of Anatomy
[3] CRC Laboratories, Department of Cancer Studies
University of Birmingham Medical School
Vincent Drive, Birmingham. B15 2TT, United Kingdom

1. ABSTRACT

Analysis of apoptosis, active and controllable cell death, has demonstrated that the size of a cell population can be regulated by changes in the cell death rate as well as in the rates of proliferation and differentiation. Factors which alter the rate of cell death, such as expression of the proto-oncogene *bcl-2*, can therefore directly affect the number of cells within a population. Bcl-2 has been shown to suppress apoptosis in response to a variety of stimuli and to act as a complementary survival signal for the random acquisition of other oncogenic mutations, such as deregulated *c-myc*.

The Epstein Barr virus (EBV) gene BHRF1 was the first of a family of *bcl-2* homologues now being identified. BHRF1 and *bcl-2* share 25% primary amino acid sequence homology. Here we show that γ radiation and several cytotoxic anticancer agents induce apoptosis in Burkitt's lymphoma (BL) cell lines, as has been found in several other systems. Using gene transfection studies we have also shown that expression of either BHRF1 or *bcl-2* in BL cell lines significantly suppresses apoptosis in response

* Present address, ICRF Laboratories, P.O.Box 123, Lincoln's Inn Fields, London WC2A 3PX.
† Corresponding author, telephone 44 1782 583032, fax 44 1782 630007.

to a variety of anticancer treatments. This has confirmed that BHRF1 is *functionally* homologous to *bcl-2* in B-cells and suggests that BHRF1 may act to prevent apoptosis during EBV infection, maximising virus particle production, as has been suggested for other human and insect viral genes. Suppression of chemotherapeutic drug induced cell death by *bcl-2* and BHRF1, as demonstrated in this cell system, results in resistance to a variety of different agents and may represent an alternative mechanism by which multidrug resistance arises during chemotherapy.

2. INTRODUCTION

Work in several diverse areas has recently established a place for apoptosis alongside proliferation and differentiation as a primary option available to the cells of higher organisms (for review see references 1–3). Apoptosis is a process of cellular self-destruction which is often controlled by specific stimuli, both inducers and suppressors, and is in many cases associated with the expression of particular genes (reviewed in references 3 and 4).

Modulation of apoptosis is of fundamental significance as a physiological control mechanism, particularly in the immune system (reviewed in references 5 and 6). During development in the thymus, T cells recognising self antigens are eliminated by apoptosis[7,8].On the other hand, more mature B cells in lymph node germinal centres which fail to interact strongly enough with foreign antigen are eliminated by apoptosis as part of the mechanism of affinity maturation in the antibody response[9,10].

Haemopoietic[11] and erythroid[12] progenitor cells undergo apoptosis when the required growth factors are removed. Investigation of apoptosis in myeloid progenitors under these conditions allowed the demonstration that the cellular oncogene *bcl-2* could act to suppress apoptosis[11,13,14]. The *bcl-2* gene was originally isolated from the breakpoint of the t(14;18) chromosomal translocation strongly associated with follicular lymphoma and has subsequently been shown to suppress apoptosis in several other cell types (reviewed by McCarthy *et al*[16]).

On the basis of limited sequence homology to *bcl-2*[17] we have suggested that the EBV gene BHRF1 may also act to suppress apoptosis[18]. The sub-cellular localisation of BHRF1 protein has also been reported to be similar to that of Bcl-2[19,20]. EBV is a typical gammaherpesvirus which infects around 95% of all human populations and is generally acquired asymptomatically during early childhood. The virus is thought to be retained throughout the host's life due to the persistent latent infection of B-lymphocytes[21,22]. EBV is associated with at least two human cancers, endemic Burkitt's lymphoma (BL), found in areas of the African malaria belt, and endemic nasopharyngeal carcinoma, found in southern China. In non-endemic areas, EBV-negative Burkitt's lymphoma predominates.

EBV is able to transform B-lymphocytes of humans and other primates producing immortalised B-lymphocytes known as Lymphoblastoid Cell Lines (LCLs). Once established within the host's cells, EBV latent genes are expressed. Biopsy cells from EBV-associated BL only express one latent gene, EBNA 1, whereas cells subsequently cultured *in vitro* can progress to express all eight latent genes i.e., EBNAs 1, 2, 3a, 3b, 3c, LP and LMPs 1 and 2 . Cell lines derived from such cells are known as group I and group III, respectively, and group III cell lines are resistant to induction of apoptosis[23] probably by virtue of EBV-induced expression of host cell *bcl-2*[24]

Several viruses infecting mammalian[25,26] and insect[27,28] cells contain genes which can suppress apoptosis. It is therefore possible that BHRF1, which is normally expressed early in the EBV lytic cycle[29,30], might act to suppress host cell apoptosis to promote the continued production of viral particles.

As well as playing a part in normal cell regulation and in viral infection, inappropriate cell survival caused by expression of genes which can suppress apoptosis might be expected to contribute to oncogenesis (reviewed in reference 18). It is likely that suppression of apoptosis by inappropriately expressed *bcl-2*, for example, is crucial to the development of follicular lymphoma[31,32]. Similarly, viral genes which can suppress apoptosis, such as v-*abl* and EBV BHRF1, could also contribute to the development of the cancers with which these viruses are associated. In this context it is particularly interesting to note that BHRF1 expression has been reported in EBV-associated B-cell lymphoma[33].

In addition to its importance in cancer development, control of apoptosis must also be considered in relation to the elimination of cancer cells during therapy. Although overwhelming cell damage leads to death by necrosis, limited treatment with ionising radiation[34] or cancer chemotherapeutic drugs such as cisplatin[35,36], induces self destruction through apoptosis (reviewed in references 18 1nd 37). This implies that genetic suppression of apoptosis by cellular or viral genes could contribute to cancer cell resistance to radiation and chemotherapy. This hypothesis is supported by the work of Tsujimoto[15], who reported that the transfection of *bcl-2* into B-cell lines conferred "stress resistance" including a reduced sensitivity to irradiation and methotrexate. Work on other cell types now makes it likely that such effects result from suppression of apoptosis[38,39]. Here we report that low levels of γ radiation and several cytotoxic drugs used in cancer therapy induce apoptosis in human B-cell lines and that transfection of BHRF1, as for its cellular homologue *bcl-2*, significantly inhibits this induction of apoptosis.

3. RESULTS

3.1. Induction of Apoptosis by Chemotherapeutic Drugs and γ-Radiation

To investigate whether the BL cell lines underwent apoptosis on treatment with cytotoxic drugs, morphological analysis was carried out both by fluorescence and electron microscopy. Morphological changes characteristic of cell death by apoptosis were seen in treated cells with both techniques and are shown unequivocally in the electron micrographs.

Untreated control Raji-BL cells and drug treated cells are shown in Figure 1 with the condensed chromatin, a characteristic marker of apoptotic cells, being clearly visible in cells treated with 10μM methotrexate (B), 1μg/ml etoposide (C) and 1μg/ml cytosine arabinoside (araC) (D). Similar apoptotic morphology was observed in drug-treated Akata-BL and Cheptage-BL (Chep-BL) cell lines (data not shown). Cells treated with araC often showed additional degradative changes indicating cell damage which could either be independent of the process of apoptosis, or could represent early onset of secondary necrosis.

A further characteristic of many apoptotic cells is the activation of a calcium dependent endonuclease resulting in fragmentation of the DNA into multiples of 180bp[40,41]. Figure 2 shows the DNA of Raji-BL cells after treatment with etoposide, methotrexate or araC, (lanes 3–5) with fragmentation being clearly visible.

The observation of apoptosis induction in response to ionising radiation which has been documented for several cell lines[34,38,42], was confirmed for these BL cell lines. Raji-BL cells were exposed to 16Gy of γ radiation and morphology was examined 24 hours after treatment. Figure 1E illustrates apoptotic morphology evident at this time point. Electrophoresis of the cellular DNA at 12 hours post treatment (see Figure 3) also showed the classical apoptotic fragmentation of the DNA.

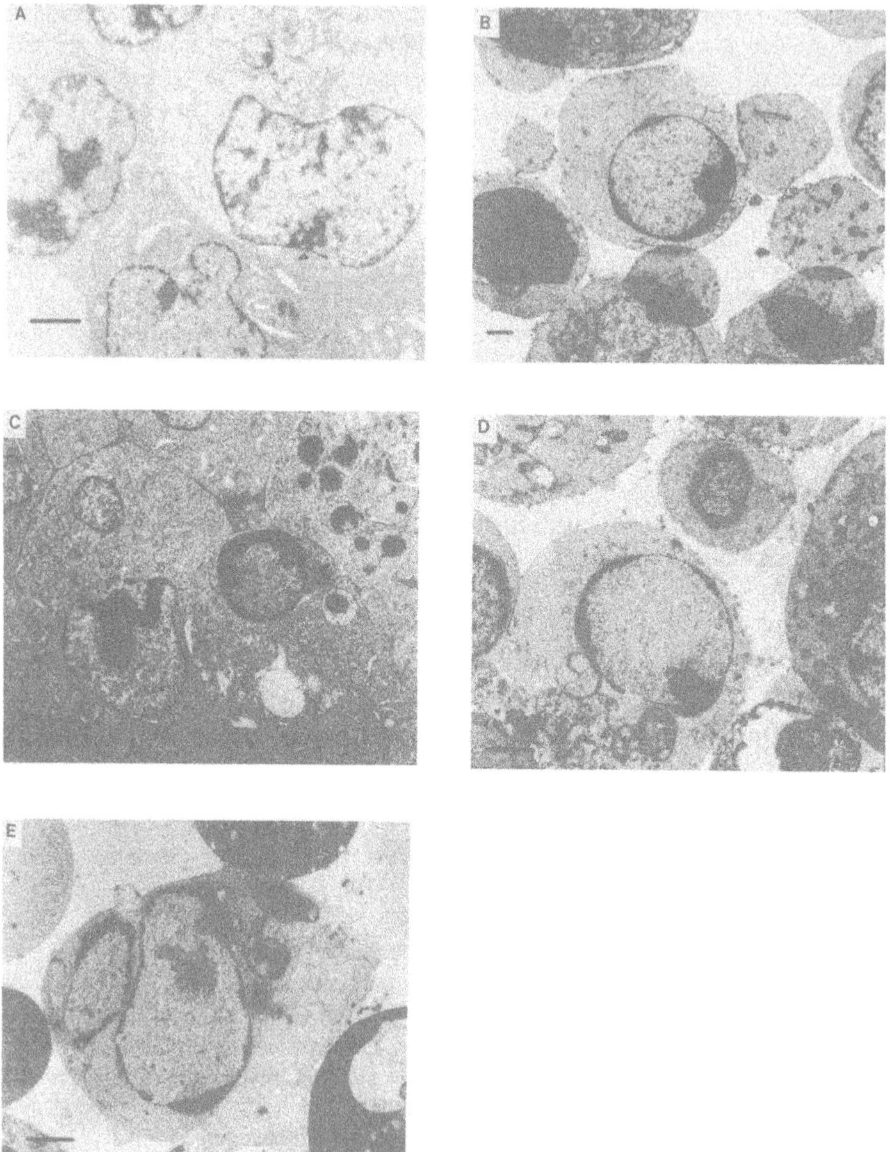

Figure 1. Electron micrographs showing induction of apoptosis in Raji-BL cells treated with cytotoxic drugs and γ radiation. Bar sizes represent 2μm. (A) Untreated control cells, note the regular appearance of the cytoplasm and normal nuclear chromatin state. Drug treated cells were exposed to (B) 10μM methotrexate, (C) 1μg/ml etoposide and (D) 1μg/ml araC, for 14 hours and subsequently grown at 37°C free of drug for 24 hours before morphological analysis. Apoptotic morphology is clearly shown in all drug treated cells by the classical condensation of the nuclear chromatin. (E) Cells treated with 16Gy of γ radiation again showing the margination of chromatin at the nuclear membrane, indicative of apoptosis, 24 hours post irradiation.

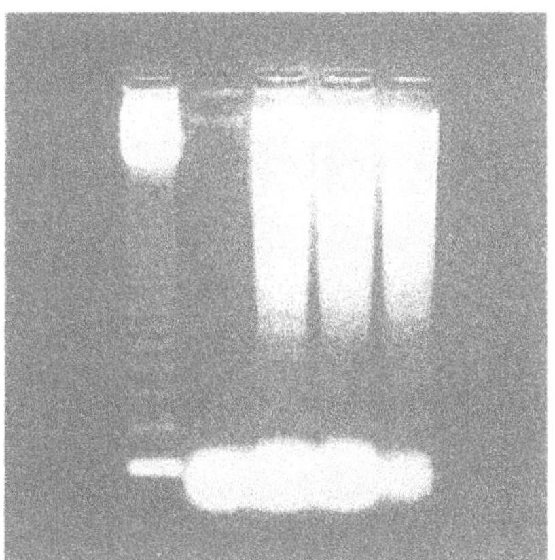

Figure 2. Raji-BL control transfectant cells. 5×10^6 cells were exposed to cytotoxic drugs for 14 hours and then washed free of the drug and cultured in normal growth medium for a further 12 hours before analysis (see materials and methods). Lane 1, 123bp ladder DNA marker (Gibco-BRL); Lane 2, untreated control; Lane 3, 1μg/ml etoposide; Lane 4, 10μM methotrexate; Lane 5, 1μg/ml araC. Note the classical DNA fragmentation evident in cells treated with the cytotoxic drugs (lanes 3–5).

3.2. Expression of Transfected *Bcl-2* Affords Protection against Chemotherapeutic Drugs and γ-Radiation

Having established that treatment of the BL cell lines with these cytotoxic agents results in apoptosis, we first examined the apoptosis-suppressing effect of *bcl-2* expression in these cells. Chep-BL *bcl-2* transfectants were analysed for protection from these cytotoxic stimuli. Figure 4A illustrates the viability of Chep-BL *bcl-2* transfectants and controls after exposure to 8Gy of γ radiation. High levels of suppression of apoptosis are seen for Chep-BL *bcl-2* transfectants when compared to controls at 24 and 48 hours post treatment ($p<0.001$).

Bcl-2 is also able to suppress apoptosis in response to cytotoxic drugs as shown in Figure 4B. Significant inhibition of cell death ($p<0.001$) is seen for Chep-BL *bcl-2* expressing transfectants compared to controls treated with either 1μg/ml etoposide or 5μM methotrexate. Chep-BL *bcl-2* expressors show a smaller inhibition of cell death in response to treatment with araC and the observations are statistically significant at the 10%, but not at the 5% level.

Chep-BL-*bcl-2* transfectants and controls were also incubated for 72 hours in the continuous presence of each drug and analysed every 24 hours to establish a time course for

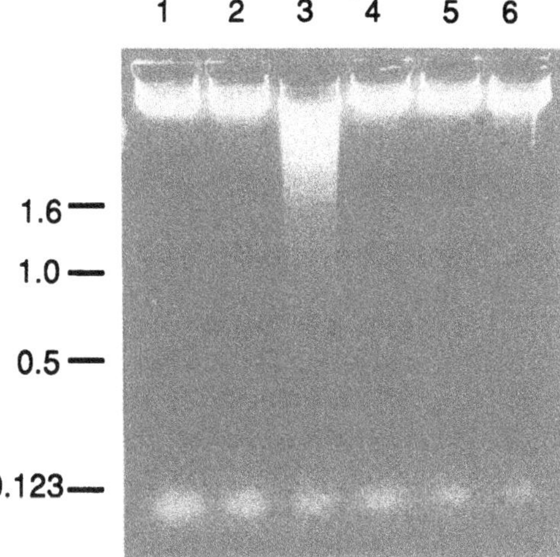

Figure 3. Gel electrophoresis of whole cell DNA from Raji-BL cells exposed to 16Gy of γ radiation. 10^6 cells were harvested 12 hours after treatment and analysed as described in materials and methods. Molecular sizes in kilobase (kb) pairs are indicated. Lane 1, control transfectants, untreated; Lane 2, control transfectants mock-treated on ice; Lane 3, control transfectants exposed to 16Gy γ radiation; Lane 4, BHRF1 transfectants untreated; Lane 5, BHRF1 transfectants mock-treated on ice; Lane 6, BHRF1 transfectants exposed to 16Gy γ radiation. Note the suppression of fragmentation in the BHRF1 transfectants when compared to the controls in response to g radiation (lanes 6 and 3 respectively).

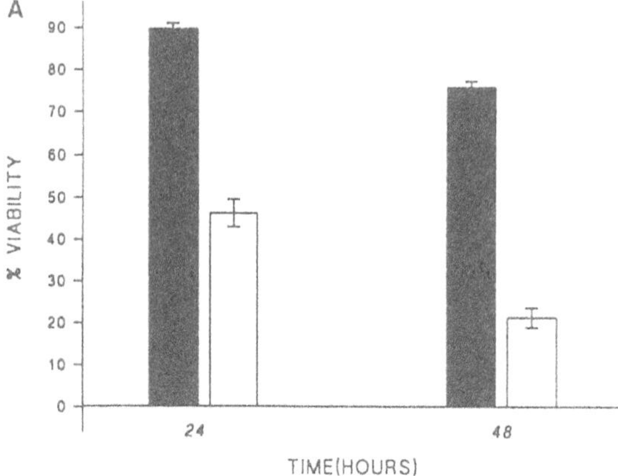

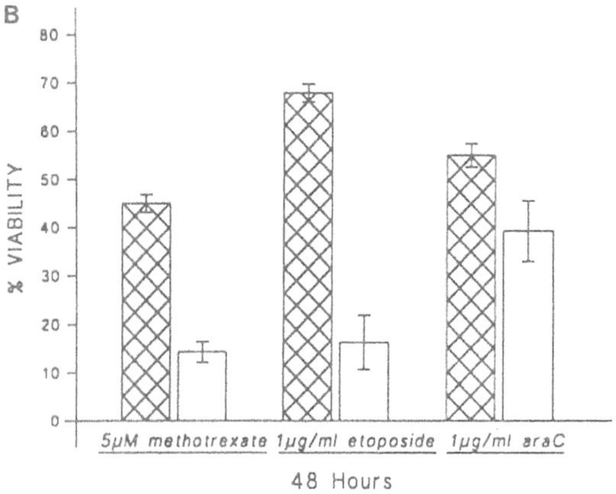

Figure 4. (A) Viability of Chep-BL cells expressing *bcl-2* and controls after exposure to 8Gy of γ radiation. Viability was determined 24 and 48 hours after treatment by Nigrosin exclusion, with the number of viable cells being expressed as a percentage of the total cell number. Significant suppression of apoptosis is again evident for *bcl-2* transfectants (closed bar) when compared to control transfectants (open bar) at 24 and 48 hours ($p<0.001$). Means and S.E. are shown for 3 replicates from a typical example of 3 separate experiments. (B) Viability of chep-BL cells expressing *bcl-2* and controls after 14 hours exposure to methotrexate, etoposide and araC and 48 hours subsequent growth. Viability was determined as above. Significant suppression of cell death is seen in Chep-BL *bcl-2* transfectants (hatched bar) when compared to control transfectants (open bar) in response to 5μM methotrexate ($p<0.001$) and 1μg/ml etoposide ($p<0.001$). A difference between bcl-2-expressing cells and controls is evident in response to 1μg/ml araC, but this is only significant at the 10% level ($p<0.1$). Means and S.E. of 3 or more replicate experiments are shown.

death. Cells expressing Bcl-2 again showed significantly greater viability when compared to controls in the presence of etoposide and methotrexate (data not shown). Two further Chep-BL-*bcl-2* clones were analysed with etoposide to confirm that the suppression of apoptosis was due to the presence of *bcl-2* and not an unknown mutation in any one clone (data not shown).

These results extend the range of apoptosis-inducing stimuli for which *bcl-2* provides a protective effect in Chep-BL cells, protection against apoptosis induced by serum withdrawal having been previously documented for these cell lines[24]. The direct demonstration of inhibition of apoptosis induction also confirms that, as found in other cell types[38], *bcl-2* expression increases cell culture viability after treatment with cytotoxic drugs by inhibiting apoptosis rather than by stimulating proliferation.

3.3. The EBV Gene BHRF1 Can Also Suppress Apoptosis

The EBV gene BHRF1 was the first gene shown to have any significant homology to *bcl-2*[17], suggesting the possibility that it too might possess the unusual and potentially highly significant property of suppressing apoptosis[18]. Raji-BL cells transfected with BHRF1 or relevant plasmid controls were exposed to 16Gy of γ radiation and their viabilities were determined at 24 and 48 hours post treatment. Inhibition of apoptosis is seen for BHRF1-expressing transfectants showing significant inhibition ($p<0.02$) of cell death when compared to controls (see Figure 5A). This suppression of cell death can also be seen by

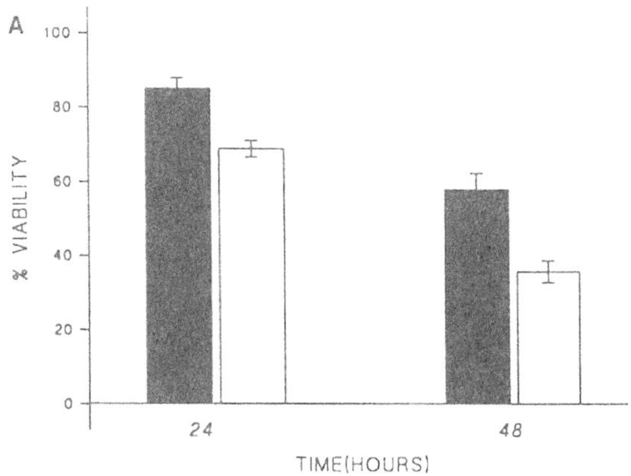

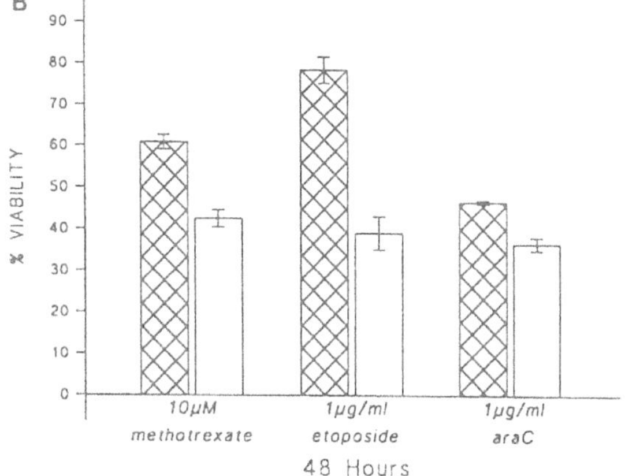

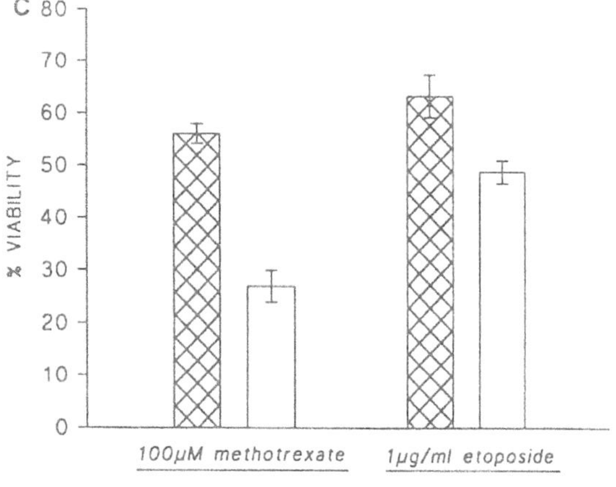

Figure 5. (A) Viability of Raji-BL cells expressing BHRF1 and controls after exposure to 16Gy of γ radiation. Significant suppression of apoptosis is seen for BHRF1 transfectants (closed bar) when compared to control transfectants (open bar) at 24 and 48 hours after treatment ($p<0.02$). Viability was determined as previously described in Figure 4A. Means and S.E. are shown for 3 replicates from a typical example of 3 separate experiments. (B) Viability of Raji-BL cells expressing BHRF1 and controls after 14 hours exposure to methotrexate, etoposide and araC and 48 hours subsequent growth. Significant suppression of apoptosis is seen in BHRF1 transfectants (hatched bar) compared to control transfectants (open bar) in response to 10μM methotrexate, 1μg/ml etoposide and 1μg/ml araC ($p<0.001$ for all 3 drugs). Viability was determined as previously described in Figure 4A. Means and S.E. are shown for 3 or more replicate experiments. (C) Viability of Akata-BL cells expressing BHRF1 and controls after 14 hours exposure to methotrexate and etoposide and 48 hours subsequent growth. Significant suppression of apoptosis is shown for BHRF1 transfectants (hatched bar) when compared to control transfectants (open bar) in response to 100μM methotrexate ($p<0.05$) and 1μg/ml etoposide ($p<0.01$). Viability was determined as described in Figure 4A. Means and S.E of 3 replicate experiments are shown.

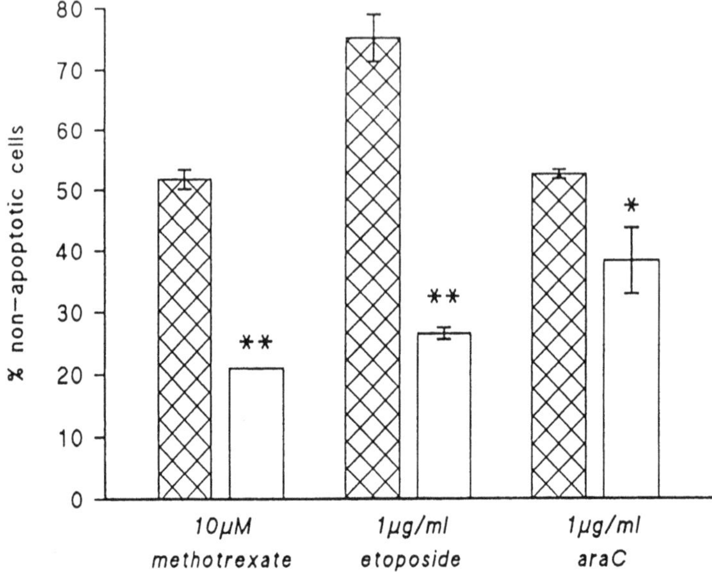

Figure 6. Percentage of morphologically normal (non-apoptotic) Raji-BL transfectants expressing BHRF1 and controls after treatment with cytotoxic drugs as determined by acridine orange staining. Significant suppression of cells with apoptotic morphology is seen for BHRF1 transfectants (hatched bar) when compared to control transfectants (open bar) in response to 10μM methotrexate ($p<0.01$), 1μg/ml etoposide ($p<0.01$) and 1μg/ml araC ($p<0.05$). Means and S.E. of 3 replicate experiments are shown.

monitoring DNA fragmentation. Figure 3 shows a qualitative delay in fragmentation in BHRF1 transfectants compared to controls analysed at 12 hours post irradiation.

On treatment with cytotoxic drugs, BHRF1 expressing Raji-BL cells show significant suppression ($p<0.001$) of apoptosis in response to all three drugs tested, as illustrated in Figure 5B. Again, greatest suppression is seen in response to etoposide, followed by methotrexate, then araC. These results were also verified by directly counting apoptotic cells stained with acridine orange (see Figure 6).

A time course of drug induced cell death was also established in these cells to examine the rates of loss of viability for each drug (see figure 7). Suppression of cell death is evident for BHRF1 transfectants when compared to controls in the presence of 10μm methotrexate ($p<0.001$) (A), 1μg/ml etoposide ($p<0.001$) (B) and 0.5μg/ml ara C ($p<0.05$) at 72 hours.

To ensure that the effect was not confined to Raji-BL cells, nor indeed the result of an undefined mutation within the Raji transfectants, BHRF1-transfected Akata clones were also studied using etoposide and methotrexate (araC could not be used with Akata clones since the parent Akata cells are insensitive to high levels of the drug (>100μg/ml) (data not shown)). Figure 5C shows significant suppression of apoptosis in response to 100μM methotrexate ($p<0.01$) and 1μg/ml etoposide ($p<0.05$), and further indicates that BHRF1 is able to prevent apoptosis in a range of genetic backgrounds.

4. DISCUSSION

It has become widely accepted that consideration of a potential role for apoptosis is essential to any complete analysis both of physiologically regulated cell populations and of

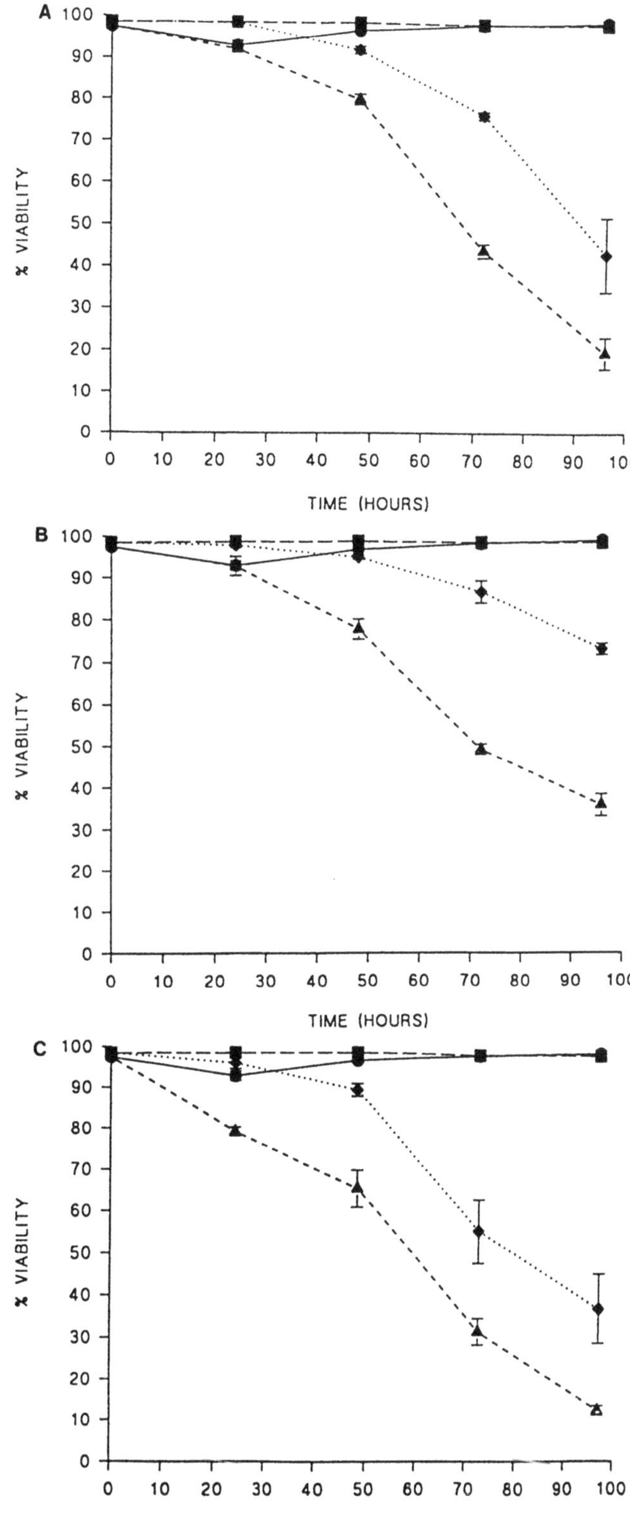

Figure 7. Viability of Raji-BL cells expressing BHRF1 and controls after constant response to methotrexate, etoposide and araC. Significant suppression of apoptosis is seen for BHRF1 transfectants (diamonds) relative to controls (triangles) in response to 1μm methotrexate (A), 1μg/ml etoposide (B), and 0.5μg/ml araC (C). Both BHRF1 transfectants (squares) and controls (circles) remain highly viable in the absence of drugs. Means and standard errors are shown ($n = 3$).

pathologically induced cell death. As a consequence, research has begun to focus on the genetic control of apoptosis[3].

The tumour suppressor p53, which is mutated in many different types of human tumour[43,44] is, at least in some cell types, required for the induction of apoptosis in response to DNA damage[44,46]. c-Myc also plays a significant role in inducing apoptosis under some circumstances[47,48] and secondary signals, such as the level of growth factor in the cells' environment, determine whether proliferation or death is selected in cells expressing deregulated c-*myc*.

Inhibition of apoptosis, on the other hand, can be produced by the expression of certain proto-oncogenes, of which *bcl-2* is the best established example (reviewed in reference 16). In addition to the EBV gene BHRF1, several other genes related to *bcl-2* have been identified in mammalian cells, and at least two of these genes can also suppress apoptosis on IL3 withdrawal[49,50]. Expression of *bcl-2* can inhibit programmed cell death in the nematode *C. elegans*[51,52]and the *C. elegans* gene *ced-9*, which normally suppresses programmed cell death in this organism, also shows limited sequence homology to *bcl-2*[52]). The emerging *bcl-2* gene family[3,53] therefore displays a striking conservation of function across a broad range of eukaryotes and is likely to play a crucial role in regulating cell population growth through modulation of apoptosis.

It has been demonstrated that apoptosis can also be an important limitation on viral infection cycles and the genetic suppression of apoptosis would be expected to have substantial advantages for the virus in preventing the premature self-destruction of the host cell. Genes which suppress apoptosis have been identified in several viruses including some infecting insect cells[27], as well as African swine fever virus[54], adenovirus[25] and Abelson murine leukaemia virus[26]. Epstein-Barr virus has at least two potential pathways for inhibiting apoptosis[18]. Expression of EBV latent proteins inhibits apoptosis on serum withdrawal[23] and this effect may be mediated by increased expression of cellular *bcl-2* induced by the viral latent gene LMP1[24]. The EBV genome itself encodes the *bcl-2* homologue BHRF1[55] which we have shown here can inhibit B-cell apoptosis induced by γ-irradiation and cytotoxic chemotherapeutic drugs, as well as that induced by serum withdrawal[19]. This is entirely consistent with BHRF1's supression of both Chinese Hamster Ovary cell apoptosis (induced by viral infection, cisplatin, etoposide or mitomycin C[56]) and apoptosis in a squamous cell carcinoma line (induced by cisplatin or serum deprivation[57]).

As for *bcl-2*, the mechanism of action of BHRF1 is not clear. BHRF1 does not appear to act through stimulation of *bcl-2* expression since direct examination by western blotting indicated that *bcl-2* levels were not significantly changed in BHRF1-transfected cells (S.H., unpublished work).

In addition to the roles played by apoptosis-suppressing genes in physiological control mechanisms and in lytic viral infection, aberrant expression of such inhibitors of apoptosis at critical stages in oncogenesis may allow developing cancerous or pre-cancerous cells to escape from the normal control mechanisms which otherwise help to prevent pathological accumulation of cells. This is likely to be particularly significant for BL cells which express high levels of c-*myc*[58] and are, at early stages of development, very sensitive to withdrawal of growth factors[23,24] and to cancer chemotherapy[59]. The deregulated expression of c-*myc* in several cancer cel lines correlates with an increased sensitivity to chemotherapeutic drugs. This reflects Myc's ability to drive cell death as well as proliferation[48] and implies that such cells require a complementary survival lesion, such as expression of *bcl-2* or the presence of survival factors, in order to suppress Myc-induced apoptosis.

The resistance to apoptosis we have observed for cells transfected either with *bcl-2* or BHRF1 was noticeably less pronounced for cells treated with araC when compared to cells treated with γ-radiation or one of the other cytotoxic drugs studied here. This suggests that cell damage may trigger apoptosis by more than one pathway, and that one or more of these pathways which is unaffected by *bcl-2* or BHRF1 expression, predominates in the cellular response to araC in these cell lines.

In general, cancer cell resistance to chemotherapy can be due to several distinct mechanisms, one of the most prominent being the multidrug resistance (MDR) phenotype[60,61]. Cells with MDR phenotype are cross resistant to a number of functionally unrelated drugs, such as the epipodophyllotoxins and the vinca alkaloids, and are able to remove these drugs from the cytoplasm by virtue of the over-expression of the MDR1 gene product P-glycoprotein, an energy dependent drug efflux pump[62].

We propose that the drug-resistance conferred by *bcl-2* or BHRF1 expression is not likely to involve P-glycoprotein since resistance to methotrexate[63] and γ-irradiation is not produced by over-expression of this protein. In addition, studies on *bcl-2*-induced resistance to 5-fluorodeoxyuridine and other inhibitors of thymidylate synthase[64] and to nitrogen mustard and camptothecin[65] have indicated that Bcl-2 does not affect events leading to DNA damage. Rather, it is the response of the cell to DNA damage, ie. induction of apoptosis, which has also been shown for *v-abl*[36]may be of substantial clinical significance in restricting the effective options for therapy.

The genetic suppression of apoptosis by cellular and viral genes may therefore play a critical role, not only in the development of some cancers, but also in determining the sensitivity of the malignant cells to therapy[66].

5. MATERIALS AND METHODS

5.1. Cell Lines

The Burkitt's lymphoma (BL) cell lines used were the EBV genome positive Latency I lines Cheptage (Chep) [67] and Akata[68] and the EBV genome positive latency III line Raji[69]. Raji-BL are unusual in that expression of LMP1 occurs without concomitant *bcl-2* up-regulation (S. H. unpublished observations).

The production of the Chep-BL *bcl-2* and SV40-cΔj control transfectants by standard techniques and analysis of expression of the transfected genes has been described previously[24], as have the Raji-BL and Akata-BL BHRF1 transfectants and pHebo vector-only controls[19]. Briefly, BL-cells were transfected with a pHebo based construct containing the BHRF1 open reading frame, or the pHebo based construct alone as a control, with selection for expressing clones carried out in the presence of 300μg/ml hygromycin. The pHebo plasmid contains the EBV-specific origin of replication (oriP) which enables the transfected plasmid to be episomally expressed producing stable transfectants in the presence of the selective antibiotic. Cell growth rates were not significantly affected by transfection.

Standard cell cultures were maintained at 37°C with 5% CO_2 in RPMI 1640 (Gibco-BRL) supplemented with 10% pre-screened fetal calf serum and 2mM L-glutamine. Transfectants were passaged in medium containing either 2.5mg/ml G418 (Sigma) (cΔj based constructs) or 300mg/ml hygromycin B (Flow) (pHebo based constructs).

5.2. Induction and Analysis of Apoptosis

BL cell lines were treated with the following drugs; methotrexate, araC and etoposide (Sigma). Stock concentrations of each drug were dissolved in DMSO, then diluted 1:10 with growth medium and filter sterilised. Serial dilution to the required 10x concentration was carried out and all drugs were kept on ice prior to addition to cell culture. DMSO was used at a final concentration of $<0.001\%$ as a control in at least 3 repeat experiments and had no detectable effect on cell viability.

BL cells, with a starting viability of > 86%, were exposed at $3x10^5$ cells/ml to the desired drug concentration for 14 hours in growth medium at 37°C and 5% CO_2. Treated and untreated controls were washed free of drug by 2 washes in warmed, gassed RPMI, and cells were resuspended in at least 3 replicate wells at $3x10^5$ cells/ml in growth medium. Viability was determined after 48 hours by the exclusion of the vital dye Nigrosin. Cell numbers were counted on a haemocytometer using a phase contrast inverted microscope.

5.3. Exposure to Ionising Radiation

Pre-washed cell suspensions of $3x10^5$ cells/ml were irradiated on ice, 8Gy (Chep-BL) or 16Gy (Raji-BL) using a Co^{60} γ source. Cells were pelleted and resuspended in 1ml cultures in fresh growth medium and incubated in a 24 well plate (Nunc) at 37°C with 5% CO_2. Viability was determined as previously described and at least 3 replicates were counted per treatment.

The statistical significance of differences in cell viability was determined using the Student t-test.

5.4. DNA Fragmentation

Fragmentation of the total genomic DNA was examined as previously[7]. Briefly, 10^6 cells were harvested and lysed in buffer containing 0.5% sodium lauryl sarkosinate and 0.5mg/ml proteinase K. RNA was digested by treatment with RNase A (0.5mg/ml) and the cell lysate was dry loaded into a 2% agarose gel containing ethidium bromide. Gels were run for 2 hours at 10V/cm at room temperature.

For more sensitive detection of fragmentation, low molecular weight DNA was separated out, based on the method of Wyllie *et al.* [70]. $5x10^6$ cells were lysed on ice in buffer containing 5mM Tris-HCl (pH 7.5), 5mM EDTA and 0.5% Triton. Low molecular weight DNA was separated by centrifugation at 27,000g for 20 mins and the resulting supernatant was treated as described above, using 5x concentrations of proteinase K and RNase A.

5.5. Fluorescence Microscopy

Equal volumes of a cell suspension and acridine orange (final concentration 25μg/ml) were mixed on a slide and the numbers of apoptotic vs. live cells were determined by counting all cells visible in a field using the x16 objective lens of a Zeiss fluorescence microscope. At least 8 separate fields were counted per replicate treatment.

5.6. Electron Microscopy

Approx. 10^6 cells were pelleted in a 1.5ml eppendorf microcentrifuge tube and fixed in 2.5% glutaraldehyde on ice for 2 hours. Pellets were post-fixed in osmium tetroxide and embedded in epoxy resin. 70nm sections were cut on an ultramicrotome and stained in uranyl acetate and Reynolds lead citrate. Sections were examined and photographed using a Joel 1200ex transmission electron microscope.

6. ACKNOWLEDGMENTS

We thank Mrs Leslie Tomkins for technical assistance with the electron microscopy and Mrs Alison Orchard for photographic assistance. This work is financially supported by the Leukaemia Research Fund (U.K.) (N.J.M., G.T.W.), the Wellcome Trust (G.T.W., S.A.H.)

and the Cancer Research Campaign (U.K.) (S.A.H., D.H., A.B.R.). D.H. is supported by a Beit fellowship.

7. REFERENCES

1. M.J. Arends and Wyllie, A., Apoptosis: mechanisms and roles in pathology. *Int. Rev. Exp. Pathol.*, **32**, 223–254 (1991).
2. G.T.Williams, Smith, C.A., McCarthy, N.J. and Grimes, E.A., Apoptosis: Final control point in cell biology. *Trends Cell. Biol.*, **2**: 263–267 (1992).
3. A.J. Hale, Smith, C.A., Sutherland, L.C., Stoneman, V.E.A., Longthorne, V.L., Culhane, A.C. and Williams, G.T., Apoptosis: molecular regulation of cell death. *Eur. J. Biochem.* 1996 (in the press).
4. C.A.Smith, Gimes, E.A., McCarthy, N.J. and Williams, G.T., Multiple gene regulation of apoptosis: Significance in immunology and oncology. *In*: L.D. Tomei and F.O. Cope (eds.), *Apoptosis. The molecular basis of cell death II*, Cold Spring Harbour Laboratory Press, Cold Spring Harbour, NY. (in press) (1994).
5. P. Golstein, Ojcius, D.M. and Young, J.D-E., Cell death mechanisms and the immune system. *Immunol. Rev.*, **121**: 29–65 (1991).
6. G.T.Williams. Apoptosis in the immune system. *J. Pathol.*, **173**: 1–4 (1994).
7. C.A. Smith, Williams, G.T., Kingston, R, Jenkinson, E.J. and Owen, J.J.T., Antibodies to the CD3/T-cell receptor complex induce death by apoptosis in immature T-cells in thymic cultures. *Nature*, **337**: 181–184 (1989).
8. Y. Shi, Sahai, B.M. and Green, D.R., Cyclosporin A inhibits activation-induced cell death in T-cell hybridomas and thymoccytes. *Nature*, **339**: 625–626 (1989).
9. Y.J.Liu, Cairns, J.A, Holder, M.J., Abbot, S.J., Jansen, K.U., Bonnefoy, J.Y., Gordon, J. and MacLennan, I.C.M.. Recombinant 25 kDa CD23 and interleukin 1α promote the survival of germinal centre B cells: evidence for bifurcation in the development of centrocytes rescued from apoptosis. *Eur. J. Immunol.* **21**, 1107–1114 (1991).
10. G.J.V.Nossal. The molecular and cellular basis of affinity maturation in the antibody response. *Cell*, **68**: 1–2 (1992).
11. G.T.Williams, Smith, C. A., Spooncer, E., Dexter, T.M and Taylor, D. R., Haemopoietic colony stimulating factors promote cell survival by suppressing apoptosis. *Nature*, **343**: 76–79 (1990).
12. M.J.Koury and Bondurant, M.C., Control of red cell prodcution: the roles of programmed cell death (apoptosis) and erythropoietin. *Transfusion*, **30**: 673–674 (1990).
13. D.L.Vaux, Cory, S. and Adams, J.M.. Bcl-2 gene promotes haemopoietic cell survival and co-operates with c-myc to immortalize pre-B cells. *Nature*, **335**:440–42 (1988).
14. D.Hockenbery, Nunez, G., Milliman, C., Schreiber, R.D. and Korsmeyer, S.J., Bcl-2 is an inner mitochondrial membrane protein that blocks programmed cell death. *Nature*, **348**: 334–336 (1990).
15. Y.Tsujimoto, Stress resistance conferred by high level of Bcl-2α protein in human B-lymphoblastoid cell. *Oncogene*, **4**: 1331–1336 (1989).
16. N.J.McCarthy, Smith, C.A. and Williams, G.T., Apoptosis in the development of the immune system: Growth factors, clonal selection and *bcl-2*. *Cancer Metastasis Rev.*, **11**: 157–178 (1992).
17. M.L.Cleary, Smith, S.D. and Sklar J., Cloning and structural analysis of cDNAs for *bcl-2* and a hybrid *bcl-2*/immunoglobulin transcript resulting from the t(14;18) translocation. *Cell*, **47**: 19–28 (1986).
18. G.T.Williams, Programmed cell death: Apoptosis and oncogenesis. *Cell*, **65**: 1097–1098 (1991).
19. S.A.Henderson, Huen, D., Rowe, M., Dawson, C., Johnson, G. and Rickinson, A., Epstein-Barr virus coded BHRF1 protein, a viral homologue of *bcl-2*, protects human B cells from programmed cell death. *Proc. Natl. Acad. Sci. U.S.A.*, **90**: 8479–8483 (1993).
20. T.Hickish, Robertson, D., Clarke, P., Hill, M., di Stefano, F., Clarke, C. and Cunnigham, D., Ultrastructural localisation of BHRF1: an Epstein-Barr virus gene product which has homology with bcl-2. *Cancer Res.* **54**, 2808–2811.
21. J.W.Gratama, Oosterveer, M.A.P., Zwann, F.E., Lepoutre, J., Klein, G. and Ernberg, I., Eradication of Epstein Barr virus by allogenic bone marrow transplantation: implcations for sites of vital latency. *Proc. Natl. Acad. Sci. USA*, **85**: 8693–8696 (1988).
22. Q.Y.Yao, Ogan, P., Rowe, M., Wood, M. and Rickinson, A.B., Epstein Barr virus infected B cells persist in the cirulation of acyclovir-treated virus carriers. *Int. J. Cancer*, **43**: 67–71 (1989).

23. C.D.Gregory, Dive, C., Henderson, S.A., Smith, C.A., Williams, G.T., Gordon, J and Rickinson, A. B., Activation of Epstein-Barr virus latent genes protects human B cells from death by apoptosis. *Nature*, **349**: 612–614 (1991).
24. S.A.Henderson, Rowe, M., Gregory, C., Croom-Carter, D., Wang, F., Longnecker, R., Kieff, E. and Rickinson, A., Induction of *bcl-2* expression by Epstein-Barr virus latent membrane protein 1 protects infected B cells from programmed cell death. *Cell*, **65**: 1107–1115 (1991).
25. E.White, Sabbatini, P. Debbas, M., Wold, W.S.M., Kusher, D. I. and Gooding L.R., The 19-Kilodalton Adenovirus E1b transforming protein inhibits programmed cell death and prevents cytolysis by tumour necrosis factor α. *Mol. Cell. Biol.*, **12**: 2570–2580.
26. J.L.Cleveland, Dean, M., Rosenberg, N., Wang, J.Y.J. and Rapp, U.R., Tyrosine kinase oncogenes abrogate interleukin 3 dependence of murine myeloid cells through signalling pathways involving *c-myc*: conditional regulation of *c-myc* transcription by temperature sensitive *v-abl*. *Mol. Cell Biol.*, **9**:5685–5695 (1989).
27. R.J.Clem, Fechheimer, M. and Miller L. K., Prevention of apoptosis by a Baculovirus gene during infection of insect cells. *Science*, **254**: 1388–1390 (1991).
28. N.E.Crook, Clem, R.J. and Miller, L.K., An apoptosis inhibiting Baculovirus gene with a zinc finger like motif. *J. Virol.*, **67**: 2168–2174 (1993).
29. M.Hummel, and Kieff, E., Epstein Barr virus RNA VIII. Viral RNA in permissively infected B95–8 cells. *J. Virol.*, **43**: 262–272 (1982).
30. M.Hummel, and Kieff, E., Mapping of polypeptides encoded by the Epstein Barr virus genome in productive infection. *Proc. Natl. Acad. Sci. U.S.A.*, **79**: 5698–5702 (1982).
31. T.J.McDonnell, Deane, N., Platt, F.M, Nunez, G., Jaeger, U., McKearn, J.P. and Korsmeyer S. J., Bcl-2-immunoglobulin transgenic mice demonstrate extended B cell survival and follicular lymphoproliferation. *Cell*: **57**: 79–88 (1989).
32. T.J.McDonnell and Korsmeyer, S. J., Progression from lymphoid hyperplasia to high-grade malignant lymphoma in mice transgenic for the t(14;18). *Nature*, **349**: 254–256 (1991).
33. J.J.Oudejans, van den Brule, A.J., Jiwa, N.M., de Bruin, P.C., Ossenkoppele, G.J., van der Valk, P., Walboomers, J.M. and Meijer, C.J., BHRF1, the Epstein-Barr virus (EBV) homologue of the Bcl-2 protooncogene, is transcribed in EBV-associated B-cell lymphomas and in reactive lymphocytes. *Blood*, **86**: 1893–1902 (1995).
34. T.Yamada, and Ohyama, H., Radiation-induced interphase death of rat thymocytes is internally programmed. *Int. J. Radiat. Biol.*, **53**: 65–75 (1988).
35. M.A.Barry, Behnke, C.A. and Eastman, A., Activation of programmed cell death (apoptosis) by cisplatin, other anticancer drugs, toxins and hyperthermia. *Biochem. Pharmacol.*, **40**: 2353–2362 (1990).
36. C.A.Evans, Owen-Lynch, P.J., Whetton, A.O. and Dive, C., Activation of the ableson tyrosine kinase activity is associated with suppression of apoptosis in haemopoietic cells. *Cancer Res.*, **53**: 1735–1738 (1993).
37. C.A.Dive and Hickman, J.A., Drug-target interactions: only the first step in a commitment to a programmed cell death. *Br. J. Cancer*, **64**: 192–196 (1991).
38. M.K.L.Collins, Marvel, J., Malde, P. and Lopez-Rivas, A., Interleukin 3 protects murine bone marrow cells from apoptosis induced by DNA damaging agents. *J. Exp. Med.*, **176**: 1043–1051 (1992).
39. T.Miyashita, and Reed, J.C., Bcl-2 oncoprotein blocks chemotherapy-induced apoptosis in a leukaemia cell line. *Blood*, **81**: 151–157 (1993).
40. A.H.Wyllie. Glucocorticoid-induced thymocyte apoptosis is associated with endogenous endonuclease activation. *Nature*, **284**: 555–556 (1980).
41. D.J.McConkey, Hartzell, P., Nicotera, P. and Orrenius, S., Calcium-activated DNA fragmentation kills immature thymocytes. *FASEB J.*, **3**: 1843–1849 (1989).
42. K.S.Sellins, and Cohen, J.J., Gene induction by gamma-irradiation leads to DNA fragmentation in lymphocytes. *J.Immunol.*, **139**: 3199–3206 (1987).
43. A.J.Levine, Momand J. and Finlay, C.A., The p53 tumour suppressor gene. *Nature*, **351**:453–456 (1991).
44. S.J.Baker, Fearon, E.R., Nigro, J.M., Hamilton, S.R., Preisinger, A.C., Jessup, J.M., vanTuinen, P., Ledbetter, D.H., Barker, D.F., Nakamura, Y., White, R. and Vogelstein, B., Chromosome 17 deletions and p53 gene mutations in colosectal carcinoma. *Science*, **244**: 217–221 (1989).
45. S.W.Lowe, Schmitt, E.M., Smith, S.W., Osborne, B.A. and Jacks, T., p53 is required for radiation induced apoptosis in mouse thymocytes. *Nature*, **362**, 847–849 (1993).
46. A.R.Clarke, Purdie, C.A., Harrison, D.J., Morris, R.G., Bird, C.C., Hooper, M.L. and Wyllie, A.H., Thymocyte apoptosis induced by p53 dependent and independent pathways. *Nature*, **362**: 849–852 (1993).

48. G.I.Evan, Wyllie, A. H., Gilbert, C. S., Littlewood, T. D., Land, H., Brooks, M, Waters, C.M. and Hancock, D. C., Induction of apoptosis in fibroblasts by *c-myc* protein. *Cell*, **63**: 119–128 (1992).
49. L.H.Boise, Gonzalez-Garcia, M., Postema, C.E., Ding, L., Lindsten, T., Turka, L.A., Mao, X., Nunez, G. and Thompson, C.B., *bcl-x*, a *bcl-2* related gene that functions as a dominant regulator of apoptotic cell death. *Cell*, **74**: 597–609 (1993).
50. Z.N.Oltvai, Milliman, C.L. & Korsmeyer, S.J., Bcl-2 heterodimerizes *in vivo* with a conserved homolog, Bax, that accelerates programmed cell death. *Cell*, **74**: 609–619 (1993).
51. D.L.Vaux, Weissman, I.L. and Kim, S.K., Prevention of programmed cell death in *Caenorhabditis elegans* by human *bcl-2*. *Science*, **258**: 1955–1957 (1992).
52. M.O.Hengartner, and Horvitz, H.R., C.elegans cell survival gene ced-9 encodes a functional homologue of the mammalian proto-oncogene bcl-2. Cell, 76: 665–676 (1994).
53. G.T.Williams, and Smith, C.A., Molecular regulation of apoptosis: Genetic controls on cell death. *Cell*, **74**: 777–779 (1993).
54. J.G.Neilan, Lu, Z., Afonso, C.L., Kutish, G.F., Sussman, M.D. and Rock, D.L., An African swine fever virus gene with similarity to the proto-oncogene *bcl-2* and the Epstein-Barr virus gene BHRF1. *J. Virol.*, **67**: 4391–4394 (1993).
55. G.R.Pearson, Luka, J., Petti, L., Sample, J., Birkenbach, M., Braun, D. and Keiff, E., Identification of an Epstein Barr virus early gene encoding a second comonent of the restricted early antigen complex. *Virology*, **160**: 151–161 (1987).
56. B.Tarodi, Subramanian, T. and Chinnadurai, G., Epstein-Barr virus BHRF1 protein protects against cell death induced by DNA-damaging agents and heterologous viral infection. *Virology*, **201**: 404–407 (1994).
57. C.W.Dawson, Eliopoulos, A.G., Dawson, J. and Young, L.S., BHRF1, a viral homologue of the BCL-2 oncogene, disturbs epithelial cell differentiation. *Oncogene*, **10**: 69–77 (1995).
58. R.Dalla-Favera, Martinotti, S., Gallo, R.C., Erikson, J. and Croce, C.M., *Science*, **219**: 963–967 (1983).
59. F.Cavalli, Chemothreapy of non-Hodgkin's lymphoma. *Bailliere's Clinical Haematology*, **4**: 157–179 (1991).
60. R.Juliano, and Ling, V., *J. Supramol. Strut.*, **4**: 521–526 (1976).
61. V.Ling. P-glycoprotein and resistance to anticancer drugs. *Cancer*, **69**: 2603–2609 (1992).
62. P.F.Juranka, Zastawny, R.L. and Ling, V., P-glycoprotein multidrug-resistance and a super family of membrane-associated transport proteins. *FASEB J.*, **3**: 2583–2592 (1989).
63. G.H.Mickisch, Merlino, G.T., Galski, H., Gottesman, M.M. and Pastan, I. Transgenic mice that express the human multidrug resistance gene in bone marrow enable a rapid identification of agents that reverse drug resistance. *Proc. Natl. Acad. Sci. USA.*, **88**: 547–551 (1991).
64. T.C.Fisher, Milner, A.E., Gregory, C.D., Jackman, A., Aherne, G.W., Hartley, J.A., Dive, C. and Hickman, J.A., *bcl-2* modulation of apoptosis induced by anticancer drugs: Resistance to thymidylate stress is independent of classical resistance pathways. *Cancer Res.*, **53**: 3321–3326 (1993).
65. M.I.Walton, Whysong, D., O'Connor, P.M., Hockenbery, D., Korsmeyer, S.J. and Kohn, K.W., Constitutive expression of human *bcl-2* modulates nitrogen mustard and camptothecin induced apoptosis. *Cancer Res.*, **53**: 1853–1861 (1993).
66. J.Lotem and Sachs, L. Regulation by *bcl2-*, *c-myc* and p53 of susceptibility to induction of apoptosis by heat shock and cancer chemotherapy compounds in differentiation competent and defective myeloid leukaemic cells. *Cell Growth Diff.*, **4**: 41–47 (1993).
67. C.M.Rooney, Gregory, C.D., Rowe, M., Finerty, S., Edwards, C., Rupani, H. and Rickinson, A.B., Endemic Burkitt's lymphoma: phenotypic analysis of Burkitt's lymphoma biopsy cell and of the derived tumour cell lines. *J.Natl. Cancer Inst.*, **77**: 681–687 (1986).
68. K.Takada and Ono. Y., Synchronous and sequential activation of latently infected Epstein-Barr virus genomes. *J.Virol.*, **63**: 445–449 (1989).
69. L.Rymo, Lindahl, T., Povey, S. and Klien, G., Anaylsis of restriction endonuclease fragments of intracellular Epstein-Barr virus type A (EBNA 2A) and type B (EBNA 2B) isolates extends to the EBNA 3 family of proteins. *Virology*, **115**: 115–124 (1981).
70. A.H.Wyllie, Morris, R.G., Smith, A.L. and Dunlop, D., Chromatin cleavage in apoptosis: Association with condensed chromatin morphology and dependence on macromolecular synthesis. *J. Pathol.*, **142**: 67–77 (1984).

10

STRUCTURE–FUNCTION ANALYSIS OF Bcl-2 FAMILY PROTEINS

Regulators of Programmed Cell Death

John C. Reed, Hongbin Zha, Christine Aime-Sempe, Shinichi Takayama, and Hong-Gang Wang

The La Jolla Cancer Research Foundation
10901 N. Torrey Pines Road
La Jolla, California 92037

ABSTRACT

The Bcl-2 protein blocks a distal step in an evolutionarily conserved pathway for programmed cell death and apoptosis. To gain better understanding of how this protein functions, we have undertaken a structure-function analysis of this protein, focusing on domains within Bcl-2 that are required for function and for interactions with other proteins. Four conserved domains are present in Bcl-2 and several of its homologs: BH1 (residues 136–155), BH2 (187–202), BH3 (93–107) and BH4 (10–30). Deletion of the BH1, BH2, or BH4 domains of Bcl-2 abolishes its ability to suppress cell death in mammalian cells and prevents homodimerization of these mutant proteins, though these mutants can still bind to the wild-type Bcl-2 protein. These mutants also fail to bind to BAG-1 and Raf-1, two proteins that we have shown can associate with protein complexes containing Bcl-2 and which cooperate with Bcl-2 to suppress cell death. Deletion of either BH1 or BH2 nullifies the ability of Bcl-2 to: (a) suppress death in mammalian cells; (b) block Bax-induced lethality in yeast; and (c) heterodimerize with Bax. In contrast, deletion of the BH4 domain of Bcl-2 nullifies anti-apoptotic function and homodimerization, but does not impair binding to the pro-apoptotic protein Bax. Taken together, the data suggest the possibility that both Bcl-2/Bcl-2 homodimerization and Bcl-2/Bax heterodimerization are necessary but insufficient for the anti-apoptotic function of the Bcl-2 protein. Homodimerization of Bcl-2 with itself involves a head-to-tail interaction, in which an N-terminal domain where BH4 resides interacts with the more distal region of Bcl-2 where BH1, BH2, and BH3 are located. In contrast, Bcl-2/Bax heterodimerization involves a tail-to-tail interaction, that requires the portion of Bcl-2 where BH1, BH2, and BH3 reside and a central region in Bax where the BH3 domain is located. The BH3 domain of Bax is also required for Bax/Bax homodimerization and pro-apoptotic function in both yeast and mammalian cells. Thus, Bcl-2 may suppress cell death at least in part by binding to Bax via the BH3 domain and thereby

Mechanisms of Lymphocyte Activation and Immune Regulation VI, Edited by Gupta and Cohen
Plenum Press, New York, 1996

preventing formation of Bax/Bax homodimers. Further studies however are required to delineate the full significance of Bcl-2/Bcl-2, Bcl-2/Bax, and Bax/Bax dimers and the biochemical mechanisms by which Bcl-2 family proteins ultimately control cell life and death.

INTRODUCTION

Programmed cell death plays an enormously important role in immune system regulation and function. During development of T-cells in the thymus and B-cells in the bone marrow, potentially autoreactive lymphocytes are removed by this cell suicide mechanism. In addition, after immune responses when inciting antigen has been cleared, cell death mechanisms help to bring the overall numbers of lymphocytes back to within normal physiological ranges. Moreover, cytolytic T-cells principally induce the destruction of virus-infected cells and tumor target cells through apoptotic mechanisms.

Defects in the regulation of programmed cell death in lymphocytes contribute significantly to several human diseases, including autoimmune disorders, immunodeficiency, and malignancy (for examples, see [1–10]). Among the various regulators of programmed cell death, the Bcl-2 protein and its homologs stand out for their ability to regulate a distal step in an evolutionarily conserved pathway for apoptosis (reviewed in [9,11]).

The human *BCL-2* gene was first discovered by virtue of its involvement in the t(14;18) chromosomal translocations commonly found in non-Hodgkin's B-cell lymphomas [12,13]. The protein encoded by the *BCL-2* gene is a potent blocker of programmed cell death [14–17]. Deregulation of the production of this protein as a result of t(14;18) translocations contributes to neoplastic B-cell expansion by preventing cell turnover rather than by accelerating rates of cell division, making *BCL-2* the first example of a human proto-oncogene that functions through effects on cell death rather than cell cycle. In fact, the indolent follicular B-cell lymphomas that typically involve *BCL-2* are characterized by mostly G_0/G_1-phase resting mature B-cells that gradually accumulate in the patient over a period of several years, with median patient survivals of ~ 7 years as opposed to the scenario with aggressive non-Hodgkin's B-cell lymphomas where the natural history of the disease entails median survivals of less than 2 years [18,19].

High levels of Bcl-2 protein production have also been detected in several types of lymphocytic and non-lymphocytic leukemia, as well as in many non-Hodgkin's B-cell lymphomas, in the absence of t(14;18) translocations, suggesting that dysregulation of *BCL-2* expression can occur through mechanisms that do not require structural alterations to the gene (for examples, see [20–23]). Of potential clinical importance, high levels of Bcl-2 protein production have been demonstrated to render lymphoma, leukemia, and solid tumor cell lines markedly resistant to the cytotoxic effects of essentially all currently available chemotherapeutic drugs and radiation, probably because chemo- and radiotherapy rely heavily on the endogenous programmed cell death pathway for triggering the apoptotic destruction of cancer cells (reviewed in [9,24]). Consistent with this idea, the presence of high levels of Bcl-2 protein has been correlated with poor response to chemotherapy, shorter disease-free survival, and shorter overall survival in some subgroups of patients with lymphomas or leukemias [21,25].

In contrast to the role that over-production of Bcl-2 plays in neoplasia and chemoresistance, studies of *bcl-2* knock-out mice have demonstrated that Bcl-2 is required for maintenance of lymphocyte survival. These *bcl-2* deficient animals experience massive apoptotic destruction of their peripheral lymphocytes a few weeks after birth, probably at about the time that endogenous glucocorticoid levels rise [26,27]. Interestingly, peripheral blood T-cells from patients with HIV-infection have been reported to contain reduced amounts of

Bcl-2 protein and to undergo spontaneous and activation-induced apoptosis in vitro with much higher frequencies than lymphocytes derived from normal individuals [28,29]. Thus, reduced levels of Bcl-2 may contribute to the decline in T-cell numbers that occurs in the setting of HIV-infection. Taken together, these associations of alterations in *BCL-2* expression with human diseases suggest a need to better understand the biochemical mechanisms of Bcl-2 protein structure and function, so that it may one day be possible to pharmacologically manipulate the activities of this protein and its homologs for therapeutic benefit.

1. THE Bcl-2 PROTEIN FAMILY

Since the discovery of Bcl-2, several homologous proteins have subsequently been identified which comprise the Bcl-2 protein family. Interestingly, some of these proteins block apoptosis when over-expressed by gene transfer methods in lymphoid or other types of cells, whereas others promote cell death [30–42]. To date, at least eight mammalian homologs of Bcl-2 have been described, including the anti-apoptotic proteins Bcl-X_L, Mcl-1, A1/Blf-1, Nr13, and the pro-apoptotic proteins Bax, Bcl-X_S, Bad, Bak and Bik. (The Bcl-X_L and Bcl-X_S protein arise through alternative mRNA slicing mechansims form the same gene [33]). In addition, four homologs of Bcl-2 have been discovered in viruses: E1b-19 kD (adenovirus), BHRF-1 (Epstein Barr Virus), LMH-5W (African Swine Fever Virus), and ORF-16 (Herpes Saimiri Virus) [43–46]. An anti-apoptotic homolog has also been identified in the nematode *C. elegans*, which has been termed *ced-9* [47]. Interestingly, several of these homologous proteins can interact with each other, thus constituting a network of homo- and heterdimers that regulate programmed cell death [32,39,48–50].

Though less is known about the in vivo roles of these other members of the Bcl-2 family for immune system regulation, immunohistochemical analysis of the patterns of expression of some of these Bcl-2 family proteins has revealed some provocative differences. For example, reciprocal patterns of Bcl-2 and Mcl-1 expression have been noted in the follicles of lymph nodes, where germinal center B-cells contain high levels of Mcl-1 but little Bcl-2, and conversely the surrounding mantle zone lymphocytes contain high levels of Bcl-2 but little Mcl-1 [51]. Reciprocal patterns of Bcl-2 and Bcl-X expression have similarly been described in the thymus, where cortical thymocytes express Bcl-X but not Bcl-2, whereas medullary thymocytes contain abundant amounts of Bcl-2 but little or no Bcl-X [52]. Presumably therefore some of these Bcl-2 family proteins play unique roles in the control of cell life and death at specific stages of lymphocyte differentiation and activation. This hypothesis has been borne out by recent studies of *bcl-X* and *bax* knock-out mice, which have demonstrated (among other things) a failure of most thymocytes to survive in the absence of *bcl-X* and a hyperplasia of thymocytes and peripheral lymphocytes in the absence of *bax* [53,54].

2. HYPOTHESIZED MECHANISMS OF Bcl-2 PROTEIN ACTION

The mechanism by which Bcl-2 and its homologs control the programmed cell death pathway remains enigmatic to date, due mostly to the lack of any significant homology between Bcl-2 family proteins and other proteins whose biochemical mechanism of action is know. The human Bcl-2 protein has a molecular mass of ~26-kDa and contains a stretch of hydrophobic amino-acids at its C-terminus that allows it to post-translationally insert into intracellular membranes, primarily the outer mitochondrial membrane, nuclear envelope, and endoplasmic reticulum [55–58]. Most other members of the Bcl-2 family also contain a transmembrane domain near the C-termini and, where examined to date, appear to reside within approximately in the same intracellular membrane compartments [59,60].

Table 1. Hypothesized mechanisms of Bcl-2 protein action

Antioxidant pathway
Ca+2 transport
Protein translocation
Proteases
Signal transduction
None of the above

Several theories have been advanced as to how Bcl-2 and its homologs control cell life and death. As summarized in Table 1, data have been presented which argue both in favor and against the possibility of an effect of Bcl-2 on an anti-oxidant pathway in cells [61–64]. It has also been suggested that Bcl-2 may regulate the homeostasis of Ca^{2+} in cells, based on experiments which have shown an ability of Bcl-2 overexpression to influence the sequestration of Ca^{2+} within the ER and (in some cases) to prevent the accumulation of Ca^{2+} in mitochondria of cells subjected to an apoptotic stimulus [65,66]. Evidence has also been represented suggesting that Bcl-2 can control the transport of proteins across biological membranes, particularly the nuclear envelope [67,68]. In this regard, electron microscopic studies have demonstrated the presence of Bcl-2 protein in association with what appear to be nuclear pore complexes [55]. It has been hypothesized that Bcl-2 may control the activity of a family of cysteine proteases with homology to the Interleukin-1β Converting Enzyme (ICE), based on genetic arguments from studies of cell death genes in the nematode *C. elegans* [69]. Indeed, recently biochemical evidence supporting this idea has been obtained through experiments which have demonstrated that over-expression of Bcl-2 can prevent the proteolytic processing and activation of the ICE homolog, CPP32/YAMA, in mammalian cells [70,71]. Finally, the association of Bcl-2 with the kinase Raf-1 and possibly with the GTPase R-Ras has raised the possibility that Bcl-2 may control a signal transduction pathway that is focused on the intracellular membrane compartments where Bcl-2 resides rather than the plasma membrane, where such enzymes are associated with growth factor receptor function [72–74]. In no case, however, has a direct cause-and-effect relation been demonstrated between Bcl-2 and these processes, and at this point we do not know whether the effects of Bcl-2 on redox state, Ca^{2+} compartmentalization, protein transport, and protease activation represent direct effects of Bcl-2 versus downstream events that are hundreds of steps removed from Bcl-2.

3. SEQUENCE COMPARISONS OF Bcl-2 FAMILY PROTEINS

In the absence of a clear function for the Bcl-2 protein, our laboratory has attempted to map functionally important domains in the Bcl-2 protein, asking what the roles of these domains are both for anti-apoptotic function and for interactions with other proteins. We have also attempt to identify novel proteins that associate with Bcl-2 or that at least enter into multiprotein complexes that contain the Bcl-2 protein. At present, at least four well-conserved domains can be recognized in the Bcl-2 protein based on sequence comparisons with the Bcl-2 proteins derived from various species (human, rat, mouse, chicken, worm) and from other Bcl-2 family proteins, as well as from functional analysis of the repurcussions of deleting or mutating these domains. Though various names for these domains can be found in the literature, we have recently proposed that these be terms BH-1, BH2, BH-3, and BH-4, where BH stands for Bcl-2 Homology Domain as originally suggested by Oltvai, et al [32,75]. For historical reasons, these fours domains are unfortunately not ordered sequentially along the protein from NH_2- to COOH-terminus.

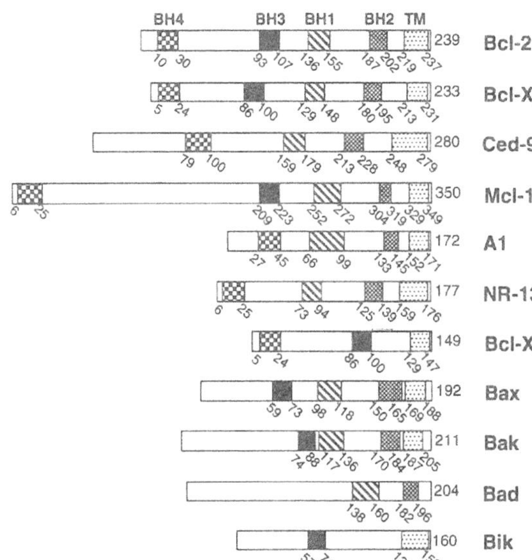

Figure 1. Structural comparisons of Bcl-2 family proteins. The domain-maps of the cellular homologs of Bcl-2 are depicited, showing the relative locations of the BH1, BH2, BH3, BH4 and TM (transmembrane) domains. The proteins represent human Bcl-2, Bcl-X_L, Bcl-X_S, Bak, Mcl-1, and Bik, as well as mouse Bax, Bad, and A1 (human homolog known as Blf-1) and *C. elegans*, ced-9.

Figure 1 depicts the structures of Bcl-2 and its cellular homologs. In the human Bcl-2 protein, the BH1, BH2, BH3, and BH4 domains reside at amino-acid positions 136–155 (BH1), 187–202 (BH2), 93–107 (BH3) and 10–30 (BH4). The transmembrane domain (TM) of Bcl-2 resides at positions 219 to 237. Of interest, the BH4 domain (also known as the A-box) [48,76,77] is not found in most pro-apoptotic Bcl-2 family proteins, including Bax, Bak, Bik, and Bad, suggesting that this domain may play a unique role in the function of the anti-apoptotic proteins such as Bcl-2, Bcl-X_L, Mcl-1, A1/Bfl-1, and ced-9. The BH4 domain however is found in the cell death promoting proteins Bcl-X_S, which suggests either that the BH4 domain is not directly involved in the anti-apoptotic function of Bcl-2 family proteins, or more likely, that the Bcl-X_S protein can function at least in part as a dominant-inhibitor of Bcl-2 and its anti-apoptotic homologs by competing for whatever proteins or protein domains that might normally interact with BH4. Also of note in Figure 1 is the observation that one of the pro-apoptotic proteins Bik contains only the BH3 domain, implying that this particular domain may be uniquely important in the promotion of apoptosis [42]. Indeed, recent deletional analysis of the Bax and Bak proteins suggest that this is the case, as will be discussed below in more detail [75,78]. Though some anti-apoptotic proteins do contain regions with homology to BH3, there exist clear sequence differences that distinguish the BH3 domains of the pro-apoptotic proteins Bax, Bak, and Bik from the anti-apoptotic proteins Bcl-2, Bcl-X_L, and Mcl-1 [75]. Moreover, it has recently been demonstrated that substituting the BH3 domain of Bax for the corresponding domain in Bcl-2 converts the Bcl-2 protein from a cell death blocker to a cell death promoter [79].

4. STRUCTURE-FUNCTION ANALYSIS OF THE Bcl-2 AND Bax PROTEINS

In an effort to better understand the functional significance of these various domains that can be recognized in Bcl-2 by sequence comparisons, we and others have deleted or mutated some of these homologous regions, asking what the effects are on anti-apoptotic function and correlating the results with the ability of Bcl-2 to form homodimers with itself or heterodimers with the pro-apoptotic protein Bax. It is important to note, however, that while we describe the interactions of Bcl-2 with itself and Bax as dimers, in fact, we do not know the stoichiometry of these protein-protein interactions.

4.1. Bcl-2/Bcl-2 Homodimerization

Deletion of the N-terminal BH4 domain of Bcl-2 as well as the downstream BH1 and BH2 domains prevents the resulting mutant Bcl-2 proteins from forming homodimers with themselves, though they can still bind the wild-type Bcl-2 protein forming mutant/wild-type heterodimers [77]. These mutants also are deficient in anti-apoptotic function in mammalian cells [80], suggesting the possibility that Bcl-2/Bcl-2 homodimerization is important for function. In this regard, recent data suggest that deletion of the BH4 domain of Bcl-2 converts the protein to a dominant inhibitor of the wild-type Bcl-2 protein, suggesting that mutant/wild-type heterodimers of Bcl-2 may be dysfunctional dimers [81]. This observation also provides insights into the potential mechanism by which the Bad protein may promote cell death, since it contains the BH1 and BH2 domains but lacks the BH4 domain (Figure 1).

The capacity of deletion mutants of Bcl-2 to retain their ability to bind the wild-type Bcl-2 protein can be explained by data which suggest that Bcl-2/Bcl-2 homodimerization involves a head-to-tail association in which sequences located in the first ~80 amino-acids of the protein where the BH4 domain resides form an interaction with sequences located in the more distal portions of the protein (~85 -> TM) where the BH1, BH2, and BH3 domain are located [48]. Thus, a Bcl-2 mutant protein in which the BH1 and/or BH2 domains have been deleted, for example, can still bind via its BH4 domain to the wild-type Bcl-2 protein or to Bcl-2 protein fragments that retain an intact distal region (~85 -> TM) where BH1, BH2, and BH3 reside. This observation may provide some insights into the mechanism by which the Bcl-X_S protein promotes apoptosis, given that it contains the BH4 domain but is lacking the BH1 and BH2 domains and can therefore bind to Bcl-2, presumably sequestering the protein in an dysfunctional Bcl-2/Bcl-X_S heterodimer [48].

To some extent the separation of the distal region of Bcl-2 into BH1, BH2, and BH3 domains may be artificial, since deletion of other segments located between these conserved domains also destroys Bcl-2's function as a suppressor of cell death [81,82]. Thus, region in Bcl-2 from approximately amino-acid 85 to the TM domain probably folds into a structure that is capable of binding the BH4-containing N-terminal portion of Bcl-2 as well as some other proteins (see below).

4.2. Bcl-2/Bax Heterodimerization

In contrast to Bcl-2/Bcl-2 homodimerization, the BH4 domain of the Bcl-2 protein is completely expendible for dimerization with Bax. Indeed, the first ~80 amino-acids of Bcl-2 can be removed without impairing heterodimerization with Bax [77]. Given that the Bax protein lacks a BH4 domain, this observation suggests that the structural features by which Bcl-2 interacts with Bax versus with itself are considerably different. Though BH4 is not required for heterodimerization with Bax, the BH1 and BH2 domains are needed (BH3 has not been tested to date) [77,83]. Moreover, single amino-acid substitutions in the BH1 and BH2 domains of Bcl-2 have been shown to impair binding to Bax and abrogate anti-apoptotic function in lymphoid cells [83]. These data argue that for Bcl-2 to suppress cell death, it must be capable of heterodimerizing with Bax. However, the finding that the N-terminal region of Bcl-2 where BH4 resides is needed for function but not for dimerization with Bax argues that Bcl-2/Bax heterodimer formation is insufficient by itself for blocking apoptosis and that the Bcl-2 protein must fulfill other functions as well. In this regard, not only are deletion mutant of Bcl-2 lacking the BH4 domain incapable of suppressing apoptosis in mammalian cells, but they also fail to rescue yeast from Bax induced lethality despite binding to Bax [48,77]. Thus, whatever the role of the N-terminal domain in suppressing Bax-mediated cell death, it appears to be evolutionarily conserved.

One idea is that BH4 domain of Bcl-2 may be required to sterrically interfere with the binding of Bax to some other death effector protein or to modulate somehow post-translational modifications of the Bax protein. It is also possible however that this domain is required for Bcl-2/Bcl-2 homodimerization or for interactions of Bcl-2 with other proteins that require the BH4 domain for their association with Bcl-2/Bax complexes. In this regard, recent data from our laboratory indicate that the association of BAG-1 and Raf-1, two proteins have have been demonstrated to associate with protein complexes containing Bcl-2 and which cooperate with Bcl-2 in co-transfection assays to suppress apoptosis [72,84], is dependent on the BH4 domain. Thus, to some extent, the N-terminal region of Bcl-2 where BH4 resides can be thought of as an effector domain that may link Bcl-2 to other proteins such as BAG-1 and Raf-1, whereas the distal portion of Bcl-2 where the BH1 and BH2 domains reside may represent a dimerization domain that serves to target Bcl-2 and its associated proteins to Bax. Alternatively, it is possible that the dependence on the BH4 domain for associations with BAG-1 and Raf-1 is an indirect consequence of the need for this domain for formation of functional Bcl-2/Bcl-2 homodimers.

4.3. Bax/Bax Homodimerization

In an effort to understand some of the structural features of the Bax protein which allow it to homodimerize with itself and heterodimerize with Bcl-2, we have tested the ability of deletion mutants of Bax lacking the BH1, BH2, or BH3 domains to interact with the wild-type Bax and Bcl-2 proteins, using yeast two-hybrid assays [75]. These studies showed that the BH1 and BH2 domains of Bax are expendible for both homodimerization with Bax and heterodimerization with Bcl-2. In addition, Bax deletion mutants lacking BH1 or BH2 are also able to bind to themselves (i.e., mutant to mutant) as well as to the wild-type Bax protein. In contrast, the BH3 domain of Bax is absolutely required for binding to both wild-type Bax and Bcl-2.

The structural features of Bax which permit it to homodimerize with itself and to heterodimerize with Bcl-2 therefore are considerably different from those required for Bcl-2 protein function in terms of dimerization events. These data suggest that despite their amino-acid sequence homology, the Bax and Bcl-2 proteins are probably extensively different in their 3-dimensional structures, but x-ray crystallographic or NMR studies are required to confirm this idea. Consistent with this idea however we have also explore the effect of deleting the NH_2-terminal domains of Bcl-2 and Bax on their dimerization with each other. As mentioned above, removal of the N-terminal first ~80 amino-acids of Bcl-2 does not impair its ability to bind to wild-type Bax. Moreover, a Bcl-2 (ΔN) mutant of this type can also form heterodimers with an N-terminal truncation mutant of Bax that lacks the first 58 amino-acids (i.e., everything upstream of BH3). Thus, unlike Bcl-2/Bcl-2 homodimerization which involves a head-to-tail interaction, Bax/Bcl-2 heterodimerization appears to occur through a tail-to-tail interaction. N-terminal truncation mutants of Bax can also homodimerize with themself, indicating that Bax/Bax homodimerization also occurs via a tail-to-tail interaction that is independent of the proximal region of the Bax protein upstream of BH3 [75].

Similar to our work with Bax, the BH3 domain of Bak was also recently shown to be necessary and sufficient for dimerization with Bcl-X_L [78]. Bak homodimerization however was not explored. Also, the pro-apoptotic protein Bik contains a BH3 domain, but lacks the BH1, BH2, or BH4 domains, suggesting that BH3 in Bik is also functionally important for

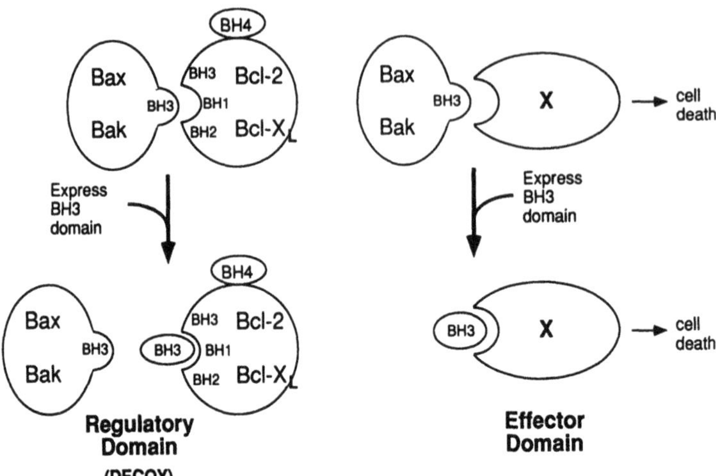

Figure 2. Alternative models for BH3 domain function. The BH3 domains of pro-apoptotic proteins such as Bax and Bak may function as regulatory domains that are involved in binding to anti-apoptotic proteins such as Bcl-2 and Bcl-X_L (left) or may engage an unidentified cell death effector protein (right). Over-expression of this domain in cells (bottom) could then either act as a decoy, occupying the binding sites on Bcl-2 and Bcl-X_L so that they cannot bind to endogenous Bax and Bak (left; bottom) or alternatively activating cell death effector proteins (right; bottom). These models are not mutually exclusive.

this protein's interactions with Bcl-2 and other anti-apoptotic Bcl-2 family proteins [42]. In gene transfer studies in Rat-1 fibroblasts, it was shown that expression of a fragment of the Bak protein that contains essentially only the BH3 domain is capable for promoting apoptosis [78], suggesting that this domain alone is sufficient for the pro-apoptotic function of this protein. However, there are at least two ways that these results can be interpreted. For example, it could be that the BH3 domain of pro-apoptotic proteins such as Bak, Bax, and Bik engages some unknown effector protein that promotes cell death. Alternatively, expression of this domain may act as a "decoy," binding to Bcl-X_L, Bcl-2, or other anti-apoptotic proteins and thereby preventing them from interacting with the endogenous wild-type Bak, Bax, or Bik proteins and thus leaving these pro-apoptotic proteins unopposed (Figure 2). In this regard, preliminary results from our laboratory indicate that mutants of Bax that have altered BH3 domains and which fail to homodimerize are incapable of promoting cell death, thus it may also be that the BH3 domain of proteins such as Bax, Bak, and Bik is required for homodimerization of these proteins and that homodimerization is critical for their pro-apoptotic function.

CONCLUSIONS

Taken together, the data currently available about Bcl-2, Bax, and other Bcl-2 family proteins fail to provide conclusive information about the functional significance of Bcl-2/Bcl-2, Bcl-2/Bax, and Bax/Bax dimers. However, several models can be imagined, most of which are not mutually exclusive (Figure 3). For example, Bax-Bax homodimers can be envisioned as the active moiety that promotes cell death, with Bcl-2 functioning essentially as a dominant-inhibitor of Bax by sequestering the protein in a heteromeric Bax/Bcl-2 complex. Alternatively, Bcl-2/Bcl-2 homodimers may actively provide "signals" for cell survival, with Bax functioning as a dominant inhibitor that prevents Bcl-2/Bcl-2 dimerization. It is also formally possible the Bcl-2/Bax heterodimers represent the most critical

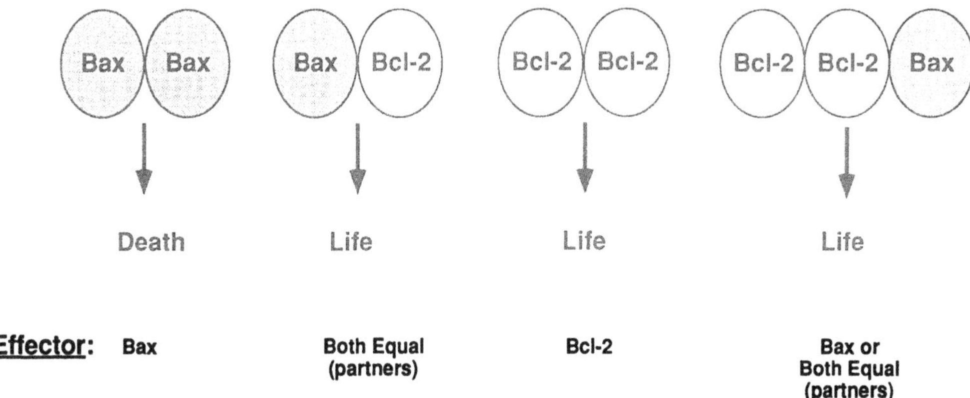

Figure 3. Potential models for Bcl-2/Bax dimerize function. (A) Bax/Bax homodimers may promote cell death. In this case, Bcl-2 serves the role of a dominant-inhibitor of Bax by preventing Bax/Bax dimerization. (B) Bax and Bcl-2 may function as equal partners, each contributing structures that in composite protect cells from apoptosis. (C) Bcl-2/Bcl-2 homodimers may provide signals for cell survival. In this case, Bax serves the role of a dominant-inhibitor of Bcl-2 that inhibits Bcl-2 function by disrupting Bcl-2/Bcl-2 homodimers. (D) One molecule of Bcl-2 may simultaneously bind via its N-terminal domain (BH4) to another molecule of Bcl-2 and via its distal domain (BH1, BH2) to Bax. In this case, both Bcl-2/Bcl-2 homodimerization and Bcl-2/Bax heterodimerization may be required for suppression of cell death.

structure, with the proteins functioning as equal partners in suppressing cell death by analogy to the α- and β-chains of many growth factors or dimeric transcription factors such as Myc/Max. Though recent *bax* knock-out experiments argue against this co-partner model, it cannot be formally excluded because of the possibility of redundancy among known and perhaps currently undiscovered pro-apoptotic members of the Bcl-2 family. Indeed, the paradoxical observation that loss of *bax* can result in excessive cell death in the developing spermatocytes of the testes raises the possibility of a need for Bax for suppression of apoptosis in some contexts [54]. Finally, because Bcl-2 can dimerize with itself through a head-to-tail interaction and yet bind to Bax via a tail-to-tail interaction, there exists the potential for Bcl-2 to simultanously bind to Bax via is distal domain where BH1 and BH2 reside and to bind another molecule of Bcl-2 via its BH4 domain-containing proximal domain, thus complexing with Bax in a 2:1 ratio (Figure 3). This last model would provide the advantage that both Bcl-2/Bax heterodimerization and Bcl-2/Bcl-2 homodimerization are simultaneously required for anti-apoptotic function, which fits well with the currently available data suggesting that both Bcl-2/Bcl-2 homodimerization and Bcl-2/Bax heterodimerization are necessary but insufficient for Bcl-2 protein function.

One caveat about the models presented in Figure 3, however, is that a recent report involving an analysis of various BH1 and BH2 domain single amino-acid substitution mutants has suggested that the Bcl-X_L protein can suppress apoptosis without necessity for binding to Bax or Bak [85]. Thus, it is possible that our current concepts about the significance of interactions of anti-apoptotic proteins such as Bcl-X_L and Bcl-2 with pro-apoptotic proteins such as Bax and Bak are erroneous. It should be borne in mind, however, that while some substitution mutants of Bcl-X_L that retain cell death-blocking activity are apparently defective in binding to Bax or Bak in vitro and could not be detectably co-immunoprecipitated with Bax or Bak from cells, it cannot be excluded that such mutants do retain at least weak affinity for Bax and Bak and that this reduced affinity is nevertheless sufficient for function. Clearly, some direct comparisons of the affinities (K_d), as well as kinetics (k_{on}; k_{off} rates), of the interactions of wild-type and mutant Bcl-X_L proteins with Bax and Bak would shead some light on this controversial issue. The effects of these substitution mutants on Bcl-X_L/Bcl-X_L homodimerization were not explored, leaving open the possibility that homodimerization is critical for the anti-apoptotic function of this protein.

Though much has been learned in recent years about the domains in Bcl-2 family proteins that are required for function and dimerization events, the fundamental biochemical mechanisms by which these proteins control cell life and death remain enigmatic. The finding that a kinase Raf-1 can participate in protein complexes that include Bcl-2 raises the possibility that at least one function of Bcl-2 may be to serve as an adaptor protein that helps to target Raf-1 to unique substrates involved in the control of programmed cell death. These substrates presumably are different from MEK or other substrates normally associated with Raf-1's role as a transducer of signals from growth factor receptors in the plasma membrane. What those substrates are is currently under investigation, and hopefully will provide new insights into the function of Bcl-2 and its homologs. In should be noted, however, that no data exist to date which demonstrate a dependence on Raf-1 for Bcl-2 function, though Raf-1 can collaborate with Bcl-2, and provide enhanced protection from cell death [72]. Nevertheless, recent findings that another Bcl-2 associated protein, BAG-1, can also bind Raf-1 and in addition can activate this kinase, presumably through a protein-protein interaction that involves binding of BAG-1 to the kinase domain of Raf-1 [86], lends further support to the notion that Bcl-2 may act (at least in part) as a docking protein around which "signal-transduction" events occur. With further investigations of the interactions of Bcl-2 and homologous proteins with Raf-1, BAG-1 and other proteins, both known and currently undiscovered, hopefully the biochemical mechanisms of Bcl-2 family proteins will be revealed. When available, there is little doubt that this information will provide new strategies for the treatment of a wide variety of human diseases, including cancer, graft-rejection, autoimmunity, and immunodeficiency disorders.

REFERENCES

1. G.H. Fisher, F.J. Rosenberg, S.E. Straus, J.K. Dale, L.A. Middelton, A.Y. Lin, W. Strober, M.J. Lenardo and J.M. Puck, Dominant interfering *fas* gene mutations impair apoptosis in a human autoimmune lymphoproliferative syndrome, *Cell* 81:935 (1995).
2. J.P. DiSanto, J.Y. Bonnefoy, J.F. Gauchat, A. Fischer and G. de Saint Basile, CD40 ligand mutations in x-linked immunodeficiency with hyper-IgM, *Nature* 361:541 (1993).
3. R.C. Allen, R.J. Armitage, M.E. Conley, H. Rosenblatt, N.A. Jenkins, N.G. Copeland, M.A. Bedell, S. Edelhoff, C.M. Disteche, D.K. Simoneaux, W.C. Fanslow, J. Belmont and M.K. Spriggs, CD40 ligand gene defects responsble for X-linked hyper-IgM syndrome, *Science* 259:990 (1993).
4. M.O. Westendorp, R. Frank, C. Ochsenbauer, K. Stricker, J. Dhein, H. Waleczak, K.-M. Debatin and P.H. Krammer, Sensitization of T cells to CD95-mediated apoptosis by HIV-1 Tat and gp120, *Nature* 375:497 (1995).
5. T.H. Finkel and N.K. Banda, Indirect mechanisms of HIV pathogenesis: how does HIV kill T cells? *Curr. Opin. Immun.* 6:605 (1994).
6. L. Meyaard, S.A. Otto, R.R. Jonker, M.J. Mijnster, R.P. Keet and F. Miedema, Programmed death of T cells in HIV-1 infection, *Science* 257:217 (1992).
7. M.L. Gougeon, S. Garcia, J. Heeney, R. Tschopp, H. Lecoeur, D. Guetard, V. Rame, C. Dauguet and L. Montagnier, Programmed cell death in AIDS-related HIV and SIV infections, *AIDS Res. Hum. Retroviruses* 9:553 (1993).
8. G.T. Williams, Programmed cell death: apoptosis and oncogenesis, *Cell* 65:1097 (1991).
9. J.C. Reed, Bcl-2 and the regulation of programmed cell death, *J. Cell Biol.* 124:1 (1994).
10. S.J. Korsmeyer, Bcl-2 initiates a new category of oncogenes: regulators of cell death, *Blood* 80:879 (1992).
11. D.L. Vaux, Toward an understanding of the molecular mechansisms of physiological cell death, *Proc. Natl. Acad. Sci. USA* 90:786 (1993).

12. Y. Tsujimoto, J. Cossman, E. Jaffe and C. Croce, Involvement of the *bcl-2* gene in human follicular lymphoma, *Science* 228:1440 (1985).
13. Y. Tsujimoto and C.M. Croce, Analysis of the structure, transcripts, and protein products of bcl-2, the gene involved in human follicular lymphoma, *Proc. Natl. Acad. Sci. USA* 83:5214 (1986).
14. D.L. Vaux, S. Cory and J.M. Adams, Bcl-2 gene promotes haemopoietic cell survival and cooperates with c-myc to immortalize pre-B cells, *Nature* 335:440 (1988).
15. D.M. Hockenbery, G. Nunez, C. Milliman, R.D. Schreiber and S.J. Korsmeyer, Bcl-2 is an inner mitochondrial membrane protein that blocks programmed cell death, *Nature* 348:334 (1990).
16. J. Reed, S. Haldar, M. Cuddy, C. Croce and D. Makover, Bcl-2-mediated tumorigenicity in a T-lymphoid cell line: Synergy with C-MYC and inhibition by Bcl-2 antisense, *Proc. Natl. Acad. Sci. USA* 87:3660 (1990).
17. J. Reed, C. Stein, S. Haldar, C. Subasinghe, C. Croce, S. Yum and J. Cohen, Antisense-mediated inhibition of *bcl-2* proto-oncogene expression and leukemic cell growth: Comparisons of phosphodiester and phosphorothioate oligodeoxynucleotides, *Cancer Res.* 50:6565 (1990).
18. D.A.G. Galton and I.C.M. MacLennan, Clinical patterns of B cell malignancy, *Clin. Hematol.* 11:561 (1982).
19. Horning SJ and Rosenberg SA, The natural history of initially untreated low-grade non-Hodgkin's lymphomas, *N. Eng. J. Med.* 311:1471 (1984).
20. M. Hanada, D. Delia, A. Aiello, E. Stadtmauer and J.C. Reed, *bcl-2* gene hypomethylation and high-level expression in B-cell chronic lymphocytic leukemia, *Blood* 82:1820 (1993).
21. L. Campos, J.-P. Roualult, O. Sabido, N. Roubi, C. Vasselon, E. Archimbaud, J.-P. Magaud and D. Guyotat, High expression of *bcl-2* protein in acute myeloid leukemia cells is associated with poor response to chemotherapy, *Blood* 81:3091 (1993).
22. D. Campana, E. Coustan-Smith, A. Manabe, M. Buschle, S.C. Raimondi, F.G. Behm, R. Ashmun, M. Aricó, A. Biondi and C.-H. Pui, Prolonged survival of B-lineage acute lymphoblastic leukemia cells is accompanied by overexpression of *bcl-2* protein, *Blood* 81:1025 (1993).
23. F. Pezzella, A.G.D. Tse, J.L. Cordell, K.A.F. Pulford, K.C. Gatter and D.Y. Mason, Expression of the *bcl-2* oncogene protein is not specific for the 14;18 chromosomal translocation, *Am. J. Pathol.* 137:225 (1990).
24. J.C. Reed, Bcl-2: Prevention of apoptosis as a mechanism of drug resistance, *Hematology/Oncology Clinics of North America* 9:451 (1995).
25. O. Hermine, C. Haioun, E. Lepage, M.-F. d'Agay, J. Briere, C. Lavignac, G. Fillet, G. Salles, J.-P. Marolleau, J. Diebold, F. Reyes and P. Gaulard, Prognostic significance of Bcl-2 protein expression in aggressive non-Hodgkin's Lymphoma, *Blood* 87:265 (1996).
26. K.-I. Nakayama, K. Nakayama, I. Negishi, K. Kuida, Y. Shinkai, M.C. Louie, L.E. Fields, P.J. Lucas, V. Stewart, F.W. Alt and D.Y. Loh, Disappearance of the lymphoid system in *bcl-2* homozygous mutant chimeric mice, *Science* 261:1584 (1993).
27. D.J. Veis, C.M. Sorenson, J.R. Shutter and S.J. Korsmeyer, Bcl-2-deficient mice demonstrate fulminant lymphoid apoptosis, polycystic kidneys, and hypopigmented hair, *Cell* 75:229 (1993).
28. A.N. Akbar, N. Borthwick, M. Salmon, W. Gombert, M. Bofill, N. Shamsadeen, D. Pilling, S. Pett, J.e. Grundy and G. Janossy, The significance of low bcl-2 expression by CD45RO T cells in normal individuals and patients with acute viral infections. The role of apoptosis in T cell memory, *J. Exp. Med.* 178:427 (1993).
29. F. Boudet, H. Lecoeur and M.-L. Gougeon, Apoptosis associated with ex vivo down-regulation of Bcl-2 and up-regulation of Fas in potential cytotoxic CDS+ T lymphocytes during HIV infection, *J. Immunol.* in press (1995).
30. K.M. Kozopas, T. Yang, H.L. Buchan, P. Zhou and R. Craig, Mcl-1, a gene expressed in programmed myeloid cell differentiation, has sequence similarity to bcl-2, *Proc. Natl. Acad. Sci. USA* 90:3516 (1993).
31. J.E. Reynolds, T. Yang, L. Qian, J.D. Jenkinson, P. Zhou, A. Eastman and R.W. Craig, Mcl-1, a member of the Bcl-2 family, delays apoptosis induced by c-Myc overexpression in Chinese hamster ovary cells, *Cancer Res.* 54:6348 (1994).
32. Z. Oltvai, C. Milliman and S.J. Korsmeyer, Bcl-2 heterodimerizes in vivo with a conserved homolog, Bax, that accelerates programmed cell death, *Cell* 74:609 (1993).
33. L.H. Boise, M. Gonzalez-Garcia, C.E. Postema, L. Ding, T. Lindsten, L.A. Turka, X. Mao, G. Nunez and C.B. Thompson, *bcl-x*, a *bcl-2*-related gene that functions as a dominant regulator of apoptotic cell death, *Cell* 74:597 (1993).
34. E.Y. Lin, A. Orlofsky, M.S. Berger and M.B. Prystowsky, Characterization of A1, a novel hemopoietic-specific early-response gene with sequence similarity to *bcl-2*, *J. Immunol.* 151:1979 (1993).

35. E.Y. Lin, A. Orlofsky, H.-G. Wang, J.C. Reed and M.B. Prystowsky, A1, a Bcl-2 family member prolongs cell survival and permits granulocyte differentiation, *Blood* 87:983 (1996).
36. S.S. Choi, I.-C. Park, J.W. Yun, Y.C. Sung, S. Hong and H. Shin, A novel Bcl-2 related gene, Bfl-1, is overexpressed in stomach cancer and preferentially expressed in bone marrow, *Oncogene* 11:1693 (1995).
37. T. Chittenden, E.A. Harrington, R. O'Connor, C. Flemington, R.J. Lutz, G.I. Evan and B.C. Guild, Induction of apoptosis by the Bcl-2 homologue Bak, *Nature* 374:733 (1995).
38. M.C. Kiefer, M.J. Brauer, V.C. Powers, J.J. Wu, s.R. Ubansky, L.D. Tomei and P.J. Barr, Modulation of apoptosis by the widely distributed Bcl-2 homologue Bak, *Nature* 374:736 (1995).
39. S.N. Farrow, J.H.M. White, I. Martinou, T. Raven, K.-T. Pun, C.J. Grinham, J.-C. Martinou and R. Brown, Cloning of a bcl-2 homologue by interaction with adenovirus E1B 19K, *Nature* 374:731 (1995).
40. G. Gillet, M. Guerin, A. Trembleau and G. Brun, A Bcl-2-related gene is activated in avian cells transformed by the Rous sarcoma virus, *EMBO J.* 14:1372 (1995).
41. E. Yang, J. Zha, J. Jockel, L.H. Boise, C.B. Thompson and S.J. Korsmeyer, Bad: a heterodimeric partner for Bcl-X_L and Bcl-2, displaces *bax* and promotes cell death, *Cell* 80:285 (1995).
42. J.M. Boyd, G.J. Gallo, B. Elangovan, A.B. Houghton, S. Malstrom, B.J. Avery, R.G. Ebb, T. Subramanian, T. Chittenden, R.J. Lutz and G. Chinnadurai, Bik, a novel death-inducing protein shares a distinct sequence motif with Bcl-2 family proteins and interacts with viral and cellular survival-promoting proteins, *Oncogene* 11:1921 (1995).
43. L. Rao, M. Debbas, P. Sabbatini, D. Hockenbery, S. Korsmeyer and E. White, The adenovirus E1A proteins induce apoptosis, which is inhibited by the E1B 19-kDa and *bcl-2* proteins, *Proc. Natl. Acad. Sci. USA* 89:7742 (1992).
44. M.L. Cleary, S.D. Smith and J. Sklar, Cloning and structural analysis of cDNAs for bcl-2 and a hybrid bcl-2/immunoglobulin transcript resulting from the t(14;18) translocation, *Cell* 47:19 (1986).
45. J.G. Neilan, Z. Lu, C.L. Afonso, G.F. Kutish, M.D. sussman and D.L. Rock, An African swine fever virus gene with similarity to the proto-oncogene *bcl-2* and the Epstein-Barr virus gene BHRF1, *J. Virol.* 67:4391 (1993).
46. C.A. Smith, A novel viral homolgue of Bcl-2 and Ced-9, *Trends Cell Biol.* 5:344 (1995).
47. M.O. Hengartner and H.R. Horvitz, *C. elegans* cell survival gene *ced-9* encodes a functional homolog of the mammlian proto-oncogene *bcl-2*, *Cell* 76:665 (1994).
48. T. Sato, M. Hanada, S. Bodrug, S. Irie, N. Iwama, L.H. Boise, C.B. Thompson, E. Golemis, L. Fong, H.-G. Wang and J.C. Reed, Interactions among members of the *bcl-2* protein family analyzed with a yeast two-hybrid system, *Proc. Natl. Acad. Sci. USA* 91:9238 (1994).
49. S.E. Bodrug, C. Aimé-Sempé, T. Sato, S. Krajewski, M. Hanada and J.C. Reed, Biochemical and functional comparisons of Mcl-1 and Bcl-2 proteins: evidence for a novel mechanism of regulating Bcl-2 family protein function, *Cell Death Differ.* 2:173 (1995).
50. T.W. Sedlak, Z.N. Oltvai, E. Yang, K. Wang, L.H. Boise, C.B. Thompson and S.J. Korsmeyer, Multiple Bcl-2 family members demonstrate selective dimerizations with Bax, *Proc. Natl. Acad. Sci. USA* 92:7834 (1995).
51. S. Krajewski, S. Bodrug, R. Gascoyne, K. Berean, M. Krajewska and J.C. Reed, Immunohistochemical analysis of mcl-1 and bcl-2 proteins in normal and neoplastic lymph nodes, *Amer. J. Pathol.* 145:515 (1994).
52. S. Krajewski, M. Krajewska, A. Shabaik, H.-G. Wang, S. Irie, L. Fong and J.C. Reed, Immunohistochemical analysis of in vivo patterns of Bcl-X expression, *Cancer Res.* 54:5501 (1994).
53. N. Motoyama, F. Wang, K.A. Roth, H. Sawa, K. Nakayama, I. Negishi, S. Senju, Q. Zhang, S. Fujii and D.Y. Loh, Massive cell death of immature hematopoietic cells and neurons in Bcl-X-deficient mice, *Science* 267:1506 (1995).
54. C.M. Knudson, K.S.K. Tung, W.G. Tourtellotte, G.A.J. Brown and S.J. Korsmeyer, Bax-deficient mice with lymphoid hyperplasia and male germ cell death, *Science* 270:96 (1995).
55. S. Krajewski, S. Tanaka, S. Takayama, M.J. Schibler, W. Fenton and J.C. Reed, Investigations of the subcellular distribution of the bcl-2 oncoprotein: residence in the nuclear envelope, endoplasmic reticulum, and outer mitochondrial membranes, *Cancer Res.* 53:4701 (1993).
56. M.D. Jacobson, J.F. Burne, M.P. King, T. Miyashita, J.C. Reed and M.C. Raff, Apoptosis and *bcl-2* protein in cells without mitochondrial DNA, *Nature* 361:365 (1993).
57. P. Monaghan, D. Robertson, T. Andrew, S. Amos, M.J.S. Dyer, D.Y. Mason and M.F. Greaves, Ultrastructural localization of *bcl-2* protein, *J. Histochem. Cytochem.* 40:1819 (1992).
58. R. Silvestrini, S. Veneroni, M.G. Daidone, E. Benini, P. Boracchi, M. Mezzetti, G. Di Fronzo, F. Rilke and U. Veronesi, The *bcl-2* protein: a prognostic indicator strongly related to p53 protein in lymph node-negative breast cancer patients, *J. Natl. Cancer Inst.* 86:499 (1994).

59. T. Yang, K.M. Kozopas and R.W. Craig, The intracellular distribution and pattern of expression of Mcl-1 overlap with, but are not identical to, those of Bcl-2, *J. Cell Biol.* 128:1173 (1995).
60. M. González-Garcia, R. Pérez-Ballestero, L. Ding, L. Duan, L.H. Boise, C.B. Thompson and G. Núñez, *bcl-XL* is the major *bcl-X* mRNA form expressed during murine development and its product localizes to mitochondria, *Development* 120:3033 (1994).
61. D.J. Kane, T.A. Sarafin, S. Auton, H. Hahn, F.B. Gralla, J.C. Valentine, T. Ord and D.E. Bredesen, Bcl-2 inhibition of neural cell death: decreased generation of reactive oxygen species, *Science* 262:1274 (1993).
62. D. Hockenbery, Z. Oltvai, X.-M. Yin, C. Milliman and S.J. Korsmeyer, Bcl-2 functions in an antioxidant pathway to prevent apoptosis, *Cell* 75:241 (1993).
63. M.D. Jacobson and M.C. Raff, Programmed cell death and Bcl-2 protection in very low oxygen, *Nature* 374:814 (1995).
64. S. Shimizu, Y. Eguchi, H. Kosaka, W. Kamiike, H. Matsuda and Y. Tsujimoto, Prevention of hypoxia-induced cell death by Bcl-2 and Bcl-X_L, *Nature* 374:811 (1995).
65. G. Baffy, T. Miyashita, J.R. Williamson and J.C. Reed, Apoptosis induced by withdrawal of Interleukin-3 [IL-3] from an IL-3-dependent hematopoietic cell line associated with repartitioning of intracellular calcium and is blocked by enforced Bcl-2 oncoprotein production, *J. Biol. Chem.* 268:6511 (1993).
66. M. Lam, G. Dubyak, L. Chen, G. Nuñez, R.L. Miesfeld and C.W. Distelhorst, Evidence that Bcl-2 represses apoptosis by regulating endoplasmic reticulum-associated Ca^{2+} fluxes, *Proc. Natl. Acad. Sci. USA* 91:6569 (1994).
67. J.J. Ryan, E. Prochownik, C.A. Gottlieb, I.J. Apel, R. Merino, G. Nuñez and M.F. Clarke, *c-myc* and *bcl-2* modulates p53 function by altering p53 subcellular trafficking during the cell cycle, *Proc. Natl. Acad. Sci. USA* 91:5878 (1994).
68. W. Meikrantz, S. Gisselbrecht, S.W. Tam and R. Schlegel, Activation of cyclin A-dependent protein kinases during apoptosis, *Proc. Natl. Acad. Sci. USA* 91:3754 (1994).
69. J. Yuan, S. Shaham, S. Ledoux, H.M. Ellis and H.R. Horvitz, The *C. elegans* cell death gene *ced-3* encodes a protein similar to mammalian interleukin-1 beta-converting enzyme, *Cell* 75:641 (1993).
70. A.M. Chinnaiyan, K. Orth, K. O'Rourke, H. Duan, G.G. Poirier and V.M. Dixit, Molecular odering of the cell death pathway: Bcl-2 and Bcl-XL function upstream of the CED-3-like apoptotic proteases, *J. Biol. Chem.* in press.:(1996).
71. C.A. Boulakia, G. Chen, F.W.H. Ng, J.G. Teodoro, P.E. Branton, D.W. Nicholson, G.G. Poirier and G.C. Shore, Bcl-2 and adenovirus E1B 19 kDA protein prevent E1A-induced processing of CPP32 and cleavage of poly(ADP-ribose) polymerase, *Oncogene* 12:29 (1996).
72. H.-G. Wang, T. Miyashita, S. Takayama, T. Sato, T. Torigoe, S. Krajewski, S. Tanaka, III Hovey,L., J. Troppmair, U.R. Rapp and J.C. Reed, Apoptosis regulation by interaction of *bcl-2* protein and Raf-1 kinase, *Oncogene* 9:2751 (1994).
73. M.J. Fernandez-Sarbia and J.R. Bischoff, Bcl-2 associates with the ras-related protein R-ras p23, *Nature* 366:274 (1993).
74. H.-G. Wang, J.A. Millan, A.D. Cox, C.J. Der, U.R. Rapp, T. Beck, H. Zha and J.C. Reed, R-ras promotes apoptosis caused by growth factor deprivation via a Bcl-2 suppressible mechanism, *J. Cell Biol.* 129:1103 (1995).
75. H. Zha, C. Aime-Sempe, T. Sato and J.C. Reed, Pro-apoptotic protein Bax heterodimerizes with Bcl-2 and homodimerizes with Bax via a novel domain (BH3) distinct from BH1 and BH2, *J. Biol. Chem.* in press (1996).
76. T. Sato, S. Irie, S. Krajewski and J.C. Reed, Cloning and sequencing of cDNA encoding rat bcl-2 protein, *Gene* 140:291 (1994).
77. M. Hanada, C. Aimé-Sempé, T. Sato and J.C. Reed, Structure-function analysis of *bcl-2* protein: identification of conserved domains important for homodimerization with *bcl-2* and heterodimerization with *bax*, *J. Biol. Chem.* 270:11962 (1995).
78. T. Chittenden, C. Flemington, A.B. Houghton, R.G. Ebb, G.J. Gallo, B. Elangovan, G. Chinnadurai and R.J. Lutz, A conserved domain in Bak, distinct from BH1 and BH2, mediates cell death and protein binding functions, *EMBO J.* 14:5589 (1995).
79. J.J. Hunter and T.G. Parslow, A peptide sequence from Bax that converts Bcl-2 into an activator of apoptosis, *J. Biol. Chem.* in press (1996).
80. C. Borner, I. Martinou, C. Mattmann, M. Irmler, E. Scharer, J.-C. Martinou and J. Tschopp, The protein bcl-2alpha does not require membrane attachment, but two conserved domains to suppress apoptosis, *J. Cell Biol.* 126:1059 (1994).
81. J.J. Hunter, B.L. Bond and T.G. Parslow, Functional dissection of the human Bcl-2 protein: sequence requirements for inhibition of apoptosis, *Mol. Cell. Biol.* in press (1996).

82. T. Subramanian, J.M. Boyd and G. Chinnadurai, Functional substitution identifies a cell survival promoting domain common to adenovirus E1B 19 kDa and Bcl-2 proteins, *Oncogene* 11:2403 (1995).
83. X.M. Yin, Z.N. Oltvai and S.J. Korsmeyer, BH1 and BH2 domains of *bcl-2* are required for inhibition of apoptosis and heterodimerization with *bax*, *Nature* 369:321 (1994).
84. S. Takayama, T. Sato, S. Krajewski, K. Kochel, S. Irie, J.A. Millan and J.C. Reed, Cloning and functional analysis of BAG-1: a novel Bcl-2 binding protein with anti-cell death activity, *Cell* 80:279 (1995).
85. E.H.-Y. Cheng, B. Levine, L.H. Boise, C.B. Thompson and J.M. Hardwick, Bax-independent inhibition of apoptosis by Bcl-X_L, *Nature* 379:554 (1996).
86. S.V. Razin and I.I. Gromova, The channels model of nuclear matrix structure, *BioEssays* 17:443 (1995).

11

ROLE OF ICE-PROTEASES IN APOPTOSIS

Vishva M. Dixit

University of Michigan
Ann Arbor, Michigan

1. INTRODUCTION

Apoptosis, or programmed cell death (PCD), is a naturally occurring process of cell suicide that plays a critical role in the development and maintenance of multicellular organisms. It is required for sculpting tissue architecture, for the development of the nervous and immune systems, as a defense against viral pathogens and protection from cancer[1–11]. Apoptosis is characterized by cellular shrinkage, nuclear condensation, membrane blebbing and fragmentation of the cell into membrane-enclosed vesicles called apoptotic bodies. The apoptotic bodies are rapidly phagocytosed by neighboring cells and macrophages without the release of cytoplasmic contents. This avoids inflammation and autoimmunization with intracellular constituents. Despite the biological importance of apoptosis, its molecular and biochemical basis remains obscure. The most important information about components of the death pathway comes from the study of the nematode worm *Caenorhabditis elegans* [12–14]. The complete cellular lineage of the hermaphrodite form of *C. elegans* has been established. Of the 1,090 cells generated during development, 131 die by a cell-intrinsic death program[7]. Genetic analysis has identified 14 genes that function in different steps of PCD in *C. elegans*. Two genes, ced-3 and ced-4, are required for these developmental deaths. If either is inactivated, none of the 131 cells die[7]. The ced-9 gene antagonizes the function of ced-3 and ced-4 by protecting cells from death[7]. In ced-9 loss-of-function mutants, cells that normally live undergo PCD early in development, resulting in embryonic lethality. Thus, ced-9 is necessary for the survival of most cells and presumably ced-9 accomplishes this by suppressing the cell death program[7]. Several recent discoveries have shed light on the function of the *C. elegans* cell-death genes and, more importantly, have shown that they have mammalian counterparts with similar functions. The machinery controlling cell viability and cell death appears to be highly conserved through evolution and is briefly reviewed below.

1.1 Ced-9 and Bcl-2

The first indication that the pathways regulating apoptosis are similar in *C. elegans* and mammals came from the discovery that the function of the ced-9 mutation can be restored by expression of the human bcl-2 gene[7]. bcl-2 was first cloned from the breakpoint

Mechanisms of Lymphocyte Activation and Immune Regulation VI, Edited by Gupta and Cohen
Plenum Press, New York, 1996

of a t (14;18) chromosomal translocation in human B-cell lymphomas [15,16]. Expression of bcl-2 suppresses apoptosis in various cell types, both *in vitro* and *in vivo*, and bcl-2 deficient mice exhibit abnormalities consistent with loss of function of a death suppressor[15,16]. The proteins encoded by ced-9 and bcl-2 are 23% identical in sequence and, furthermore, the human bcl-2 gene can function in *C. elegans* to suppress apoptosis[17]. It is now clear that bcl-2 is a member of a large family of cellular and viral proteins, many of which are membrane bound on the cytosolic surfaces of the endoplasmic reticulum, nuclear envelope and mitochondria[15,16]. Some members of the family, such as bcl-2 and bcl-x, function as death suppressors, whereas others, such as Bax, antagonize this death suppressing function[16].

1.2. Ced-3 and ICE

Two other genes, ced-3 and ced-4, are essential for cell death to occur[13]. The ced-4 gene encodes a novel 63 kDa protein with a potential calcium binding domain but, to date, the mechanism of its function is completely obscure and, as yet, has no known mammalian homologue. On the other hand, sequence analysis of ced-3 has revealed it to be homologous to the mammalian cysteine protease interleukin-1β-converting enzyme (ICE), providing further evidence of evolutionary conservation of apoptotic pathways[14]. A distinguishing feature of ICE and ICE-like thiol proteases is an absolute specificity for specific aspartic acid residues[10]. ICE is expressed as a 45 kDa precursor that is cleaved to yield two subunits of 10 kDa and 20 kDa, both of which are required for activity[10]. ICE was originally identified because of its function in processing interleukin-1β, a proinflammatory cytokine[18]. First isolated from monocytic cells, ICE was shown to cleave the 31 kDa IL-1β precursor (pro-IL-1β) at Asp116-Ala117 to generate the 17.5 kDa mature form of IL-1β.

Inhibitor studies confirmed ICE to be a cysteine protease, with Cys 285 serving as the catalytic residue[18]. Mutations of this residue result in complete loss of enzyme activity[18]. The crystal structure of ICE indicates that the catalytically active form of the enzyme is a tetramer consisting of a $(p20\text{-}p10)_2$ homodimer[11]. The active site spans both p20 and p10 subunits[11], which explains the requirement for both subunits for catalysis. In the crystal structure, covalent binding of the ICE inhibitor Ac-Tyr-Val-Ala-Asp-chloromethyl ketone (Ac-YVAD-CMK) at Cys 285 confirmed ICE to be a cysteine protease[11].

Several lines of evidence suggest that ICE or its homologues have a direct role in the regulation of apoptosis:

a. Overexpression of the gene encoding ICE in Rat-1 cells induces apoptosis. Mutation of the catalytic Cys residue completely abolishes its ability to induce apoptosis[19].
b. Microinjection of ICE cDNA expression vectors into chicken dorsal root ganglion neurons also results in cell death, while microinjection of CrmA, a serpin-like inhibitor of ICE, prevents neuronal cell death induced by deprivation of nerve growth factor. Further, the pro-apoptotic effects of ICE can be prevented by cotransfection with bcl-2, a mammalian cell death suppressor gene[9–11].
c. In Rat-1 cells, CrmA inhibits apoptosis induced by serum depletion, presumably by inactivating ICE or a related protease[14].

Despite the above stated evidence suggesting a role for ICE in apoptosis, ICE-deficient mice display normal developmental and *ex vivo* apoptosis (29), suggesting that ICE itself does not play a prominent role in mammalian cell death, whereas related family members might.

2. ICE RELATED PROTEINS

To date, six homologues of ced-3 and ICE have been characterized and include Nedd-2/ICH1,[21,22] Yama/CPP32/Apopain,[23–25] Tx/ICH2/ICE rel-II,[26–28] ICE rel-III,[26] Mch2,[29] and ICE-LAP3/Mch3/CMH-1[30–32]. All members contain an active QACRG pentapeptide and their ectopic expression induces apoptosis.

Phylogenetic analysis of the ICE/CED-3 gene family reveals three subfamilies. Yama, ICE-LAP3 and Mch2 are most closely related to *C. elegans* ced-3 and comprise the ced-3 subfamily. ICE and the ICE-related genes, ICE rel II and ICE rel III, form the ICE subfamily, while ICH1 and its mouse homologue, NEDD-2, form the NEDD-2 subfamily[30].

An important advance came with the finding that in a cell free system of apoptosis there occurred the rapid cleavage of the DNA repair enzyme poly (ADP-ribose) polymerase (PARP) into signature apoptotic fragments[31]. Although the cleavage followed an aspartic acid residue, purified ICE failed to cleave PARP[31] suggesting that the activity (termed prICE) was distinct from ICE. However, it soon became evident that Yama, unlike ICE, rapidly cleaved the death substrate PARP into characteristic fragments[23,25]. Active Yama was biochemically purified using the Ac-DEVD-CHO tetrapeptide aldehyde corresponding to the amino acids of the PARP cleavage site. Using this peptide, it was possible to deplete apoptotic cell extracts of Yama and Yama-like enzymes and demonstrate an attenuation in apoptotic potential *in vitro*. Apoptotic activity could be restored by adding back purified Yama to the depleted extracts[25]. Overall, these observations suggested that Yama could function as the PARP cleavage enzyme (prICE) and thereby potentially represent the mammalian equivalent of ced-3. However, it was soon discovered that both Mch2 and ICE-LAP3 cleave PARP efficiently and both are members of the Yama subfamily of ICE-like proteases that are most related to ced-3[29,31,32].

Additionally, ICE-LAP3 was inhibited by the tetrapeptide aldehyde inhibitor, Ac-DEVD-CHO with kinetics similar to that for Yama inhibition[31,32]. Therefore, the PARP cleavage activity may be represented by more than one ICE-like protease, suggesting a degree of redundancy in the cell death pathway. To further implicate individual ICE/ced-3 family members in the cell death pathway, we and others have attempted to show activation of the zymogen form to active heterodimeric enzyme upon receipt of a death stimulus[33,34]. Yama, for example, was found to be activated from its 32kDa precursor form to processed p17 and p12 subunits following exposure to apoptotic signals including engagement of CD95 (Fas/Apo-1) and treatment with the broad spectrum protein kinase inhibitor, staurosporine[33].

An important area of investigation presently is the characterization of substrates that are cleaved during apoptosis. PARP cleavage, per se, is probably not critical for apoptotic death since PARP knockout mice develop normally[35]. More recently identified substrates for ICE-like proteases include the nuclear lamins[36,37], U1–70kDa snRP[38,39], D4-GDI[40], SREBPs[41], and protein kinase C δ[42]. It is unclear which of these cleavages play a pivotal role in apoptosis.

Prior studies have shown that enucleated cells exposed to various death stimuli undergo apoptosis consistent with the cell death machinery being present outside the nucleus[43,44]. It is unclear how the cytoplasmic cell death machinery gains access to the nucleus and other organelle compartments. This, undoubtedly will be the focus of future research.

3. REFERENCES

1. M.C. Raff, Social controls on cell survival and cell death. *Nature* 356:397–400 (1992).
2. C.B.Thompson, Apoptosis in the Pathogenesis and Treatment of Disease. *Science* 267:1456–1462 (1995).

3. J.J. Cohen, Apoptosis. *Immunology Today* 14:126–130 (1993).
4. G. Majno, and I.Joris, Apoptosis, Oncosis, and Necrosis. *American Journal of Pathology* 146:3–15 (1995).
5. M.C. Peitsch, H.G. Mannherz, and J. Tschopp, The apoptosis endonucleases: cleaning up after cell death? *Trends in Cell Biology* 4:37–41 (1994).
6. D.L. Vaux, G. Haecker, and A. Strasser, An Evolutionary Perspective on Apoptosis. *Cell* 76:777–779 (1994).
7. M.O. Hengartner, Programmed cell death in invertebrates. *Current Opinion in Genetics and Development* 6:34–38 (1996).
8. G.I. Evan, L. Brown, M. Whyte, and E. Harrington, Apoptosis and cell cycle. *Current Opinion in Cell Biology* 7:825–836 (1995).
9. E. White, Life, death, and the pursuit of apoptosis. *Genes and Development* 10:1–15 (1996).
10. S.J. Martin, and D.R. Green, Protease activation during apoptosis: death by a thousand cuts? *Cell* 82:349–352 (1995).
11. S. Kumar, and N.L. Harvey, Role of multiple cellular proteases in the execution of programmed cell death. *FEBS Letters* 375:169–173 (1995).
12. H.M. Ellis, and H.R. Horvitz, Genetic control of programmed cell death in the nematode *Caenorhabditis elegans*. *Cell* 44:817–829 (1986).
13. J.Y. Yuan, and H.R. Horvitz, The Caenorhabditis elegans genes ced-3 and ced-4 act cell autonomously to cause programmed cell death. *Dev Biol* 138:33–41 (1990).
14. J. Yuan, S. Shaham, S. Ledoux, H.M. Ellis, and H.R. Horvitz, The C. elegans cell death gene *ced-3* encodes a protein similar to mammalian Interleukin-1ß-Converting Enzyme. *Cell* 75:641–652 (1993).
15. J.C. Reed, Bcl-2 and the Regulation of Programmed Cell Death. *J Cell Biol* 124:1–6 (1994).
16. G. Nunez, and M.F. Clarke, The bcl-2 family of proteins: regulators of cell death and survival. *Trends in Cell Biology* 4:399–403 (1994).
17. M.O. Hengartner, R.E. Ellis, and H.R. Horvitz, Caenorhabditis elegans gene ced-9 protects cells from programmed cell death. *Nature* 356:494–9 (1992).
18. N.A. Thornberry, H.G. Bull, J.R. Calaycay, K.T. Chapman, A.D. Howard, M.J. Kostura, et al., A novel heterodimeric cysteine protease is required for interleukin-1ß processing in monocytes. *Nature* 356:768–774 (1992).
19. M. Miura, H, Zhu, R. Rotello, E.A. Hartwieg, and J. Yuan, Induction of Apoptosis in Fibroblasts by IL-1ß Converting Enzyme, a Mammalian Homolog of the C. elegans Cell Death Gene *ced-3*. *Cell* 75:653–660 (1993).
20. P. Li , H. Allen, S. Banerjee, S. Franklin, L. Herzog, C. Johnston C, et al.: Mice deficient in IL-1ß-converting enzyme are defective in production of mature IL-1ß and resistant to endotoxic shock. *Cell* 80:401–411 (1995).
21. S. Kumar, M. Kinoshita, M. Noda, N.G. Copeland, and N.A. Jenkins, Induction of apoptosis by the mouse *Nedd2* gene, which encodes a protein similar to the product of the *Caenorhabditis elegans* cell death gene *ced-3* and the mammalian IL-1ß-converting enzyme. *Genes and Development* 8:1613–1626 (1994).
22. L. Wang, M. Miura, L. Bergeron, H. Zhu, and J. Yuan, Ich-1, an ICE/ced-3-related gene, encodes both positive and negative regulators of programmed cell death. *Cell* 78:739–750 (1994).
23. M. Tewari, L.T. Quan, K. O'Rourke, S. Desnoyers, Z. Zeng, D.R. Beidler , et al., Yama/CPP32ß, a mammalian homolog of CED-3, is a CrmA-inhibitable protease that cleaves the death substrate poly(ADP-ribose) polymerase. *Cell* 81:801–809 (1995).
24. T. Fernandes-Alnemri, G. Litwack, E.S. and E.S. Alnemri, CPP32, a novel human apoptotic protein with homology to Caenorhabditis elegans cell death protein Ced-3 and mammalian interleukin-1ß-converting enzyme. *J Biol Chem* 269:30761–30764 (1994).
25. D.W. Nicholson, A.Ali, N.A. Thornberry, J.P. Vaillancourt, C.K. Ding, M. Gallant, et al., Identification and inhibition of the ICE/CED-3 protease necessary for mammalian apoptosis. *Nature* 376:37–43 (1995).
26. N.A. Munday, J.P. Vaillancourt , A. Ali, F.J. Casano, D.K. Miller, S.M. Molineaux, et al., Molecular cloning and pro-apoptotic activity of ICErel II and ICErel III, members of the ICE/CED-3 family of cysteine proteases. *J Biol Chem* 270:15870–6 (1995).
27. C. Faucheu, A. Diu, A.W. Chan , A.M. Blanchet, C. Miossec, F. Herve, et al.: A novel human protease similar to the interleukin-1 beta converting enzyme induces apoptosis in transfected cells. *Embo J* 14:1914–22 (1995).
28. J. Kamens, M. Paskind, M. Hugunin, R.V. Talanian, H. Allen, D. Banach, et al.: Identification and characterization of ICH-2, a novel member of the interleukin-1 beta-converting enzyme family of cysteine proteases. *J Biol Chem* 270:15250–6 (1995).

29. T. Fernandes-Alnemri, G. Litwack, and E.S. Alnemri, Mch2, a new member of the apoptotic Ced-3/Ice cysteine protease gene family. *Cancer Res* 55:2737–42 (1995).
30. H. Duan, A.M. Chinnaiyan, P.L. Hudson, J.P. Wing, W. He, and V.M. Dixit, ICE-LAP3, a Novel Mammalian Homolog of the Caenorhabditis elegans Cell Death Protein CED-3 is Activated During Fas- and Tumor Necrosis Factor-Induced Apoptosis. *J Biol Chem* 271:35013–35 (1996).
31. T. Fernandes-Alnemri, A. Takahashi, R. Armstrong, J. Krebs, L. Fritz, K.J. Tomaselli, et al.: Mch3, a Novel Human Apoptotic Cysteine Protease Highly Related to CPP32. *Cancer Res* 55:6045–6052 (1995).
32. J.A. Lippke, Y. Gu, C. Sarnecki, P.R. Caron, M.S.-S. Su, Identification and characterization of CPP32/Mch2 homolog 1, a novel cysteine protease similar to CPP32. *J Biol Chem* 271:1825–1828 (1996).
33. A.M. Chinnaiyan, K. Orth, K. O'Rourke, H. Duan, G.G. Poirier, and V.M. Dixit, Molecular ordering of the cell death pathway: bcl-2 and bcl-xL function upstream of the CED-3-like apoptotic proteases. *J Biol Chem* 271:4573–4576 (1996).
34. J. Schlegel, I. Peters, S. Orrenius, D.K. Miller, N.A. Thornberry, T.-T. Yamin, et al., CPP32/Apopain is the key interleukin-1β converting enzyme-like protease involved in Fas-mediated apoptosis. *J Biol Chem* 271:1841–1844 (1996).
35. Z.-Q. Wang, B. Auer, L. Stingl, H. Berghammer, D. Haidacher, M. Schweiger M, et al.: Mice lacking ADPRT and poly(ADP-ribosyl)ation develop normally but are susceptible to skin disease. *Genes and Dev* 9:509–520 (1995).
36. Y.A. Lazebnik, A. Takahashi, R. Moir, R. Goldman, G.G. Poirier, S.H. Kaufmann, et al.: Studies of the lamin proteinase reveal multiple parallel biochemical pathways during apoptotic execution. *Proc Natl Acad of Sci USA* 92:9042–9046 (1995).
37. N. Neamati, A. Fernandez, S. Wright, J. Kiefer, and D.J. McConkey, Degradation of lamin B1 precedes oligonucleosomal DNA fragmentation in apoptotic thymocytes and isolated thymocyte nuclei. *J Immunol* 154:1693–1700 (1994).
38. M. Tewari, D.R. Beidler, and V.M. Dixit , CrmA-inhibitable cleavage of the 70 kDa protein component of the U1 small nuclear ribonucleoprotein during Fas- and tumor necrosis factor-induced apoptosis. *J. Biol. Chem.* 270:18738–18741 (1995).
39. L.A. Casciola-Rosen, D.K. Miller, G.J. Anhalt, and A. Rosen, Specific cleavage of the 70-kDa protein component of small nuclear ribonucleoprotein is a characteristic biochemical feature of apoptotic cell death. *J. Biol. Chem.* 269:30757–30760 (1994).
40. S. Na, R.H. Chuang, T.G. Turi, J.H. Hanke, G.M. Bokoch, and D.E. Danley, D4-GDI, a substrate for CPP32, is proteolyzed during Fas-induced apoptosis. *J Biol Chem* In Press (1996).
41. X. Wang, N.G. Zelenski, J. Yang, J. Sakai, M.S. Brown, and J.L. Goldstein, Cleavage of the sterol regulatory element binding proteins (SREBPs) by CPP32 during apoptosis. *EMBO J* 15: (1996)
42. Y. Emoto, Y. Manome, G. Meinhardt , T. Ghayur, W.W. Wong, R. Kamen, et al., Proteolytic activation of protein kinase C δ an ICE-like protease in apoptotic cells. *EMBO J* 14:6148–6156 (1995).
43. M.D. Jacobson, J.F. Burne, and M.C. Raff, Programmed cell death and Bcl-2 protection in the absence of a nucleus. *EMBO J* 13:1899–910 (1994).
44. K. Schulze-Osthoff, H. Walczak, W. Droge, and P.H. Krammer, Cell Nucleus and DNA Fragmentation are not Required for Apoptosis. *J Cell Biol* 127:15–20 (1994).

12

Fas-MEDIATED APOPTOSIS

Shigekazu Nagata

Osaka University Medical School and
Osaka Bioscience Institute
Suita, Osaka 565, Japan

SUMMARY

Homeostasis in vertebrates is tightly regulated by cell death as well as by cell proliferation. The death of cells during embryogenesis, metamorphosis, endocrine-dependent tissue atrophy, and normal tissue turnover is "programmed cell death", mediated by a process called "apoptosis". Cytotoxic T lymphocytes and natural killer cells kill the target cells by inducing apoptosis. Apoptosis can be distinguished from necrosis, which occurs as a result of injury, complement attack, severe hypoxia and hyperthermia. Morphological and biochemical analyses of the apoptotic cell death process indicated that apoptosis is accompanied by condensation of cytoplasm, loss of plasma membrane microvilli, segmentation of nucleus, and extensive degradation of chromosomal DNA into oligomers of 180 bp. Cellular proliferation and differentiation are mediated by a family of proteins called cytokines. Our studies on the Fas ligand and Fas have indicated that apoptosis is also mediated by a cytokine and its receptor in some cases. Here, I summarize the current status of the Fas death factor system.

1. Fas AND Fas LIGAND

In 1989, Yonehara *et al*[1] established a mouse monoclonal antibody of IgM-type (anti-Fas antibody) which has a cytolytic activity on various human cell lines. Molecular cloning of the antigen recognized by the antibody revealed that Fas, also called CD95 or APO-1, is a type I membrane protein consisting of 270 amino acids with Mr of 45 kDa[2]. Its amino acid sequence indicated that Fas belongs to the tumor necrosis factor (TNF)/nerve growth factor (NGF) receptor family which includes two different TNF receptors (type I and II), the low affinity NGF receptor, CD40, CD27 and CD30[3]. When mouse cell lines transformed with human Fas were treated with the anti-human Fas antibody, they quickly died. The death process was accompanied by apoptotic morphological changes and DNA fragmentation, indicating that Fas can transduce an apoptotic signal into cells, and the anti-Fas antibody works as an agonist[2]. Mouse Fas cDNA was then isolated by cross-hybridization with human Fas cDNA, and Northern hybridization analysis indicated that the

Mechanisms of Lymphocyte Activation and Immune Regulation VI, Edited by Gupta and Cohen
Plenum Press, New York, 1996

Fas mRNA is expressed in various cell lines and tissues[4]. In particular, abundant Fas expression was found in the thymus, liver, lung and kidney. Interferon (IFN)-γ or TNF induced the Fas gene expression in macrophages and T cells.

The fact that Fas is a member of the TNF/NGF receptor family suggested that Fas is a receptor of an unidentified cytokine. In 1993, Rouvier *et al.* have found that a cytotoxic T cell line (d10S) can kill the Fas-expressing cells but not the cells which do not express Fas, suggesting that d10S cells express the Fas ligand (FasL)[5]. In fact, the FasL expression could be detected on the cell surface of this cell line using the biotinylated Fas-Fc molecule which is a fusion protein between the Fas extracellular region and the Fc portion of human immunoglobulin[6]. The d10S cells were subjected to repeated sorting with a cell sorter (FACS) to enrich the cells expressing FasL. FasL was purified as a protein of 40 kDa using affinity column of Fas-Fc from the solubilized membrane fraction of a d10S subline overexpressing FasL. Subsequent molecular cloning of FasL has revealed that it is a type II-membrane protein consisting of 270 amino acids[7]. FasL has a significant homology with members of the TNF family which include TNFα, α and β chains of lymphotoxin, CD40 ligand, CD27 ligand and CD30 ligand. FasL is predominantly expressed in activated T and natural killer (NK) cells. It is also constitutively expressed in other tissues such as the stroma of the retina and the Sertoli cells of the testis[8]. TNFα, a prototype of the TNF family, is synthesized as a type II membrane protein, but it is cleaved into a soluble form by a matrix metalloproteinase[9]. Similarly, when FasL is synthesized endogenously in activated T cells or exogenously in COS cells transfected with human FasL expression plasmid, a large amount of FasL activity was found in the culture supernatant of the cells[10,11]. This secretion of FasL from the cells was inhibited by an inhibitor of metalloproteinase, indicating that the membrane-bound FasL is cleaved off by a metalloproteinase to become a soluble form. The soluble as well as the membrane-bound FasL are active to induce apoptosis by binding to its receptor Fas.

II. FasL-INDUCED APOPTOSIS

Recombinant FasL comprising of the extracellular region of human FasL is trimer[10]. Oligomeric anti-Fas antibody (IgM or IgG_3 class) but not its dimeric form (IgG_1 or $F(ab)_2$ fragment) induces apoptosis[1,12,13], suggesting that the Fas receptor must be oligomerized to transduce the signal. Since FasL is a homotrimer, it is likely that binding of FasL induces trimerization of the Fas receptor, which then transduces the signal. A domain of about 80 amino acids in the Fas cytoplasmic region (a death domain) is responsible for the signal transduction, and a similar domain was found in type I (p55) TNF receptor[14] which also can transduce the apoptotic signal[15]. The Fas-mediated apoptosis occurs in enucleated cells[16], and are independent from RNA or protein synthesis[1,2]. Recently, two molecules (FADD/MORT1 and RIP) which bind to the Fas cytoplasmic region were identified by yeast two hybrid system[17,18]. Both of them contain a death domain-like sequence through which they bind to the death domain of the Fas cytoplasmic region. FADD/MORT1 and RIP can induce apoptosis when they are overexpressed in the cells, suggesting they are responsible to mediate the Fas-induced death signal. In particular, a specific binding of FADD to the activated trimerized Fas was recently demonstrated[19].

Similar to the apoptosis in *C. elegans*, the Fas-mediated apoptosis is executed by members of the ICE (interleukin-1β-converting enzyme)/Ced-3 family[20] and can be inhibited by an oncogene product Bcl-2/Ced-9[21,22]. The members of the ICE superfamily were recently divided into three subgroups, ICE-, CPP32- and ICH-1-like proteases. The Fas engagement sequentially activates ICE- and CPP32-like proteases to proceed apoptosis[23]. The cytosolic fraction from the Fas-activated cells quickly induces apoptosis in nuclei from living cells in a cell-free system, and causes DNA degradation[24]. When the cytosolic fraction from growing, non-apoptotic cells was supplemented with the recombinant active CPP32, the extracts now induced apoptosis, confirming that the CPP32-like protease activated during Fas-mediated apoptosis is responsible to proceed the process[23]. However, many questions such as how the activation of ICE is triggered, how Bcl-2 blocks the Fas-mediated apoptosis and what kinds of molecules are responsible for DNA degradation and membrane blebbing, remain to be studied.

III. PHYSIOLOGICAL ROLES OF THE Fas SYSTEM

FasL is expressed in activated T cells and NK cells[5,6,11,25,26], and works as an effector of cytotoxic T lymphocytes (CTL) and NK cells to remove the cells infected by virus, or cancerous cells[3]. Thus, CTL and NK cells have two major mechanisms to kill the target cells. In one mechanism, perforins released from the effector cells create pores in the target cells, through which granzymes, serine proteases, go into the cells to induce the cell death. The other mechanism is the Fas/FasL system, in which FasL expressed on the cell surface of the effector cells or its soluble form binds to the Fas receptor on the target cells to induce apoptosis.

Genetic and molecular analyses of the Fas and FasL genes have revealed that mouse mutations of lymphoproliferation (*lpr*) and generalized lymphoproliferative disease (*gld*) are loss-of-function mutations in the Fas and FasL genes, respectively[27–30]. Mice carrying the *lpr* or *gld* mutation develop lymphadenopathy and splenomegaly by accumulating $CD4^-CD8^-$ cells of the T cell origin, and develop autoimmune disease in some strains of mice. These mutant mice were immunologically and biochemically analyzed, and the results of these analyses indicated that the Fas system is involved in peripheral clonal deletion, and the activation-induced suicide of T cells to down-regulate the immune reaction[3,31]. When the activated T cells accomplished their task, they should be smoothly removed. Otherwise, the organisms would be filled up with the activated T cells. The thymic clonal deletion or negative selection, the other death process which occurs during the development of T cells, seems to be not mediated by the Fas system[32]. The Fas-null mice have elevated levels of immunoglobulins of various classes which include anti-ssDNA and anti-dsDNA antibodies, suggesting an involvement of the Fas system in deletion of activated or autoreactive B lymphocytes[32]. In fact, immunization of mice with antigens rapidly induced the Fas expression in germinal centers[33]. Although these results suggest that FasL-expressing T cells kill the Fas-expressing activated B cells in the germinal centers, its precise mechanism remains to be studied. Human patients carrying a defect in the Fas gene have been also identified[34,35]. These patients showed phenotypes similar to those of *lpr*-mice, including lymphadenopathy, splenomegaly and autoimmune disease.

In addition to the thymocytes and lymphocytes, Fas mRNA is abundantly expressed in the liver, heart, lung and ovary[4]. The Fas-null mice generated by gene targeting showed lymphadenopathy and splenomegaly, which were much more accelerated and pronounced than those in *lpr* mice[32]. Moreover, the Fas-null mice showed hyperplasia in the liver with increased number of hepatocytes carrying enlarged nuclei, suggesting an involvement of the Fas system in turn-over of senescent hepatocytes[34].

IV. PATHOLOGICAL ROLES OF THE Fas SYSTEM

There are two types of diseases which can be caused by disregulation of the Fas system[3]. The first one is the lymphoproliferation and hepatomegaly caused by malfunction of the system as described above. The lymphocytes and hepatocytes accumulated in these animals and patients are not tumorigenic. However, one more mutation in the oncogene or tumor suppressor gene may transform the cells tumorigenic. In this regard, the death factor and its receptor, such as FasL and Fas, may be regarded as tumor suppressor genes.

The other type of diseases is probably caused by the exaggeration of the death factor system. When an agonistic anti-Fas antibody was injected into mice to activate the Fas system *in vivo*, the mice were quickly killed by the liver failure with a symptom similar to human fulminant hepatitis[35]. Human fulminant hepatitis is known to be caused by abnormally activated T cells. And the transformation of hepatocytes with HBV (hepatitis B virus) or HCV (hepatitis C virus) causes the up-regulation of the Fas expression[36]. These results suggest that CTL recognizing the virus antigen expressed on the cell surface of transformed hepatocytes are activated through the T cell receptor, and kill the hepatocytes by Fas/Fas ligand system. If this killing process works in order, it would benefit the organisms. However, when this system is exaggerated, it may causes serious diseases leading to the fulminant hepatitis. It is possible that other CTL-induced autoimunne diseases such as graft-versus-host disease, AIDS and insulitis may also be mediated by the Fas system.

Various cancer patients produce TNFα in a soluble form, and it works as a cachectin to induce systemic tissue damages. We have recently developed an ELISA system to quantify the soluble form of FasL in human serum. Surveying the sera of human patients carrying various hematological disorders showed that the patients of NK lymphoma or large granular lymphocytes leukemia (LGL) of NK or T cell type[11] carry a significant amount of the soluble FasL in their serum. The leukemic cells themselves were found to express the functional FasL on their surface. These patients often show systemic tissue damages such as hepatitis and neutropenia. Since hepatocytes and neutrophils are particularly sensitive to the Fas-mediated apoptosis, it is possible that the systemic tissue damages observed in these patients are due to the FasL in the serum or FasL expressed in the circulating leukemic cells.

PERSPECTIVES

Our project, which was aimed to elucidate how a monoclonal antibody (anti-Fas antibody) kills cells, has led to the identification of a death factor and its receptor. Subsequent unexpected findings of mutations of Fas and FasL genes in *lpr* and *gld* mice pointed out the importance of this death factor system in mammalian homeostasis, specifically in the life and death of lymphocytes. Subsequent findings of FasL in CTL and NK cells have settled the long-time debate on the effector (killer) molecules of these cytotoxic cells. Human diseases caused by a loss-of function mutation of the Fas system have been identified. It is likely that the exaggeration of the Fas system would cause tissue damages in various human diseases. If the involvement of the Fas system in human diseases are proved, neutralizing antibodies against Fas or FasL, or inhibitors of the Fas-mediated apoptosis would have potential uses for clinical application.

ACKNOWLEDGMENTS

The works described here was carried out in Osaka Bioscience Institute, I thank all members in my laboratory. This work was supported in part by Grants-in-Aid from the Ministry of Education, Science and Culture of Japan, and performed in part through Special Coordination Funds of the Science and Technology Agency of the Japanese Government.

REFERENCES

1. S. Yonehara, A. Ishii, and M. Yonehara. A cell-killing monoclonal antibody (anti-Fas) to a cell surface antigen co-downregulated with the receptor of tumor necrosis factor. *J. Exp. Med.* 169: 1747 (1989).
2. N. Itoh, S. Yonehara, A. Ishii, M. Yonehara, S. Mizushima, M. Sameshima, A. Hase, Y. Seto, and S. Nagata. The polypeptide encoded by the cDNA for human cell surface antigen Fas can mediate apoptosis. *Cell* 66: 233 (1991).
3. S. Nagata, and P. Golstein. The Fas death factor. *Science* 267: 1449 (1995).
4. R. Watanabe-Fukunaga, C.I. Brannan, N. Itoh, S. Yonehara, N.G. Copeland, N.A. Jenkins, and S. Nagata. The cDNA structure, expression, and chromosomal assignment of the mouse Fas antigen. *J. Immunol.* 148: 1274 (1992).
5. E. Rouvier, M.-F. Luciani, and P. Golstein. Fas involvement in Ca^{2+}-independent T cell-mediated cytotoxicity. *J. Exp. Med.* 177: 195 (1993).
6. T. Suda, and S. Nagata. Purification and characterization of the Fas ligand that induces apoptosis. *J. Exp. Med.* 179: 873 (1994).
7. T. Suda, T. Takahashi, P. Golstein, and S. Nagata. Molecular cloning and expression of the Fas ligand: a novel member of the tumor necrosis factor family. *Cell* 75: 1169 (1993).
8. T. Griffith, T. Brunner, S. Fletcher, D. Green, and T. Ferguson. Fas ligand-induced apoptosis as a mechanism of immune privilege. *Science* 270: 1189 (1995).
9. D. Pennica, G.E. Nedwin, J.S. Hayflick, P.H. Seeburg, R. Derynck, M.A. Palladino, W.J. Kohr, B.B. Aggarwal, and D.V. Goeddel. Human tumour necrosis factor: precursor structure, expression and homology to lymphotoxin. *Nature* 312: 724 (1984).
10. M. Tanaka, T. Suda, T. Takahashi, and S. Nagata. Expression of the functional soluble form of human Fas ligand in activated lymphocytes. *EMBO J.* 14: 1129 (1995).
11. M. Tanaka, T. Suda, K. Haze, N. Nakamura, K. Sato, F. Kimura, K. Motoyoshi, M. Mizuki, S. Tagawa, S. Ohga, K. Hatake, A. Drummond, and S. Nagata. Fas ligand in human serum. *Nature Medicine* 2: 317 (1996).
12. B.C. Trauth, C. Klas, A.M.J. Peters, S. Matzuku, P. Möller, W. Falk, K.-M. Debatin, and P.H. Krammer. Monoclonal antibody-mediated tumor regression by induction of apoptosis. *Science* 245: 301 (1989).
13. J. Dhein, P.T. Daniel, B.C. Trauth, A. Oehm, P. Möller, and P.H. Krammer. Induction of apoptosis by monoclonal antibody anti-APO-1 class switch variants is dependent on cross-linking of APO-1 cell surface antigens. *J. Immunol.* 149: 3166 (1992).
14. N. Itoh, and S. Nagata. A novel protein domain required for apoptosis: mutational analysis of human Fas antigen. *J. Biol. Chem.* 268: 10932 (1993).
15. L.A. Tartaglia, T.M. Ayres, G.H.W. Wong, and D.V. Goeddel. A novel domain within the 55 kd TNF receptor signals cell death. *Cell* 74: 845 (1993).
16. K. Schulze-Osthoff, H. Walczak, W. Dröge, and P.H. Krammer. Cell nucleus and DNA fragmentation are not required for apoptosis. *J. Cell Biol.* 127: 15 (1994).
17. A.M. Chinnaiyan, K. O'Rourke, M. Tewari, and V.M. Dixit. FADD, a novel death domain-containing protein, interacts with the death domain of Fas and initiates apoptosis. *Cell* 81: 505 (1995).
18. B.Z. Stanger, P. Leder, T.-H. Lee, E. Kim, and B. Seed. RIP: A novel protein containing a death domain that interacts with FAS/APO-1 (CD95) in yeast and causes cell death. *Cell* 81: 513 (1995).
19. F. Kischkel, S. Hellbardt, I. Behrmann, M. Germer, M. Pawlita, P. Krammer, and M. Peter. Cytotoxicity-dependent APO-1 (Fas/CD95)-associated proteins from a death-inducing signaling complex (DISC) with the receptor. *EMBO J.* 14: 5579 (1995).
20. M. Enari, H. Hug, and S. Nagata. Involvement of an ICE-like protease in Fas-mediated apoptosis. *Nature* 375: 78 (1995).
21. N. Itoh, Y. Tsujimoto, and S. Nagata. Effect of bcl-2 on Fas antigen-mediated cell death. *J. Immunol.* 151: 621 (1993).
22. I. Rodriguez, K. Matsuura, K. Khatib, J. Reed, S. Nagata, and P. Vassalli. A bcl-2 transgene expressed in hepatocytes protects mice from fulminant liver destruction but not from rapid death induced by anti-Fas antibody injection. *J. Exp. Med.*, in press (1996).
23. M. Enari, R.V. Talanian, W.W. Wong, and S. Nagata. Sequential activation of ICE-like and CPP32-like proteases during Fas-mediated apoptosis. *Nature*, in press (1996).
24. M. Enari, A. Hase, and S. Nagata. Apoptosis by a cytosolic extract from Fas-activated cells. *EMBO J.* 14: 5201 (1995).
25. T. Suda, T. Okazaki, Y. Naito, T. Yokota, N. Arai, S. Ozaki, K. Nakao, and S. Nagata. Expression of the Fas ligand in T-cell-lineage. *J. Immunol.* 154: 3806 (1995).

26. H. Arase, N. Arase, and T. Saito. Fas-mediated cytotoxicity by freshly isolated natural killer cells. *J. Exp. Med.* 181: 1235 (1995).
27. R. Watanabe-Fukunaga, C.I. Brannan, N.G. Copeland, N.A. Jenkins, and S. Nagata. Lymphoproliferation disorder in mice explained by defects in Fas antigen that mediates apoptosis. *Nature* 356: 314 (1992).
28. M. Adachi, R. Watanabe-Fukunaga, and S. Nagata. Aberrant transcription caused by the insertion of an early transposable element in an intron of the Fas antigen gene of *lpr* mice. *Proc. Natl. Acad. Sci. USA* 90: 1756 (1993).
29. T. Takahashi, M. Tanaka, C.I. Brannan, N.A. Jenkins, N.G. Copeland, T. Suda, and S. Nagata. Generalized lymphoproliferative disease in mice, caused by a point mutation in the Fas ligand. *Cell* 76: 969 (1994).
30. S. Nagata, and T. Suda. Fas and Fas ligand: *lpr* and *gld* mutations. *Immunol. Today* 16: 39 (1995).
31. D. Watanabe, T. Suda, H. Hashimoto, and S. Nagata. Constitutive activation of the Fas ligand gene in mouse lymphoproliferative disorders. *EMBO J.* 14: 12 (1995).
32. M. Adachi, S. Suematsu, T. Suda, D. Watanabe, H. Fukuyama, J. Ogasawara, T. Tanaka, N. Yoshida, and S. Nagata. Enhanced and accelerated lymphoproliferation in Fas-null mice. *Proc. Natl. Acad. Sci. USA* 93: 2137 (1996).
33. D. Watanabe, T. Suda, and S. Nagata. Expression of Fas in B cells of the mouse germinal center and Fas-dependent killing of activated B cells. *Int. Immunology* 7: 1949 (1995).
34. M. Adachi, S. Suematsu, T. Kondo, J. Ogasawara, T. Tanaka, N. Yoshida, and S. Nagata. Targeted mutation in the Fas gene causes hyperplasia in the peripheral lymphoid organs and liver. *Nature Genetics* 11: 294 (1995).
35. J. Ogasawara, R. Watanabe-Fukunaga, M. Adachi, A. Matsuzawa, T. Kasugai, Y. Kitamura, N. Itoh, T. Suda, and S. Nagata. Lethal effect of the anti-Fas antibody in mice. *Nature* 364: 806 (1993).
36. N. Hiramatsu, N. Hayashi, K. Katayama, K. Mochizuki, Y. Kawanishi, A. Kasahara, H. Fusamoto, and T. Kamada. Immunohistochemical detection of Fas antigen in liver tissue of patients with chronic hepatitis C. *Hepatology* 19: 1354 (1994).

Fas SPLICING VARIANTS AND THEIR EFFECT ON APOPTOSIS

Giovina Ruberti,* Isabella Cascino, Giuliana Papoff, and Adriana Eramo

Department of Immunobiology
Institute of Cell Biology
National Research Council
Rome, Italy

1. INTRODUCTION

Higher vertebrates frequently contain multigene families of related ligands and their receptors, often with overlapping specificities. Presumably such an organization allows for greater flexibility in the timing and tissue distribution of these molecules. A further level of complexity is introduced by the fact that variants of the same growth factor or receptor can be encoded as alternative transcripts of the same gene. Such variants may remain membrane associated or may be efficiently secreted.

Alternative splicing of pre-mRNA is a widely used mechanism to increase the coding potential of genes[1]. In most cases alternative splicing gives rise to protein isoforms sharing extensive regions of identity, and varying only in specific domains, thus allowing for the fine modulation of protein function. Alternative splicing is also used to quantitatively regulate gene expression, by an on/off switch for gene function. This has been well documented for a number of Drosophila genes[2], including *Sex lethal* [3] and *transformer*,[4] which are both important regulatory genes in sexual differentiation.

Interestingly several members of the tumor necrosis factor receptor (TNFR) family and of the respective ligand family[5], including Fas/Apo-1(CD95) and FasL, have been shown to be expressed as both membrane-bound and soluble molecules[6].

There is growing evidence that not all Fas positive cells are susceptible to apoptosis induction suggesting that Fas expression is not sufficient to induce a "death signal" and that Fas-mediated apoptosis must be tightly regulated. Multiple pathways originate upon TNFR1[7] and Fas/Apo-1[8–12] crosslinking and both signals are becoming more and more complex. Moreover Fas/Apo-1 has also been reported to generate co-stimulatory signals in lymphocytes[13]. It should be pointed out that engagement of TNFR1 may also result in the

*Address correspondence to Giovina Ruberti, Department of Immunobiology, Institute of Cell Biology, National Research Council, Viale Marx, 43; 00137 Roma-Italy. Tel. +39–6–86090294; Fax +39–6–8273287; e.mail: giovina@biocell.irmkant.rm.cnr.it.

Mechanisms of Lymphocyte Activation and Immune Regulation VI, Edited by Gupta and Cohen
Plenum Press, New York, 1996

generation of non apoptotic signals, among others antiviral activity and cellular proliferation in several systems[14]. Fas-mediated apoptosis, such as that involving TNFR1, undoubtedly involves a delicate balance of receptor/ligand interactions and many of these may be modulated by protein variants.

In this article we summarize our data on human Fas splicing variants that code for soluble and for membrane bound isoforms. The role of these variants in the physiological and pathological modulation of apoptosis and consequently in the regulation of the immune responses is discussed.

2. SEVERAL HUMAN FAS SPLICING VARIANT mRNAs ENCODE SOLUBLE PROTEINS

Our group and others have reported that normal human activated lymphocytes express, in addition to the full length mRNA, less abundant shorter Fas mRNA species that encode soluble Fas proteins[15–18]. The genomic intron/exon organization of the regions surrounding the deleted sequences demonstrated that the transcripts derive by alternative splicing of the Fas gene. A more descriptive nomenclature of the Fas variants has been proposed based on the Fas/Apo-1 gene structure[19,20]: FasExo6Del (previously called FasTMDel)[15,16]; FasExo3,4Del (previously called FasDel2)[16]; FasExo3,4,6Del (previously called FasDel3)[16]; FasExo4Del[17]; FasExo4,6Del[17] and FasExo4,7Del[18]. With the exception of FasExo6Del, which is characterized by an in frame deletion of exon 6, in all of these variants the deletions result in a different reading frame with premature termination codons. Thus, the variants should code for smaller mature Fas proteins with a C-terminal end of 21 or 38 amino acids that differ from those of the Fas membrane-bound form[16]. Interestingly in a patient with a human autoimmune lymphoproliferative syndrome (ALPS)[21,22] a mutation in the 5' splice site of intron 3 probably affecting its function has been reported[22]. This patient expresses FasExo3,4Del in addition to a new splicing variant with an in frame deletion of exon 3 and to normal sized Fas mRNA.

Several of the variants depicted in Fig. 1 retain the hydrophobic leader peptide but lack the hydrophobic transmembrane domain. This suggested that they might be expressed as soluble forms and secreted. In fact it has been shown, in transfected cells, by immunofluorescence and enzyme-linked immunosorbent assays that these mRNA variants can be translated as proteins and, more importantly, that they are secreted in the extra-cellular fluid[15–17]. This initially presented some problems for the splicing variants with deletions in the extracytoplasmic region because several Fas monoclonal antibodies (i.e. CH-11 (Upstate Biotechnology Inc., Lake Placid, NY) and DX2[8] reacted with Fas and FasExo6Del but did not react with the other isoforms[16]. After a rather extensive search three antibodies, M24[26], M1[26] and FasN18 (Santa Cruz, Biotechnology,CA) that recognize all variants were identified[17]. These antibodies should recognize epitopes at the N-terminal region, in fact all variants share only a small region (exons 1 and 2) corresponding to the N-terminal 49 aminoacids of the mature protein. With a sandwich enzyme-linked immunosorbent assay using M24 and FasN18 antibodies the Fas proteins were found to be present in the supernatants at a concentration two orders of magnitude higher than intracellularly, suggesting that the Fas isoforms code for Fas soluble proteins and that these accumulate in the medium[17].

By using in vitro apoptosis inhibition assays it has been shown that the Fas isoforms block Fas-mediated apoptosis induced by agonistic antibody[15,16] and, more importantly, by its natural ligand[17] suggesting a potential regulatory function of these proteins in apoptosis modulation.

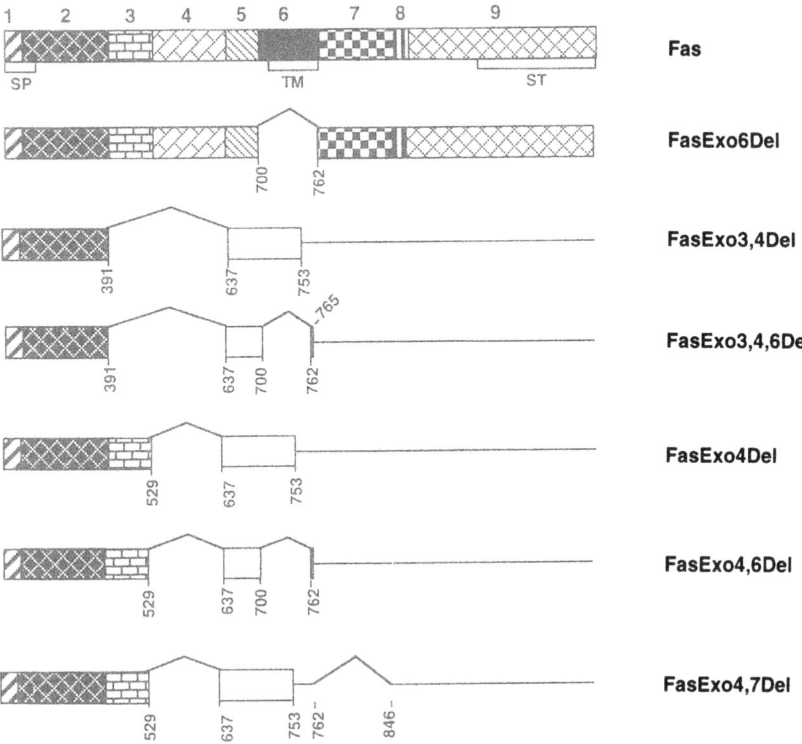

Figure 1. Schematic representation of human Fas cDNAs. The coding regions are represented as boxes. Sequences generated by a different reading frame are indicated as open boxes[16]. SP indicates signal peptide; TM indicates the transmembrane region, and ST represents the signal transducing domain[23,24]. Lines indicate untranslated regions. Numbering is according to Itoh et al.[25]. The coding regions corresponding to exons 1 to 9 are indicated.

Although Fas/Apo-1 is widely expressed in several tissues and cell types, its role in the regulation of the immune system has been more extensively investigated[6]. Fas/Apo-1 and FasL are involved in the down-regulation of immune reactions by controlling the clonal expansion and the accumulation of lymphocytes following antigen challenge in lymphoid organs[27]. Both Fas/Apo-1 and FasL are expressed in activated mature T cells and prolonged activation induce susceptibility to Fas-mediated apoptosis[28–35]. In addition Fas/Apo-1 is expressed on several hematopoietic and non-hematopoietic tumor cells[36–39]. The Fas variants also are known to be expressed not only in peripheral mononuclear cells (PBMC) but also in tumor cells. For example FasExo6Del expression has been reported in human hepatoma[40] and osteosarcoma cells [41]. However a correlation between the expression pattern of the different sFas in human cells and sensitivity to apoptosis remains to be determined. The role of sFas in both physiological and pathological conditions, in particular in the regulation of apoptosis in the immune system, need further investigation. Interestingly, preliminary data from our group show marked qualitative and quantitative variations in the contribution of each variant to the "soluble Fas-pool" in PBMC from different individuals. It will be important to test whether these variations are associated with a differential regulation of apoptosis. The qualitative and quantitative variation in the mRNA expression of these variants in a heterogeneous population may reflect either a differential expression of the variants in particular cell subsets or a regulation by particular stimuli. Interestingly intra-hepatic T lymphocytes in the mouse have been shown to possess different sensitivity to apoptosis according to the mRNA expression of a soluble Fas isoform (Fas β)[42]. Moreover sera of patients with systemic lupus erithematosus (SLE)[15] and sera of patients with different high- and low-grade malignant B- and T-cell leukemias and lymphomas[43] have been reported to have an increased level of FasTMDel (FasExo6Del). The cells producing FasExo6Del that accumulate in the sera have not been characterized. Moreover other studies in SLE patients have failed to confirm an increase of FasExo6Del in the sera[44]. Direct proof of a role of sFas in those diseases remains to be established.

The mechanism of the inhibitory effect is still unknown. However as for the majority of soluble receptors at least two hypotheses can be considered. One might involve extracellular competition for cognate ligand and the second suppression of receptor function by heterotrimer formation. A simple competition of sFas molecules with Fas for binding is unlikely for Ab-induced apoptosis since sFas isoforms, with the exception of FasExo6Del, are not recognized by the CH-11 agonistic IgM Ab[16]. However it may be valid for FasL-induced apoptosis. In this case, this would imply different mechanisms for apoptosis inhibition in the two *in vitro* systems. Regarding the second model a necessary requirement for Fas-induced apoptosis is supposed to be the trimerization of the Fas molecules, as suggested by the homology to TNFα and TNFβ[45–49], the fact that IgM anti-Fas antibodies have agonistic activities[50] and the report that FasL forms homotrimers[51]. Based on these data it can be postulated that sFas forms are still able to trimerize with Fas but that they are not able to form active trimers because of the lack of other domains including the death domain. As a consequence, the signal provided by FasL or by Fas antibody is prevented from reaching the inside of the cell and results in inhibition of apoptosis. The fact that all variants exhibit a marked inhibition points toward the N-terminal domain as being responsible for this effect[17]. This "inactive heterotrimer" model (depicted in Fig. 2), even though more appealing, is at the present only a working hypothesis that may help to organize further experimentation. A physical interaction between Fas and FasL with the N-terminal 49 amino acids Fas region remains to be proven. A possible analogy with sFas apoptosis inhibition can be found in the inhibition of tyrosine kinase activity of the epidermal growth factor receptor (EGFR). This was found to be regulated by a truncated receptor not by simple competition for available EGF but by specific association with the EGFR[52].

3. A TRUNCATED Fas RECEPTOR LACKING THE INTRACYTOPLASMIC SIGNALING DOMAIN DERIVED BY ALTERNATIVE SPLICING

In HUT78, a Fas positive lymphoma T cell line, following treatment with anti-Fas IgM antibody most (about 90%), but not all, cells undergo apoptosis. HUT78 cells were grown in the presence of Fas antibody and spontaneously occuring Fas resistant cell clones were selected and further characterized. They all expressed Fas, were resistant to agonistic antibody and to FasL induced apoptosis. Importantly, exogenous ceramide exposure was able to trigger apoptosis in these cells, indicating that they express a functional apoptotic machinery, including the relevant target for ceramide[53]. To investigate the mechanisms of apoptosis resistance the Fas transcripts were analyzed. While the parental HUT78 cell line contained only the full length sequence, the resistant clones were found to express two transcripts, one corresponding to the expected entire open reading frame, the other showing a deletion of 25 bp corresponding to exon 8. The splicing out of exon 8 brings as a consequence a downstream change of the reading frame with a premature termination codon and elimination of the whole intracytoplasmic death domain[23,24] (Fig. 3). FasExo8Del should thus code for a truncated Fas molecule that lacks the intracellular death-signaling domain.

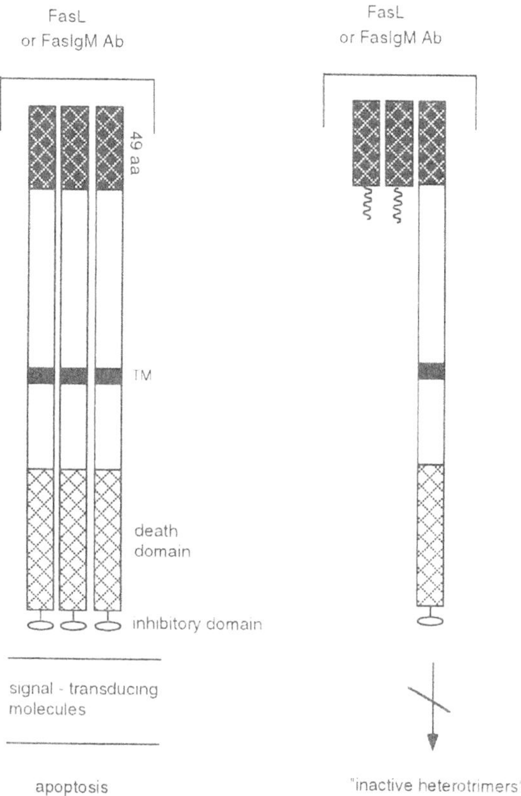

Figure 2. Hypothetical model of apoptosis inhibition.

The mutation involved was identified as a deletion-insertion in the intron7/Exon8 region of the Fas gene.

Moreover this mutation affects the phenotype in a dominant negative fashion, i.e. in the presence of the normal receptor[54]. Interestingly similar dominant negative mutants have been described in patients with ALPS[21,22]. These human Fas gene mutations, similar to the spontaneous loss-of-function mutations of Fas and FasL in *lpr* and *gld* mice[55] have been associated with T cell apoptosis defects, and to the disease progression characterized by accumulation of activated lymphocytes in tissues and autoimmune manifestation.

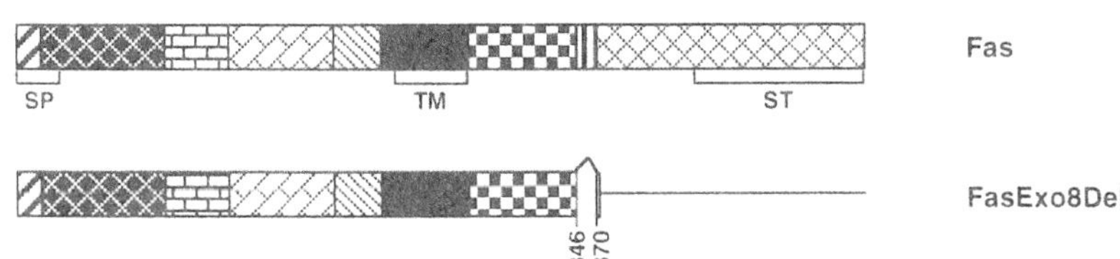

Figure 3. Schematic representation of the open reading frame of Fas and of FasExo8Del.

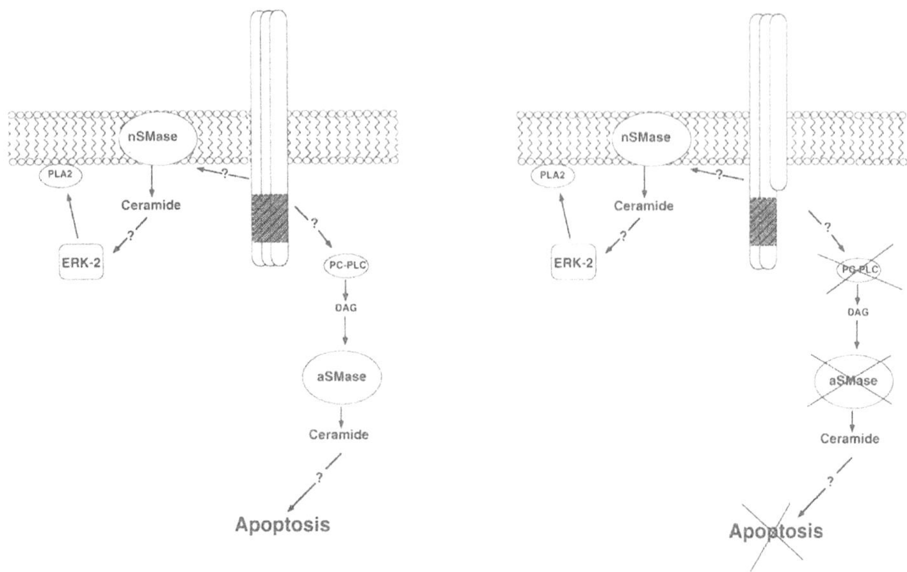

Figure 4. Model of apoptosis resistance.

We postulate that assembly of Fas and Fas truncated receptors interferes with Fas induction signaling (Fig. 4).

Heterotrimers may render the receptor unable to couple with relevant death domain-binding effectors, or unable to "displace" death domain-binding effectors. In fact by using the Yeast two-hybrid method, a number of proteins that have regions of homology with the death domain have been cloned. They physically associate with Fas and if overexpressed induce cell death [56–59]. Moreover sequence comparison revealed homology of the death domain with a protein, reaper, which has been associated with apoptosis in Drosophila[60,61]. These data underline the importance of the death domain structure in apoptosis. However it is also becoming clear that the death domain is a specific protein structure that facilitates protein aggregation and that is also present in proteins not involved in apoptosis[62,63]. Among the proteins that associate with Fas and may be involved in the death signal is emerging FADD/MORT1[56,57]. Crosslinking of CD95 by agonistic antibodies initiated *in vivo* association of receptors with four proteins designated CAP[64]. CAP1 and CAP2 were identified as serine phosporylated FADD/MORT1 proteins. Interestingly in cells transfected with a truncated CD95 lacking the 57 C-terminal amino acids no association of CAP1–4 was observed [64]. Relatively little is known concerning the early biochemical changes involved in Fas/Apo-1 signaling and these are also still controversial. These include sphingomyelin hydrolysis and ceramide production[8–12]. Both a neutral and an acidic sphingomyelinase, as in the TNFR1 crosslinking contribute to the sphingomyelin breakdown. It is noteworthy that Fas crosslinking in the resistant clones fails to activate phosphatidylcholine-specific phospholipase C and the acidic sphingomyelinase[12](Fig. 4).This suggests that the inability of the receptor to couple with relevant transducers may silence a specific pathway and that this may be involved in the propagation of the apoptotic signal.

Mechanisms of Fas-mediated apoptosis resistance, not involving mutations of the Fas gene have been described in tumor cells. These include a defective expression of hematopoietic cell protein tyrosine phosphatase (HCP) in lymphoid cells[65]; high expression of FAP-1, a protein tyrosine phosphatase that associates with Fas[66]; low expression of bax-α, a bcl-2 family member [67]. Using PCR we found that the sensitive HUT78 cell line as well as the resistant cell clones express high levels of both FAP-1 and HCP mRNA suggesting that these phosphatases may modulate Fas-mediated apoptosis in some cell lines but is unlikely to be necessarly involved in Fas signaling (G. Papoff and G. Ruberti unpublished results). Regarding the inhibitory role of the large bcl-2 protein family, there are several reports dealing with this issue. Their role in Fas-mediated apoptosis is still controversial.

We think that several steps of Fas signaling are emerging but many, however, have not consistently been found in all cells. This suggest that Fas signaling is very complex and that the identification and characterization of dominant negative mutants may contribute to the understanding of apoptotic signaling. We think that this approach may allow the identification of: 1) the steps that are necessary for apoptosis induction following Fas/FasL interaction; 2) some mechanisms of apoptosis resistance; 3) some targets to interfere with this pathway.

ACKNOWLEDGMENTS

We are very grateful to R. Tosi and R. Butler for the stimulating discussion and critical reading of the manuscript. This work was partially supported by grant n. 8206–12 AIDS, from National Institute of Health, by a grant from Associazione Italiana per la Ricerca sul Cancro and by the EEC Human Capital and Mobility Program grant CHRX-CT94–0537.

REFERENCES

1. C.W.J. Smith, J. G. Patton, and B. Nadal-Ginard, Alternative splicing in the control of gene expression, *Annu. Rev. Genet.* 23:527 (1989).
2. P.M. Bingham, T. -B. Chou, I. Mims, and Z. Zachar, On/Off regulation of gene expression at the level of splicing, *Trends Genet.* 4:134 (1988).
3. L.R. Bell, E.M. Maine, P. Schedl, and T.W. Cline, Sex-lethal, a drosophila sex determination switch gene, inhibits sex-specific RNA splicing and sequence similarity to RNA binding proteins, *Cell* 55:1037 (1988).
4. R.T. Boggs, P. Gregor, S. Idriss, J.M. Belote, and M. Mc Keown, Regulation of sexual differentiation in *D. melanogaster* via alternative splicing of RNA frome the transformer gene, *Cell* 50:739 (1987).
5. C.A. Smith, T. Farrah, and R.G. Goodwin, The TNF receptor superfamily of cellular and viral proteins: activation, costimulation and death, *Cell* 76:959 (1994).
6. S. Nagata, and P. Golstein, The Fas death factor, *Science* 267:1449 (1995).
7. P. Vandenabeele, W. Declercq, R. Beyaert, and W. Fiers, Two tumour necrosis factor receptors: structure and function, *TIBS* 5:392 (1995).
8. M.G. Cifone, R. De Maria, P. Roncaioli, M.R. Rippo, M. Azuma, L.L. Lanier, A. Santoni, R. and Testi, R, Apoptotic signaling through CD95 (Fas/Apo-1) activates and acidic sphingomyelinase, *J. Exp. Med.* 180:1547 (1994).
9. B.M. Gill, H. Nishikata, G. Chan, T.L. Delovitch, and A. Ochi, Fas antigen and shingomyelin-ceramide turnover-mediated signaling: role in life and death of T lymphocytes, *Immunol. Rev.* 142:113 (1995)
10. E. Gulbins, R. Bissonette, A. Mahboubi, S. Martin, W. Nishioka, T. Brunner, G. Baier, G. Baier-Bitterlich, C. Byrd, F. Lang, R. Kolesnick, A. Altman, and D. Green, Fas-induced apoptosis is mediated via a ceramide-initiated RAS signaling pathway, Immunity 2:341 (1995).
11. C. Tepper, S. Jayadev, B. Liu, A. Bielawska, R. Wolff, S. Yonehara, Y.A. Hannun, and M.F. Seldin, Role of ceramide as an endogenous mediator of Fas-induced cytotoxicity, *Proc. Natl. Acad. Sci. USA* 92:8443 (1995).
12. M.G. Cifone, P. Roncaioli, R. De Maria, G. Camarda, A. Santoni, G. Ruberti, and R. Testi, Multiple pathways originate at the Fas/Apo-1 (CD95) receptor: sequential involvement of phosphatidylcholine-specific phospholipase C and acidic sphingomyelinase in the propagation of the apoptotic signal, *EMBO J.* 14:5859 (1995).

13. M.R. Alderson, R.J. Armitage, E. Maraskowsky, T.W. Tough, E. Roux, K. Schooley, F. Ramsdell, and D.H. Lynch, Fas transduces activation signals in normal human T lymphocytes, *J. Exp. Med.* 178: 2231 (1993).
14. R.A. Heller, and M. Kronke,Tumor necrosis factor-mediated signaling pathways, *J. Cell Biol.* 126:5 (1994).
15. J. Cheng, T. Zhou, C. Liu, J.P. Shapiro, M.J. Brauer, M.C. Kiefer, P.J. Barr, and J.D. Mountz, Protection from Fas-mediated apoptosis by a soluble form of the Fas molecule, *Science* 263:1759 (1994).
16. I. Cascino, G. Fiucci, G. Papoff, and G. Ruberti, Three functional soluble forms of the human apoptosis-inducing Fas molecule are produced by alternative splicing, *J. Immunol.* 154:2706 (1995).
17. G. Papoff, I. Cascino, A. Eramo, G. Starace, D.H. Lynch, and G. Ruberti, An N-terminal domain shared by Fas/Apo-1 (CD95) soluble variants prevents cell death in vitro, (submitted).
18. C. Liu, J. Cheng, and J.D. Mountz, Differential expression of human Fas mRNA species upon peripheral blood mononuclear cell activation, *Biochem. J.* 310:957 (1995)
19. I. Behrmann, H. Walczak, and P.H. Krammer, Structure of the human APO-1 gene, *Eur. J. Immunol.* 24:3057 (1994).
20. J. Cheng, L. Changdan, W.J. Koopman, and J.D. Mountz, Characterization of the human Fas gene. Exon/Intron organization and promoter region, *J. Immunol.* 154:1239 (1995).
21. F. Rieux-Laucat, F. Le Deist, C. Hivroz, I.A.G. Roberts, K.M. Debatin, A. Fischer, and J.P. de Villartay, Mutations in Fas associated with human lymphoproliferative syndrome and autoimmunity, *Science* 268:1347 (1995).
22. G.H. Fisher, F.J. Rosenberg, S.E. Straus, J.K. Dale, L.A. Middelton, A.Y. Lin, W. Strober, M.J. Lenardo, J.M. and Puck, Dominant interfering Fas gene mutations impair apoptosis in a human autoimmune lymphoproliferative syndrome, *Cell* 81:935 (1995).
23. N. Itoh, and S. Nagata, A novel protein domain required for apoptosis, *J. Biol. Chem.* 268:10932 (1993).
24. L.A. Tartaglia, T.M. Ayres, G.H.W. Wong, and D.V. Goeddel, A novel domain within the 55 Kd TNF receptor signals cell death, *Cell* 74:845 (1993).
25. N. Itoh, S. Yonehara, A. Ishii, M. Yonehara, S. Mizushima,M. Sameshima, A. Hase, Y. Seto, and S. Nagata, The polypeptide encoded by the cDNA for human cell surface antigen Fas can mediate apoptosis, *Cell* 66:233 (1991).
26. M.R. Alderson, T.W. Tough, S. Braddy, T. Davis-Smith, E. Roux, K. Schooley, R.E. Miller, and D.H. Lynch, Regulation of apoptosis and T cell activation by Fas-specific monoclonal antibodies. *Int. Immunol.* 6:1799 (1994).
27. P.H. Krammer, et al., The role of APO-1 mediated apoptosis in the immune system, *Immunol. Rev.* 142:175 (1994).
28. L.B. Owen-Schaub, S. Yonehara, W.L. Crump, and E.A. Grimm, DNA fragmentation and cell death is selectively triggered in activated human lymphocytes by Fas antigen engagement,*Cellular Immunol.* 140:197 (1992).
29. C. Klas, K.-M. Debatin, R.R. Jonker, and P.H. Krammer, Activation interferes with the APO-1 patway in mature human T cells, *Int. Immunol.* 5:625 (1993).
30. A. Anel, M. Buferne, C. Boyer, A.-M. Schmitt-Verhulst, and P. Golstein, T cell receptor-induced Fas ligand expression in cytotoxic T lymphocyte clones is blocked by protein tyrosine kinase inhibitors and Cyclosporin A, *Eur. J. Immunol.* 24:2469 (1994).
31. F. Vignaux, E. Vivier, B. Malissen,V. Depraetere, S. Nagata, and P. Golstein, TCR/CD3 coupling to Fas-based cytotoxicity, *J. Exp. Med.* 181:781 (1995).
32. J. Dhein, H. Walczak, C. Baumier, K.-M. Debatin, and P.H. Krammer, Autocrine T-cell suicide mediated by APO-1/(Fas/CD95), *Nature* 373: 438 (1995).
33. T. Brunner, R.J. Mogil, D. La Face, N.J. Yoo, A. Mahboubi, F. Echeverri, S.J. Martin, W.R. Force, D.H. Lynch, C.F. Ware, and D.R.Green, Cell- autonomous Fas (CD95)/Fas-ligand interaction mediates activation- induced apoptosis in T cell hybridomas, *Nature* 373:441 (1995).
34. S.-T. Ju, D.J. Panka, H. Cui, R.M. Ettinger, D.H. El-Khatib, B. Sherr, Z. Stanger, and A. Marshak-Rothstein, Fas (CD95)/FasL interactions required for programmed cell death after T-cell activation, *Nature* 373:444 (1995).
35. M.R. Alderson, T.W. Tough, T. Davis-Smith, S. Braddy, B. Falk, K.A. Schooley, R.G. Goodwin, C.A. Smith, F. Ramsdell, and D.H. Lynch, Fas ligand mediates activation-induced cell death in human T lymphocytes, *J. Exp. Med.* 181:71 (1995).
36. F. Leithauser, J. Dhein, G. Mechtersheimer, K. Koretz, S. Brunderlein, C. Henne, A. Schmidt, K.-M. Debatin, P.H. Krammer, and P. Moller, Constitutive and induced expression of APO-1, a new member of the nerve growth factor/tumor necrosis factor receptor superfamily, in normal and neoplastic cells, *Lab. Invest.* 69:415 (1993).

37. L.B. Owen-Schaub, R. Radinsky, E. Kruzel, K. Berry, and S. Yonehara, Anti-Fas mediated apoptosis in nonhematopoietic tumors: neither Fas/Apo-1 nor bcl-2 expression is predictive of biological responsiveness, *Cancer Res* 54:1580 (1994).
38. M.Y. Mapara, R. Bargou, C. Zugck, H. Dohner, F. Ustaoglu, R.R. Jonker, P.H. Krammer, and B. Dorken, APO-1 mediated apoptosis or proliferation in human chronic B lymphocytic leukemia: correlation with bcl-2 oncogene expression, *Eur. J. Immunol.* 23:702 (1993).
39. K.-M. Debatin, C.K. Goldman, T.A. Waldmann and P.H. Krammer, APO-1 induced apoptosis of leukemia cells from patients with adult T-cell leukemia, *Blood* 81:2972 (1993).
40. G. Natoli, A. Ianni, A. Costanzo, G. De Petrillo, I. Ilari, P. Chirillo, C. Balsano, and M. Levrero, Resistance to Fas-mediated apoptosis in human hepatoma cells, *Oncogene* 11:1157 (1995).
41. L.B. Owen-Schaub, L.S. Angelo, R. Radinsky, C.F. Ware, T.G. Gesner, and D.P. Bartos, Soluble Fas/Apo-1 in tumor cells: a potential regulator of apoptosis, *Cancer Letters* 94:1 (1995).
42. D.P.M. Hughes, and I.N. Crispe, A Naturally occurring soluble isoforms of murine Fas generated by alternative splicing, *J. Exp. Med.* 182:1395 (1995).
43. E. Knipping, K.-M. Debatin, K. Stricker, B. Heilig, A. Eder, and P.H. Krammer, Identification of soluble APO-1 in supernatants of human B- and T-cell lines and increased serum levels in B- and T-cell leukemias, *Blood* 85:1562 (1995).
44. N. Goel, D.T. Ulrich, E. St Clair, J.A. Fleming, D.H. Lynch, M.F. Seldin, Lack of correlation of serum soluble Fas levels and autoimmune disease, *Arthritis Rheum.* 38:1738 (1995).
45. R.A. Smith, and C. Baglioni, The active form of tumor necrosis factor is a trimer, *J. Biol. Chem.* 262:6951 (1987).
46. M.J. Eck, and S.R. Sprang, The structure of tumor necrosis factor at 2.6 A resolution, *J. Biol. Chem.* 264:17595 (1989).
47. E.Y. Jones, D.I. Stuart , and N.P.C. Walker, Structure of tumor necrosis factor, *Nature* 338:225 (1989).
48. M.J. Eck, M. Ultsch, E. Rinderknecht, A.M. deVos, and S.R. Spring, The structure of human lymphotoxin (tumor necrosis factor beta) at 1.9- A resolution, *J. Biol. Chem.* 267:2119 (1992).
49. D. Banner, A. D'Arcy, W. Janes, R. Gentz, H.-J. Schoenfeld, C. Broger, H. Loetscher, and W. Lesslauer, Crystal structure of the soluble human 55kd TNF receptor-human TNFβ complex: implications for TNF receptor activation, *Cell* 73:431 (1993).
50. S. Yonehara, A. Ishii, and M. Yonehara, A cell killing monoclonal antibody (anti-Fas) to a cell surface antigen co-downregulated with the receptor of TNF, *J. Exp. Med.* 169:1747 (1989).
51. M. Tanaka, T. Suda, T. Takahashi, and S. Nagata, Expression of the functional soluble form of human Fas ligand in activated lymphocytes, *EMBO J.* 14:1129 (1995).
52. A. Basu, M. Raghmath, S. Bishayee, and M. Das, Inhibition of tyrosine kinase activity of the epidermal growth factor (EGF) receptor by a truncated receptor form that binds to EGF: role for interreceptor interaction in kinase regulation, *Mol. Cell. Biol.* 9:671 (1989).
53. L.M. Obeid, C.M. Linardic, L.A. Karolak, and Y.A. Hannun, Programmed cell death induced by ceramide, *Science* 259:1769 (1993).
54. I. Cascino, G. Papoff, R. De Maria, R. Testi, and G. Ruberti, Fas/Apo- 1/CD95 receptor lacking the intracytoplasmic signaling domain protects tumor cells from Fas-mediated apoptosis, *J. Immunol.* 156:13 (1996).
55. S. Nagata, and T. Suda, Fas and Fas ligand: *lpr* and *gld* mutations, *Immunol. Today* 16:39 (1995).
56. A.M. Chinnaiyan, K. O'Rourke, M. Tewari, and V.M. Dixit, FADD, a novel death domain-containing protein, interacts with the death domain of Fas and initiates apoptosis, *Cell* 81:505 (1995).
57. M.P. Boldin, E.E. Varfolomeev, Z. Pancer, I.L. Mett, J.H. Camonis, and D. Wallach, A novel protein that interact with the death domain of Fas/Apo-1 contains a sequence motif related to the death domain, *J. Biol. Chem.* 270:7795 (1995).
58. H. Hsu, J. Xiong, and D.V. Goeddel, The TNF receptor 1-associated protein TRADD signals cell death and NF-kB activation, *Cell* 81:495 (1995).
59. B.Z. Stanger, P. Leder, T.-H. Lee, E. Kim, and B. Seed, RIP: a novel protein containing a death domain that interacts with Fas/Apo-1 (CD95) in Yeast and causes cell death, *Cell* 81:513 (1995).
60. K. White, M.E. Grether, J.M. Abrams, L. Young, K. Farrell, and H. Steller, Genetic control of programmed cell death in Drosophila, *Science* 264:677 (1994).
61. P. Golstein, D. Marguet, and V. Depraetere, Homology between Reaper and the cell death domains of Fas and TNFR1, *Cell* 81:185 (1995).
62. E. Feinstein, A. Kimchi, D. Wallach, M. Boldin, and E. Varfolomeev, The death domain: a module shared by proteins with diverse cellular functions, *TIBS* 20:342 (1995).
63. K. Hofman, and J. Tschopp, The death domain motif in Fas (Apo-1) and TNF receptor is present in proteins involved in apoptosis and axonal guidance, *FEBS Lett.* 371:321 (1995).

64. F.C. Kischkel, S. Hellbardt, I. Behrmann, M. Germer, M. Pawlita, P.H. Krammer, and M.E. Peter, Cytotoxicity-dependent APO-1 (Fas/CD95)- associated proteins form a death-inducing signaling complex (DISC) with the receptor, *EMBO J.* 14:5579 (1995).
65. X. Su, T. Zhou, Z. Wang, P. Yang, R.S. Jope, J.D. Mountz, Defective expression of hematopoietic cell protein tyrosine phosphatase (HCP) in lymphoid cells blocks Fas-mediated apoptosis, *Immunity* 2:353 (1995).
66. T. Sato, S. Irie, S. Kitada, and J.C. Reed, FAP-1: a protein tyrosine phosphatase that associates with Fas, *Science* 268:411 (1995).
67. R.C. Bargou, P.T. Daniel, M.Y. Mapara, K. Bommert, C. Wagener, B. Kallinich, H.D. Royer, and B. Dorken, Expression of the bcl-2 gene family in normal and malignant breast tissue: low bax-α expression in tumor cells correlates with resistance towards apoptosis, *Int. J. Cancer* 60:854 (1995).

THE ROLE OF FasL AND TNF IN THE HOMEOSTATIC REGULATION OF IMMUNE RESPONSES

David H. Lynch

Immunex Corp.
51 University Street
Seattle, Washington 98101

The Fas/APO-1 protein (CD95) is expressed on a wide variety of cells and tissues, including monocytes, myelocytes, activated or transformed lymphocytes, thymus, liver, ovary and heart [1]. It is a member of the TNF receptor/nerve growth factor receptor family of cell-surface receptors[2–4]. Early studies had indicated that ligation of this receptor in mAb resulted in the death of certain transformed cell lines via a process known as apoptosis[5,6]. This process is characterized by cytoplasmic condensation, plasma membrane convolution, nuclear condensation and DNA fragmentation[7].

One of the seminal early findings was that the gene encoding this receptor mapped to the *lpr* locus in mice[3], and that the autoimmune disease process observed in *lpr/lpr* mice was due to a mutation in the gene that results in aberrant processing of Fas mRNA[8,9]. It is noteworthy that an identical autoimmune process also occurs in mice homozygous for the *gld* (generalized lymphoproliferative disease) gene, although this gene is clearly distinct from the *lpr* gene (reviewed in 10). Indeed, it had been speculated that the *gld* gene encoded a mutant form of the ligand for Fas[11], a prediction that was ultimately demonstrated to be true using both functional studies[12] and molecular cloning and mapping[13,14]. Thus, the data indicated that Fas/FasL interactions likely played an important role in the homeostatic regulation of immune responses *in vivo*.

Our efforts to understand the mechanisms by which Fas-mediated signals mediated their biologic effects were substantially aided by the generation of a novel panel of Fas-specific mAb. Several of these mAb were found to be agonistic when crosslinked by immobilization on a solid-phase support, but not when used in soluble form. Further analysis of the properties of the mAb demonstrated that some, but not all, of our Fas-specific mAb effectively inhibited apoptosis induced by ligation of Fas with the prototypic Fas-specific mAb CH-11[15]. These reagents were also found to act as potent antagonists of apoptosis induced by the cognate ligand for Fas (FasL), and gave us an important tool to use to elucidate the biological role(s) of Fas/FasL interactions. In this regard, it had been known for some years that agents that lead to T cell activation (e.g., mAb to either CD3 or TcR-α/β, PHA, SEB, phorbol ester and calcium ionophore, soluble antigen) also resulted in substantial levels

Mechanisms of Lymphocyte Activation and Immune Regulation VI, Edited by Gupta and Cohen
Plenum Press, New York, 1996

of T cell death. The striking aspect of the death of T cells after activation was that it occurred through an apoptotic process characterized by cytoplasmic and nuclear condensation and DNA fragmentation. This led us to address the question of whether activation-induced cell death (AICD) of T cells was mediated by Fas/FasL interactions. To answer this question, cultures of cloned human CD4⁺ T cells were stimulated with a variety of agents (including PMA and ionomycin, PHA, SEB and plate-bound CD3 antibody) either in the absence or presence of inhibitors of Fas-mediated apoptosis. In all cases, the AICD induced in cultures was effectively inhibited by either antagonistic Fas-specific mAb or by the Fas.Fc fusion protein, but not by the TNFR.Fc fusion protein[16]. This led us to conclude that FasL/Fas interactions were the principal, if not the only, interactions mediating AICD. Further, the data indicated that TNF/TNFR interactions played no discernable role in this process. However, as will be discussed below, this last conclusion was, in fact, incorrect.

Although the data clearly demonstrated that the process of AICD was mediated by FasL/Fas interactions, populations of activated T cells from Fas-deficient *lpr/lpr* mice could also be demonstrated to undergo AICD under some culture conditions[17]. Careful analysis of the conditions under which the experiment was performed indicated a crucial methodological difference in the manner in which these studies were conducted as compared to earlier studies demonstrating the apparent sole role for FasL/Fas interactions in AICD: the studies in which the role of FasL/Fas interactions in AICD were found to play the sole role used cells that had been in culture for only 24hrs, whereas the studies demonstrating CD3-induced AICD using T cells from *lpr/lpr* mice were cultured for 48hrs. Could this seeminly minor difference in the culture period allow us to detect an additional ligand/receptor interaction pathway that played a role in AICD? A careful kinetic analysis of populations of heterogeneous murine T cells after 24 hr of stimulation by solid-phase CD3 mAb clearly demonstrated that all of the AICD detected was mediated by FasL/Fas interactions. After 48 hr of culture, however, it was also clear that a significant proportion of AICD was also being mediated by TNF/TNFR interactions. Thus, partial inhibition of AICD detected at 48hr of culture could be inhibited by inclusion of TNFR.Fc. Partial inhibition of AICD was also observed in cultures of cells into which the Fas.Fc fusion protein had been added, and virtually complete inhibition was observed in cultures in which a combination of Fas.Fc and TNFR.Fc fusion proteins were added. Thus, the data indicate that both FasL and TNF can play a role in AICD, but they do so with different kinetics. Interestingly, data obtained using purified populations of $CD4^+$ and $CD8^+$ T cells demonstrated that the principal apoptosis-inducing receptor/ligand interaction for $CD4^+$ T cells was via Fas and FasL, whereas for $CD8^+$ T cells it was TNFR and TNF.

It should be noted that there are two distinct receptors for TNF. The 55kD receptor (TNFRI) contains a stretch of amino acids in the intracelluar portion of the molecule that is homologous to a stretch of amino acids in the intracellar portion of Fas[18, 19]. Interestingly, these sequences appear to be critically required for the induction of cell death following ligation of these receptors. The 75 kD receptor (TNFRII) does not contain a homologous sequence, and has generally been considered not to play a role in the induction of cell death upon ligation[20]. Using mice genetically deficient in the ability to express either TNFRI or TNFRII, or genetically deficient in the expresession of both TNFRI and TNFRII we made the rather unexpected finding was that it was the p75 TNFR (which does not contain the intracellular "death domain" homology with Fas) that was mediating the apoptotic process[17]. However, this conclusion should not be construed to mean that apoptosis is not induced by ligation of the TNFR (p55). Instead, binding studies demonstrated that although T cells express significant levels of TNFR (p75), they express little or no TNFR (p55).

The results of these studies have lead us to hypothesize that AICD (whether mediated by Fas/FasL or TNF/TNFR interactions) play an important role in the clonal downsizing of normal immune responses. Thus, envision a site of a T cell-mediated immune response containing $CD4^+$ and/or $CD8^+$ T cells which are actively engaged in the recognition and elimination of their cognate target cells. In such a setting the continuous exposure of the effector T cells to antigen on the target cells would result not only in the chronic stimulation required to induce susceptibility of T cells to undergo apoptosis, but also in the persistent expression of FasL and TNF. Early in the course of the effector cell response the relative ratio of T cells to target cells is undoubtedly low, and thus the likelihood of T cell-T cell interactions is also liable to be low. As target cells are eliminated, however, the effector:target cell ratio is envisioned as steadily increasing, and thus the opportunity for contact between the T cells also increases. As a result of such interactions (especially when essentially all the target cells have been eliminated) a significant proportion of the T cells may undergo apoptosis, thereby decreasing the absolute number of T cells at the end of an immune response.

There are two important predictions of this model: first, that defects in the pathways leading to apoptosis should lead to an accumulation of cells in the periphery; and second, that pathologic conditions which result in continuous simulation of T cells should lead to a progressive deletion of peripheral T cells. Support for the first prediction clearly exists in the form of the *lpr* or the *gld* genes (which are mutations in Fas and FasL, respectively). In mice homozygous for either of these genes, the Fas/FasL interaction pathway in mediating AICD is inoperative, and thus clonal downsizing of immune responses is impaired. As a result of the impaired elimination of lymphoid cells that have "clonally expanded" in response to antigenic stimulation the total number of lymphoid cells would be expected to continuously increase. Indeed, this is exactly the situation observed in peripheral lymphoid tissues in *lpr/lpr* and *gld/gld* mice (reviewed in 10).

Support for the second prediction, that pathologic conditions that result in continuous stimulation of T cells also result in progressive deletion of peripheral lymphocytes, is gaining support with our increasing understanding of HIV-induced pathologies. As early as 1991, Ameisen and Capron proposed that the progressive depletion of $CD4^+$ T cells in HIV was due to inappropriate induction of apoptosis in these cells[21]. Recently Katsikis et al., have demonstrated that peripheral blood T cells from HIV-infected individuals show an increased susceptibility to undergo apoptosis subsequent to ligation of Fas, but not other members of the TNFR family such as CD27, CD30, CD40, 4–1BB, TNFR-I, or TNFR-RP[22]. $CD4^+$ T cells were found to be more susceptible to Fas-mediated apoptosis than $CD8^+$ T cells and the degree of apoptosis induced was higher in T cells from symptomatic HIV^+ individuals than from asymptomatic HIV^+ individuals. Indeed, an inverse correlation of Fas-mediated apoptosis with the peripheral blood $CD4^+$ T cell counts was observed.

The potential involvement of Fas/FasL interactions in mediating the progressive depletion of $CD4^+$ T cells is supported by the recent findings of Badley *et al.*, who demonstrated that HIV infection of monocytic cells not only increases the expression of Fas, but also induces the *de novo* expression of FasL[23]. Further, such HIV-infected macrophages were shown to induce apoptotic death of both Jurkat T cells and peripheral blood T lymphocytes in a Fas-dependent manner. Thus, the data strongly suggest that expression of FasL by macrophages following HIV infection may participate in the progressive depletion of lymphocytes that ultimately results in the immunodeficiency state that defines AIDS.

In conclusion, it is clear that both FasL and TNF play significant roles in the homeostatic regulation of normal immune responses. The data also suggest that these molecules may play important roles in certain pathological conditions arising from either too little or too much activation via these pathways. The potential use of inhibitors of the pathways induced via these receptor-ligand interactions (either at the cell surface or intracellularly) represent important areas for further research with profound implications in both health and disease.

REFERENCES

1. Watanabe-Fukunaga R, Brannan CI, Itoh N, Yonehara S, Copeland NG, Jenkins NA, Nagata S: The cDNA structure, expression, and chromosomal assignment of the mouse Fas antigen. J Immunol 148:1274, 1992
2. Itoh N, Yonehara S, Ishii A, Yonehara M, Mizushima S-I, Sameshima M, Hase A, Seto Y, Nagata S: The polypeptide encoded by the cDNA for human cell surface antigen Fas can mediate apoptosis. Cell 66:233, 1991
3. Watanabe-Fukunaga R, Brannan CI, Copeland NG, Jenkins NA, Nagata S: Lymphoproliferation disorder in mice explained by defects in Fas antigen that mediates apoptosis. Nature 356:314, 1992
4. Smith CA, Farrah T, Goodwin RG: The TNF receptor superfamily of cellular and viral proteins: activation, costimulation, and death. Cell 76:959, 1994
5. Yonehara S, Ishii A, Yonehara M: A cell-killing monoclonal antibody (anti-Fas) to a cell surface antigen co-downregulated with the receptor of tumor necrosis factor. J Exp Med 169:1747, 1989
6. Trauth BC, Klas C, Peters AMJ, Matzku S, Möller P, Falk W, Debatin K-M, Krammer PH: Monoclonal antibody-mediated tumor regression by induction of apoptosis. Science 245:301, 1989
7. Wylie AH, Kerr JFR, Currie AR: Cell death: The significance of apoptosis. Int Rev Cytol 68:251, 1980
8. Kobayashi S, Hirano T, Kakinuma M, Uede T: Transcriptional repression and differential splicing of Fas mRNA by early transposon (ETn) insertion in autoimmune lpr mice. Biochem Biophys Res Commun 191:617, 1993
9. Adachi M, Watanabe-Fukunaga R, Nagata S: Aberrant transcription caused by the insertion of an early transposable element in an intron of the Fas antigen gene of *lpr* mice. Proc Natl Acad Sci USA 90:1756, 1993
10. Cohen PL, Eisenberg RA: *Lpr* and *gld*: Single gene models of systemic autoimmunity and lymphoproliferative disease. Ann Rev Immunol 9:243, 1991
11. Allen RD, Marshall JD, Roths JB, Sidman CD: Differences defined by bone marrow transplantation suggest that *lpr* and *gld* are mutations of genes encoding an interacting pair of molecules. J Exp Med 172:1367, 1990
12. Ramsdell F, Seaman MS, Miller RE, Tough TW, Alderson MR, Lynch DH: *gld/gld* mice are unable to express a functional ligand for Fas. Eur J Immunol 24:928, 1994
13. Lynch DH, Watson ML, Alderson MR, Baum PR, Miller RE, Tough T, Gibson M, Davis-Smith T, Smith CA, Hunter K, Bhat D, Din W, Goodwin RG, Seldin MF: The mouse Fas-ligand gene is mutated in *gld* mice and is part of a TNF family gene cluster. Immunity 1:131, 1994
14. Takahashi T, Tanaka M, Brannan CI, Jenkins NA, Copeland NG, Suda T, Nagata S: Generalized lymphoproliferative disease in mice, caused by a point mutation in the Fas ligand. Cell 76:969, 1994
15. Alderson MR, Tough TW, Braddy S, Davis-Smith T, Roux E, Schooley K, Miller RE, Lynch DH: Regulation of apoptosis and T cell activation by Fas-specific mAb. Int Immunol 6:1799, 1994
16. Alderson MR, Tough TW, Davis-Smith T, Braddy S, Falk B, Schooley KA, Goodwin RG, Smith CA, Ramsdell F, Lynch DH: Fas ligand mediates activation-induced cell death in human T lymphocytes. J Exp Med 181:71, 1995
17. Zheng LX, Fisher G, Miller RE, Peschon J, Lynch DH, Lenardo MJ: Induction of apoptosis in mature T cells by tumour necrosis factor. Nature 377:348, 1995
18. Itoh N, Nagata S: A novel protein domain required for apoptosis. Mutational analysis of human Fas antigen. J Biol Chem 268:10932, 1993
19. Tartaglia LA, Ayres TM, Wong GHW, Goeddel DV: A novel domain within the 55 kd TNF receptor signals cell death. Cell 74:845, 1993
20. Tartaglia LA, Rothe M, Hu YF, Goeddel DV: Tumor necrosis factor's cytotoxic activity is signaled by the p55 TNF receptor. Cell 73:213, 1993
21. Ameisen JC: Programmed cell death and AIDS: from hypothesis to experiment. Immunol Today 13:388, 1992
22. Katsikis PD, Wunderlich ES, Smith CA, Herzenberg LA, Herzenberg LA: Fas antigen stimulation induces marked apoptosis of T lymphocytes in human immunodeficiency virus-infected individuals. J Exp Med 181:2029, 1995
23. Badley AD, McElhinny JA, Leibson PA, Lynch DH, Alderson MR, Paya CV: Upregulation of Fas-ligand expression by human immunodeficiency virus in human macrophages mediates apoptosis of uninfected T lymphocytes. J Virol 70:199, 1996

15

SIGNALS FOR SURVIVAL AND APOPTOSIS IN NORMAL AND NEOPLASTIC B LYMPHOCYTES

John Gordon,* Christopher D. Gregory, Gillian Grafton, and John D. Pound

Department of Immunology
The Medical School
Birmingham, United Kingdom

SUMMARY

B lymphocytes are subject to selection within germinal centers following somatic hyper-mutation on immunoglobulin variable region genes based on their ability to bind antigen with high affinity. Non-selected cells die by apoptosis. Tumors with features of germinal center B cells include follicular center cell lymphoma and Burkitt's lymphoma. We have used the latter extensively as a neoplastic model of germinal center cells and have compared directly the behaviour of cell lines dervied from biopsy material with that of the normal counterparts. Here we describe some of our findings in the two systems with regard to signals regulating survival and apoptosis.

1. INTRODUCTION

Germinal centers (GC) of secondary lymphoid tissues are sites of extensive B cell death and selection. They develop from primary follicles as B lymphocytes proliferate and differentiate in response to T-dependent antigens. Following an initial rapid expansion phase of the antigen-specific B cell, somatic mutation on immunoglobulin (Ig) variable region genes occurs. This may yield changes in the affinity of the B cell's antigen receptors. Cells which have mutated their surface Ig such as to produce a higher affinity binding can be positively selected by competing for native antigen held on follicular dendritic cells in the form of immune complexes. Those B cells which fail to display such an advantage (the majority) will not be selected but instead die by apoptosis. The GC is colonized by so-called

* Address correspondence to: John Gordon, Department of Immunology, The Medical School, Vincent Drive, Birmingham B15 2TT, U.K. **Tel**: 44–121 414 4034; **Fax**: 44–121 414 3599; **e-mail**: j.gordon@bham.ac.uk

Mechanisms of Lymphocyte Activation and Immune Regulation VI, Edited by Gupta and Cohen
Plenum Press, New York, 1996

"large tingible body macrophages": these are highly efficient scavenger cell which are there to facilitate the demise and ensure the clearance of B cells destined to undergo programmed cell death [see refs. 1 and 2].

There are two tumours of B cells which appear to represent a neoplastic expansion of a clone frozen at the GC stage of differentiation. One of these is the most common of the non-Hodgkin lymphoma, that is: Follicular Centre Cell (FCC) Lymphoma. This lymphoma carries a hallmark *t*14:18 translocation and this led to the discovery of the *bcl*-2 protooncogene which is constitutively expressed in FCC lymphoma due to its inappropriate involvement with the Ig locus. It was subsequently shown that the Bcl-2 protein can promote survival in hemopoietic cells by actively suppressing programmed cell death. In marked contrast to its constitutive expression in FCC lymphoma Bcl-2 protein is absent from normal GC B cells strongly implicating the *bcl*-2 gene directly in the oncogenic progression of this tumor.

The other tumor displaying a phenotype consistent with cells of GC origin is Burkitt lymphoma (BL). This lymphoma is characterized by a so-called "starry sky" histology which reflects the high rate of *in situ* apoptosis occurring and the concomitant appearance of large tingible body macrophages to deal with the dying neoplastic cells.

Despite the spontaneous high rate of apoptosis, cell lines can be generated from BL which initially demand a feeder requirement from fibroblasts. With careful weaning, stroma-independent lines can be established which appear to remain faithful to the original biopsy phenotype in that they carry all the surface markers associated with B cells of GC origin. They also retain their propensity to undergo apoptosis. Thus, in addition to a variable degree of spontaneous apoptosis ensuing in these lines, high rate programmed death can be readily induced with a number of external stimuli. These triggers include: serum deprivation; thermal shock; exposure to calcium ionophore; and cross-linking of surface immunoglobulin. BL lines displaying this tendency fall into the so-called "group I" category which have remained biopsy-like. At the other extreme, group III BL lines which have adopted a lymphoblastoid phenotype in culture due to the expression of latent genes from the endogenous Epstein-Barr virus (EBV) - carried by the majority of BL in endemic regions - are now resistant to potential apoptotic signals. This resistance is accompanied by a turn on of *bcl*-2, again through the influence of EBV latent genes. Introduction of *bcl*-2 by transfection into the otherwise Bcl-2-negative group I lines directly renders them resistant to apoptotic stimuli. Group I BL lines thereby provide useful models for investigating mechanisms regulating apoptosis in neoplastic B cells [3]. Their behaviour can be compared directly with that of their normal counterparts isolated from secondary lymphoid tissues.

2. SIGNALS REGULATING SURVIVAL IN GERMINAL CENTER B CELLS

2.1. Antigen Receptor

The B cell antigen receptor (BCR) couples in GC cells to a survival pathway. Apoptosis is a spontaneous process in freshly-isolated GC B cells from human tonsils.

Cross-linking of BCR by polyvalent anti-Ig antibodies held on a solid support provides a transient rescue signal to GC cells from programmed death allowing them to survive for approximately 24 hours longer than they would do otherwise [1]. This interaction *in vitro* is believed to reflect rescue of high affinity mutants by antigen held on FDC *in vivo*.

As in peripheral B cells, BCR of GC cells is coupled to the PI-hydrolytic pathway of signalling via tyrosine kinases [4]. Phorbol ester and calcium ionophore together not only promote efficient rescue in GC B cells from programmed death but also drive high rate proliferation for several days. As would be anticipated, triggering through BCR provokes the release of intracellular Ca^{2+} in GC B cells and more recently was shown to efficiently open a Ca^{2+} gate to provide a sustained entry of extracellular Ca^{2+} [5].

2.2. CD20

CD20 is a tetraspanning pan-B surface membrane glycoprotein with both its amino- and carboxy-termini located within the cytoplasm of the cell. It has been suggested to function as a Ca^{2+} ion channel in resting B cells. We have recently described how the tethering of monoclonal antibody (mAb) to CD20 provides a rescue signal for GC B cells which is of a similar magnitude and duration to that delivered on cross-linking of BCR [5]. Somewhat surprisingly, this rescue was not accompanied by changes in intracellular Ca^{2+} levels through either liberation of intracellular Ca^{2+} stores or the opening of a Ca^{2+} gate. Neither did tethering of CD20 influence changes in intracellular Ca^{2+} mediated via BCR. Despite operating through different signal transduction pathways, rescue mediated via CD20 and BCR was not cooperative with neither additive nor synergistic effects being seen. Both CD20 and BCR were found to be unable to either drive directly or prime GC B cells for proliferation - this was in marked contrast to their actions on resting B cells. The ability of CD20 mAb to effect rescue in GC B cells without influencing Ca^{2+} levels indicate that it functions in survival independently of any role it may have as a Ca^{2+} gate. The possibility that there exists an extracellular ligand for CD20 should therefore be considered.

2.3. CD19/CD21

CD19 and CD21 can associate at the B cell surface as part of a macromolecular signal transduction complex. It was independently shown that mAb to CD19 and selective mAb to CD21 could effect transient rescue in a proportion of GC B cells. Cells rescued via these pathways tended to adopt a plasmablast morphology although clearly not developing toward full plasma cells. While there is as yet no ligand identified for CD19, CD21 has several. It is the receptor for the C3d component of complement and although it has not been assessed *in vitro*, the possibility that C3d-coated onto immune complexes held on FDC might bind and influence GC B cell development can be considered. Interferon-α can also bind to CD21 and has been demonstrated to provide some rescue from apoptosis of GC cells *in vitro*. A more recently identified counterstructure to CD21 is CD23 and soluble recombinant forms of the latter work in synergy with IL1 to promote survival of GC B cells. Interestingly, FDC in the apical aspect of the GC contain large amounts of CD23.

2.4. CD40

The 50 kDa pan-B glycoprotein CD40 and its corresponding counterstructure, CD40L, are undoubtedly central players in GC development. There are now several reviews available on this subject [for example, ref.2] but to summarize: mAb to CD40 can suppress apoptosis in all GC B cells for up to 48 hours; administration into mice undergoing a TD-response of reagents that block the CD40-CD40L interaction leads to the dissolution of the GC; the absence of either CD40L (humans and mice) or CD40 (mice) leads to an absence of GC formation and an inability to mount secondary, isotype-switched responses to TD antigens. Although recent data indicate that TH cells in GC provide rescue signals to GC B cells additional to those delivered via induced CD40L, there is no doubting the crucial role played by the CD40-CD40L pairing in the development of the GC response and, hence, B cell memory.

2.5. Interleukin-2

Initial studies failed to note an effect of IL2 on the GC B cell population, the majority of cells still dying rapidly in its presence. However, it later became clear that a minor subset of GC B cells could be maintained in IL2, actively proliferating for several weeks [6]. This was shown to be a selective effect of IL2 on the small number of cells in the GC B cell fraction which are $CD5^{+ve}$. The precise nature and role of this subset remains unclear although several possibilities have been put forward [6].

2.6. Other Signals

There is one report to suggest that CD38, on its binding by an appropriate mAb, delivers a signal for GC B cell rescue [7]. While this was not reproduced by others (M. Holder and J. Gordon, unpublished observations) the observation is of interest given the homology described between CD38 and cADP ribose-synthase, an enzyme whose product inside the cell is involved in regulating Ca^{2+} levels.

IL10 has been described as directly suppressing apoptosis in GC B cells from human spleen. We have failed to note a direct protective effect of IL10 on tonsillar GC B cells but IL10 was recently shown to be an essential component of all cytokine combinations which were able to maintain CD40-dependent proliferation of GC B cell populations [J. Pound and J. Gordon, submitted]. In this study, it was found that to support GC cell survival and growth over several weeks, there was an absolute requirement for stroma, a multimeric CD40 signal, and cytokines. Cells with a centroblast phenotype were maintained and there was no evidence for differentiation towards plasma cells.

3. SIGNALS FOR APOPTOSIS IN BURKITT LYMPHOMA CELLS

3.1. Transforming Growth Factor-β

Transforming growth factor (TGF)-β has been found to be growth inhibitory for a large number of different cell types including some BL lines. We have now shown that for group I BL lines such growth inhibition is a prelude to their entry into apoptosis under the influence of this cytokine. Of particular interest was the observation that TGF-β could synergise with anti-Ig to drive apoptosis in group I lines [9].

In all group I lines tested, anti-Ig and the cytokine showed cooperativity in their ability to kill. This was related to them imposing an arrest on the cells in the G1 phase of cell cycle. As with GC B cells, apoptosis in group I BL lines has been shown to be effectively rescued by signals delivered through CD40. While CD40-dependent rescue was also evident in response to synergistic combinations of TGF-β and anti-Ig, it was both incomplete and short-lived. This raises interesting possibilities as tumor B cell surface Ig through its expression of a unique idiotype provides a tumor-specific target for immunotherapy.

3.2. Role of Ca^{2+} in Anti-Ig-driven Apoptosis of Burkitt Lymphoma Cells

Clearly it is important to understand the mechanisms through which surface Ig delivers its death signal to BL cells. Where studied, BCR has been found to couple to increases in cytosolic free Ca^{2+} levels at every stage of B cell differentiation. The question we have asked is whether such an increase in BL cells is, in turn, coupled to apoptosis.

In a study of over 20 mAb, we found that the capacity of an anti-Ig antibody to drive apoptosis correlated closely with its ability to raise levels of cytosolic free Ca^{2+} in group I

BL cells. From this, and dose response studies, there was a threshold concentration of Ca^{2+} at around 100 nM below which apoptosis did not proceed. Programmed cell death in group I BL cells could be induced directly, by-passing receptor-mediated events, using thapsigargin which releases Ca^{2+} from intracellular endoplasmic reticulum stores: again, the threshold concentration of free Ca^{2+} required for this receptor-independent killing to occur was 100 nM.

BCR in group I BL cells was shown to be linked to tyrosine kinases. Thus anti-Ig stimulation led to tyrosine phosphorylation of a relatively large number of substrates. Tyrphostins - which specifically inhibit selective classes of protein tyrosine kinases - were found to block both BCR-triggered Ca^{2+}-mobilization and also BCR-mediated apoptosis in BL cells. The accumulated data indicate that the elevation of cytosolic free Ca^{2+} is a key component of the pathway leading to programmed death in group I BL cells and that this is coupled to BCR via tyrosine kinases [G. Grafton, C.D. Gregory, M. Goodall and J. Gordon; in preparation].

4. PERSPECTIVES

The continued study of GC B cells and BL cell lines should throw light on our understanding of mechanisms of apoptosis and survival not only in B cells and their associated tumors but in cells generally. Areas of further study which are warranted and are underway include a detailed analysis of gene products implicated in regulating the cell cycle and programmed cell death. We have already confirmed for both normal GC B cells and BL counterparts that *bcl*-2 expression confers protection from apoptosis. Interestingly though, the ligand for CD40 (CD40L) can confer survival on both cell types in the absence of Bcl-2 induction [10]. This has been shown to relate to its capacity to maintain cells in active cycle. Indeed, we have found a consistent correlation between cell-cycle stage and entry into apoptosis. Other *bcl*-2 family members are now being studied in detail along with p53, c-Myc and the *Rb* gene product.

ACKNOWLEDGMENTS

This work has been supported with funding from the Medical Research Council (U.K.), the Leukaemia Research Fund and the EC BIOTECH Program.

REFERENCES

1. Liu, Y-J., Joshua, D.E., Williams, G.T., Smith, C.A., Gordon, J. and MacLennan, I.C.M. Mechanisms of antigen-driven selection in germinal centres. Nature, 1989, 342: 929–933.
2. Gordon J. CD40 and its ligand: central players in B lymphocyte survival, growth, and differentiation. Blood Rev. 1995. 9:53–56.
3. Gregory, C.D., Dive, C., Henderson, S., Smith, C.A., Williams, G.T., Gordon, J. and Rickinson, A.B. Activation of Epstein-Barr virus latent genes protects human B cells from death by apoptosis. Nature 1991, 349: 612–614.
4. Knox, K.A. and Gordon, J. Signal transduction pathways involved in the regulation of survival in normal and neoplastic B lymphocytes. 1995. In: Apoptosis and the Immune Response (ed. C.D. Gregory) pp313–339. Wiley-Liss, New York.
5. Holder, M., Grafton, G., MacDonald, I., Finney M. and Gordon, J. Engagement of CD20 suppresses apoptosis in germinal center B cells. Eur. J. Immunol. 1995, 25: 3160–3164.
6. Holder, M.J., Abbot, S.D., Milner, A.E., Gregory, C.D., Casamyor, M., Johnson, G.D., MacLennan, I.C.M. and Gordon, J. Interleukin-2 expands and maintains IgM-plasmablasts froma CD5-positive subset

containe dwithin the gemrinal center cell-enriched fraction of human tonsil. Int. Immunol. 1993. 5: 1059–1066.

7. Zupo, S., Rugari, E., Dono, M., Taborelli, G., Malavesi, F. and Ferrarini, M. CD38 signaling by agonistic monoclonal antibody prevents apoptosis of human germinal center B cells. Eur. J. Immunol. 1994. 24:1218–1222.
8. Levy, Y. and Brouet, J-C. Interleukin-10 prevents spontaneous death of germinal center B cells by induction of the Bcl-2 protein. J. Clin. Invest. 1994. 93: 424–450.
9. MacDonald, I., Wang, H., Grand, R., Armitage, R.J., Fanslow, W.C., Gregory, C.D. and Gordon, J. Transforming growth factor-β1 cooperates with anti-Ig for the induction of apoptosis in group I (biopsy-like) Burkitt lymphoma cell lines. Blood (in press).
10. Holder, M.J., Wang, H., Milner, A.E., Casamayor, M. et al. Suppression of apoptosis in normal and neoplastic human B lymphocytes by CD40 ligand is independent of Bcl-2 induction. Eur. J. Immunol. 1993, 23: 2368–2371.

16

REGULATION OF B CELL APOPTOSIS

Robert F. Ashman,[1,2] David Peckham,[1] and Laura L. Stunz[1]

[1] Iowa City DVA Medical Center
[2] Department of Medicine and Department of Microbiology
University of Iowa College of Medicine
Iowa City, Iowa 52242

Only 1% of thymocytes have hypodiploid nuclei, an index of apoptosis. Yet there is evidence that 99% of thymocytes die by apoptosis, and are destroyed so quickly by phagocytes they are hardly apparent histologically. So when we see 1% of spleen B (or T) cells have hypodiploid nuclei, it tells us little about the apoptotic turnover of these cells. Despite evidence that germinal center B cells enter apoptosis when activated through their antigen receptor in the absence of costimulation[1], and that terminal differentiation also leads to apoptosis, life span estimates for the majority of mature B cells run to several months[2,3]. So we were surprised to find that small dense B cells from the spleens of specific pathogen-free mice spontaneously undergo apoptosis *in vitro*[4] so fast that about 75% of small dense B cells have hypodiploid nuclei by 64h. The rate is linear with no lag period to indicate an induction process. Varying fetal calf serum concentration from 1 to 10%, varying the B cell (or T cell) concentration from 0.1 to 10 x 10^6/ml (to rule out autocrine mechanisms), and purifying the cells (adherence to BSA-coated plastic, exposure to anti-thy 1.2 plus complement, centrifugation through Percoll) had no major effect on the spontaneous apoptosis rate[5](and our unpublished data). Neither did inclusion of 10% normal mouse serum, or coculture with other spleen cell types[5]. Cycloheximide (CHX) and other protein synthesis inhibitors accelerated apoptosis in B and T cells[4,5] while blocking apoptosis in thymocytes[5,6]. We appeared to be observing an apoptosis program "hard-wired" into the B cell, very different from that of the thymocyte, requiring no stimulation and no "death protein" synthesis.

How can spontaneous apoptosis at this rate *in vitro* be reconciled with a much longer life *in vivo*, and the near-absence of hypodiploid nuclei in fresh suspensions of small dense B cells? One or more (maybe all) of the following explanations must hold: 1) *in vivo*, the apoptosis rate is controlled by apoptosis inhibitors; 2) *in vivo*, apoptotic B cells are rapidly disposed of by phagocytes, as described for thymocytes[7]; and 3) apoptotic cells rapidly depart from lymphoid organs.

There is no proof yet for the attractive concept that inhibitors of apoptosis restrain B cell apoptosis in certain safe zones *in vivo*. That such inhibition is possible is shown by the fact that the apoptosis rate *in vitro* is decreased by IL-4, γ-interferon, low dose cycloheximide (CHX-L = 0.1 μg/ml), and agents that activate protein kinase C. The most powerful single

Mechanisms of Lymphocyte Activation and Immune Regulation VI, Edited by Gupta and Cohen
Plenum Press, New York, 1996

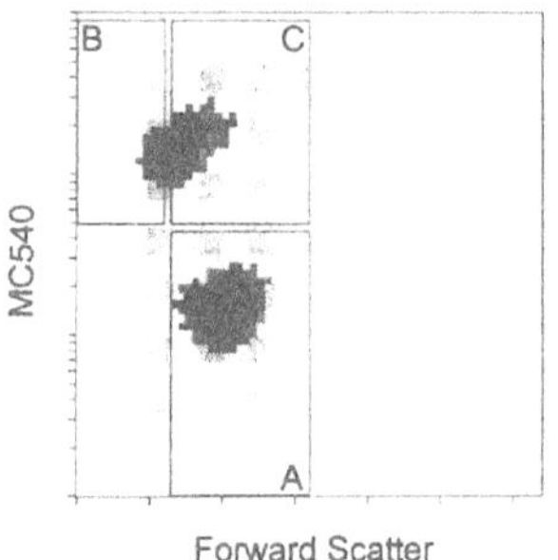

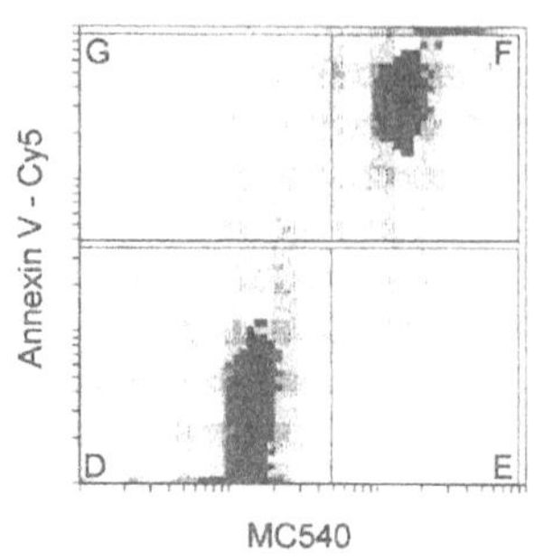

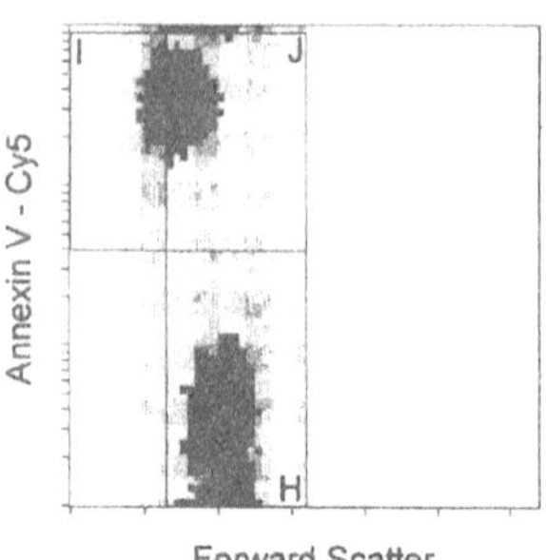

Figure 1. During B cell apoptosis, the increased binding of MC540, decreased forward scatter, and increased exposure of phosphatidyl serine affected the same cells at the same time. The transition was abrupt, with few cells showing intermediate properties. Percoll purified small dense B cells were cultured for 4 hr in 10 μg/ml cycloheximide (CHX) to accelerate apoptosis. B cells were then stained with merocyanine (MC) 540 and annexin V-Cy5 (a phosphatidyl-serine-binding protein), and analyzed by 3-parameter FACS analysis.

apoptosis inhibitor discovered so far is dextran sulfate (DS), a comitogen for B cells[8,9] suggesting a hitherto unknown function for polysulfated proteoglycans that abound in lymphoid organs[10]. Most agents that slow the appearance of hypodiploid nuclei, also slow an earlier event for which we have coined the term "membrane unpacking". En route to apoptosis, B cells abruptly lose the normal phospholipid asymmetry of their plasma membrane. The tight packing of phospholipid head groups is also lost, allowing lipophilic dyes like merocyanine (MC) 540 to bind the outer lamina[11]. This change results from the appearance in the outer lamina of phosphatidyl serine, carrying unsaturated fatty acids. By flow cytometry we observed that MC540 binding, phosphatidyl serine expression (detected by its ligand Annexin-V-Cy5), and cell shrinkage (decreased forward scatter) are concurrent events in individual cells (Fig. 1). Cell sorting experiments have shown that hypodiploid nuclei appear only in cells which have undergone membrane unpacking several hours earlier[11]. Why DNA cleavage and membrane unpacking are so tightly coregulated is an intriguing mystery.

How can we demonstrate the existence of a rapid disposal mechanism *in vivo* absent *in vitro*? *In vitro* 10 μg/ml cycloheximide accelerated B cell apoptosis such that 75% of nuclei were hypodiploid by 16 hr, instead of the normal 30%. When we injected 5 mg of

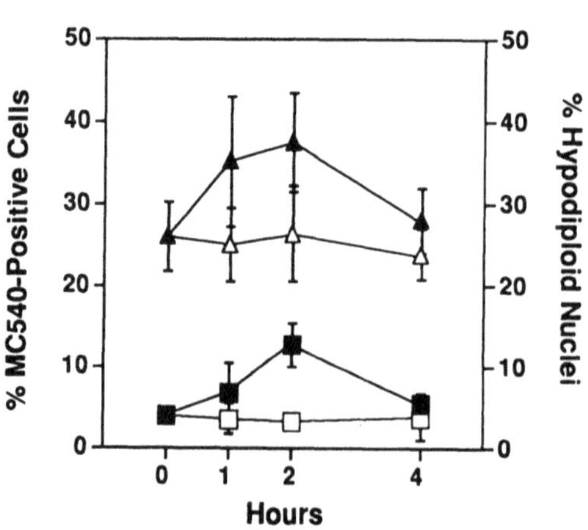

Figure 2. *In vivo* splenocyte apoptosis following parenteral cycloheximide. Mice were injected at time 0 with 5 mg CHX in 0.5 ml of 10% ethanol or with ethanol vehicle alone. At the indicated times, spleen cells were suspended without purification and immediately assayed for MC540-positive cells (triangles) and for hypodiploid nuclei (squares). Black symbols indicate data from mice that received CHX. Open symbols indicate data from mice that received ethanol vehicle only. Mean of three experiments ± SEM. CHX injection induced an increase in apoptotic cells by both assays, peaking at 2h. Between 2h and 4h, the excess apoptotic cells disappeared, testifying to the existence of a rapid disposal mechanism as adept at clearing cells with unpacked membranes as cells with hypodiploid nuclei.

cycloheximide (CHX) intraperitoneally into a group of mice, we observed a wave of apoptosis peaking at 2 hr, and returning to baseline by 4 hr (Fig. 2). Spleen B cells evidently responded to CHX by accelerating apoptosis *in vivo* as well as *in vitro*, but *in vivo* the excess apoptotic cells rapidly disappeared. This rapid disposal was just as effective for cells with unpacked membranes as for cells with hypodiploid nuclei, suggesting that recognition of the unpacked membrane may have triggered the destruction of apoptotic cells before they cleaved their DNA. Perhaps it is critical that nucleosomes not be released before phagocytosis, as they have been shown to act as polyclonal activators[12].

Can we achieve apoptosis protection *in vitro* complete enough to be compatible with an *in vivo* life span estimate of several months? Even dextran sulfate, by itself, allows 15% of B cells to enter apoptosis by 64h (Fig. 3). But combinations of protective agents afforded virtually complete protection, reducing the proportion of apoptotic cells at 64 hr to the level seen with fresh cells. The coregulation of membrane unpacking and hypodiploid nuclei was striking. How much of the total cell death and cell loss *in vitro* is attributable to apoptosis? When plasma membrane impermeability to propidium iodide was used as an index of cell viability, apoptosis prevention preserved both viability and cell recovery (Fig. 4). In other words, apoptosis accounts for virtually all B cell death and loss under our *in vitro* conditions, at least until 64 hr.

How well does apoptosis prevention preserve the ability of cells to respond to polyclonal activation? Flow cytometry with acridine orange provides a means of assessing apoptosis, the G0-to-G1 transition and the G1-to-S transition from the same histograms. An

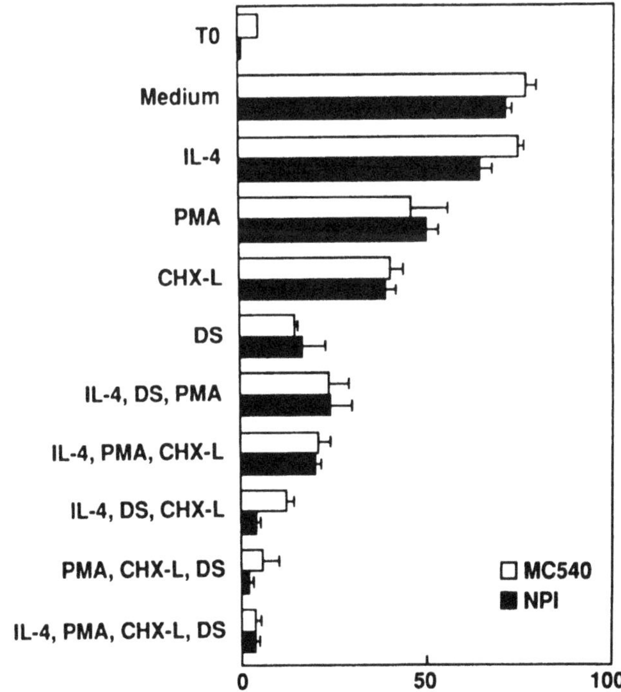

Figure 3. B cells were assayed for membrane unpacking (□) and hypodiploid nuclei (■) as fresh cells (To) or after 64 hr in medium alone, or single apoptosis protective agents or combinations of these agents: IL-4 (25 U/ml), PMA 100 ng/ml, CHX-L 0.1 μg/ml, dextran sulfate (DS),10 μg/ml). Brackets indicate 1 SEM for 3 experiments. PMA, CHX-L, DS and IL-4 kept apoptotic cells at T0 levels for 64h. Note that CHX at 0.1 μg/ml is protective whereas 10 μg/ml accelerates apoptosis.

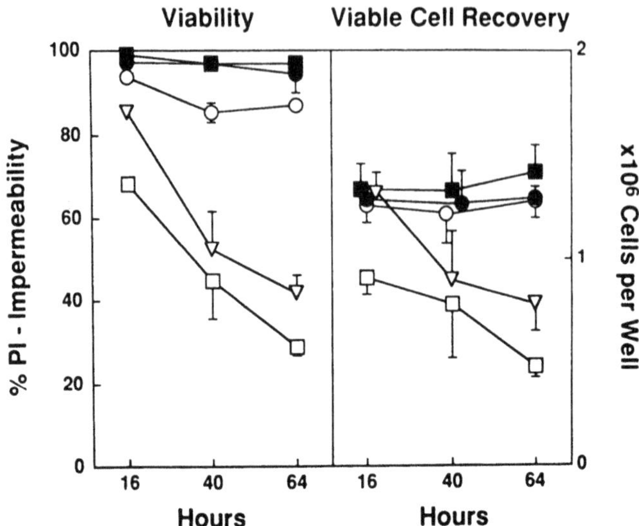

Figure 4. Recovered B cells were counted and viability was measured by propidium iodide permeability after culture up to 64 hr in medium alone (□), 25 U/ml IL-4 (∇), 10 μg/ml DS (○), a combination of 10 μg/ml DS, 100 ng/ml PMA, and 0.1 μg/ml cycloheximide (●), and the same combination + IL-4 (■). Brackets represent 1 SEM for 3 experiments. To = freshly prepared B cells. Adding CHX-L and PMA ± IL-4 to DS preserved PI impermeability and cell recovery for 64 hr.

experiment was conducted where some of the B cells were cultured 64 hr with or without the apoptosis-protective cocktail from Fig. 3 before exposure to LPS for an additional 72 hr (Fig. 5). Although only 25% as many B cells survived for 64h in the absence of apoptosis inhibitors (Fig. 4), there was no major difference in the response of the surviving cells to LPS, measured as % of cells exiting G0 or entering S phase. The apoptosis inhibitors (principally DS) allowed a few cells to enter G1, but most stayed in G0. Yet over 90% of B cells entered cycle when LPS and a cocktail of apoptosis protective agents were present simultaneously, regardless of the cells' previous history.

What accounts for this striking effect? Douglas Green has proposed the "Two Signal/Death or Survival" model for defining the relationship of activation to apoptosis[13], specifically to account for the enigmatic observation that many polyclonal activators (anti-CD3 in T cells and anti-Ig in B cells), and especially elevation of the activation-associated oncogene c-myc[14], can increase apoptosis instead of (or as well as) proliferation in certain circumstances. Fig. 6 presents a version of this model modified to include the spontaneous apoptosis pathway. A stimulus (signal 1) driving cells from G0 also puts them into a "vulnerable period" where the cell must quickly choose between S-phase entry and apoptosis. So far, the various forms of signal 2 (including the bcl-2 family, DS, IL-4, etc.), which shunt activated cells either into S phase or into apoptosis, also appear to have similar effects on the much slower progress of G0 cells into apoptosis. By closing the apoptosis option, the signal 2 apoptosis inhibitors enhanced S phase entry only while cells were being actively driven into the vulnerable period (Fig. 5).

A second example of activation-induced apoptosis which illustrates the usefulness of measuring *in vitro* apoptosis rates derives from a collaboration with Dr. Fred Finkelman. He has observed that when parenteral anti-δ antibody is given to mice, spectacular direct B cell activation ensues[15]. When effective T help was eliminated with anti-CD4, B cell replacement from the bone marrow was prevented by anti IL-7, and B cell opsonization was

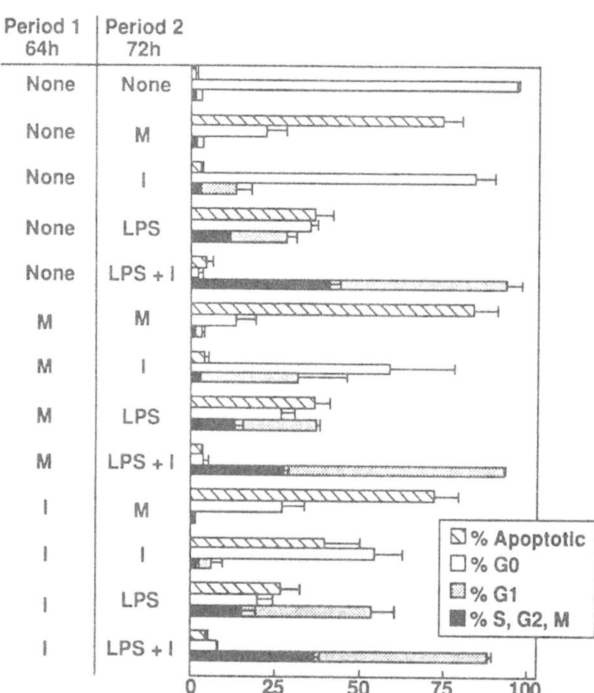

Figure 5. Fresh B cells were assayed by acridine orange flow cytometry to distinguish four categories: G1 (▩) and S, G2 and M (■) sharing the same bar; G0 □ or apoptotic (< 1 x DNA) ◪. Other B cells were cultured 72 hr (period 2) in medium alone (M), a combination of inhibitors (I), 25 μg/ml LPS from (Sigma), and LPS + I. I contains 25 U/ml IL-4, 100 ng/ml PMA, 0.1 μg/ml CHX(L), and 10 μg/ml DS. Other B cells were treated the same way, but only after a 64 hr preincubation (period 1), either in medium (M), or in the inhibitors (I). Brackets indicate 1 SEM for 3 experiments. The percent of cells surviving to 64h that were driven into cycle by LPS between 64 and 136h was similar whether these cells were previously exposed to inhibitors or not. Apoptosis inhibitors present *together with* LPS permitted a much greater proportion of cells to enter cycle.

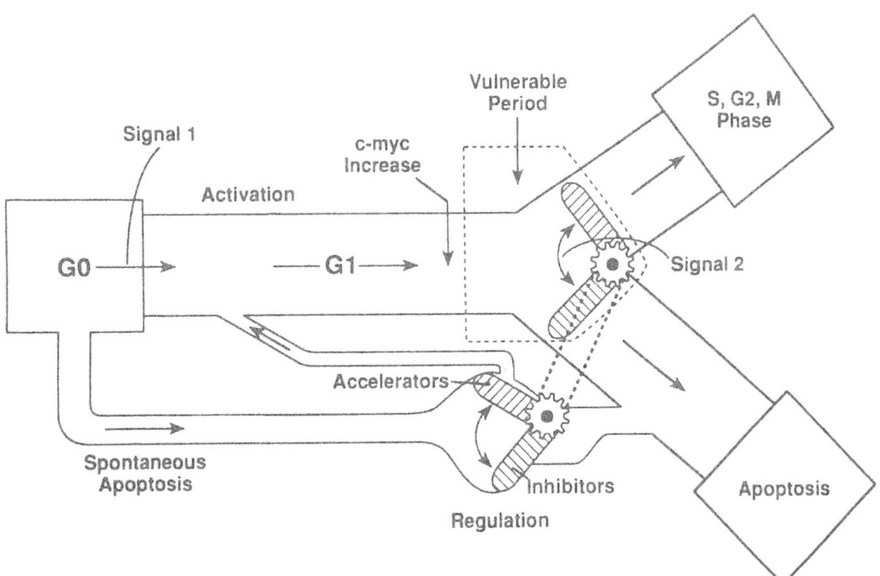

Figure 6. Relationship between B cell activation and apoptosis. Modified from the "Two Signal: Death/Survival" model of Green (13), this model makes the following predictions, in accord with current data: 1) polyclonal activators (Signal 1) or overexpression of c-myc can increase lymphocyte traffic into either cell cycle or apoptosis or both, depending on the prevailing "signal-2" conditions; 2) G0 cells may enter apoptosis by a spontaneous pathway regulated by the same "signal 2" molecules operating in the activation pathway. In certain *in vivo* locations, the combined action of apoptosis inhibitors could block this apoptosis pathway almost completely. 3) Apoptosis inhibitors (signal 2) present together with polyclonal activators (signal 1) will cause the maximal number of cells to enter cycle.

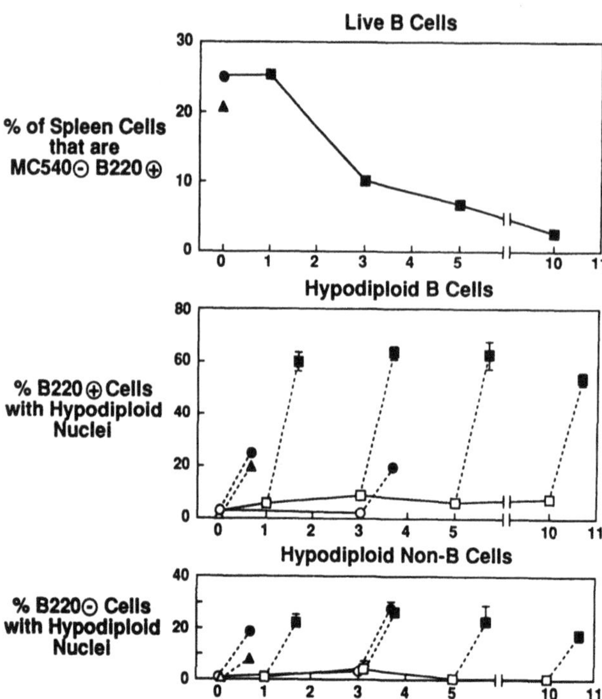

Figure 7. B cell activation by anti-δ *in vivo* leads to accelerated apoptosis and B cell depletion. Anti-B220-Cy5 was used to identify spleen B cells without purification so they could be fixed and exposed to PI and RNAse for hypodiploid cells (2nd and 3rd panel). There were 3 treatment groups: 1) fresh cells (day 0) from untreated animals (triangles); 2) spleen cells from mice given anti-IL-7 every 2 to 3 days beginning 2 weeks before "day 0" to prevent recruitment of new B cells into the spleen, GK 1.5 anti-CD4 on days -4 and 0 to deplete T help, and 2.4 G2 anti-FcγRII on day 0 and the day before harvest to prevent opsonization (circles), 3) spleen cells from mice who received all the above antibodies plus 100 μg anti-δ I.P. on day 0 (squares). Panel 1 shows the disappearance of B cells. Data obtained immediately ex vivo are open symbols with solid lines in panels 2 and 3. Data revealing spontaneous apoptosis rates obtained after 16 hr culture in medium alone are shown as black symbols on dotted lines. Since anti-δ should only affect B cells, non-B cells provide a control (3rd panel). In a T-help-depleted environment, with influx of new B cells from the marrow prevented, anti-δ activation of B cells caused massive B cell loss by apoptosis, shown both by the 5-fold increase in B cells with hypodiploid nuclei and by the increased spontaneous B cell apoptosis rate in vitro. Non-B cells were unaffected.

inhibited by antibody to Fcγ receptors, this wave of activation was followed by disappearance of mature B cells from lymphoid organs[16]. Fig. 7 presents a recreation of this basic experiment modified to include assessment of hypodiploid $B220^+$ cells. The decline of viable $B220^+$ cells as a percentage of total spleen cells underestimated the loss, as total spleen cells also declined. Mice treated with anti-δ, anti-CD4, anti-FcR and anti-IL-7 showed an increase in the percentage of hypodiploid spleen B cells from 1% into the 6–9% range, but also a striking acceleration of their *in vitro* "spontaneous" apoptosis rates, evident on day 1 and still present on day 10. B220-negative cells from the same mice were not affected. Likewise mice subjected to the same treatment protocol without anti-δ showed no acceleration of B cell apoptosis *in vitro* and no increase in hypodiploid cells *in vivo*. This experiment illustrates the ability of signal 1 to accelerate both proliferation and apoptosis in the absence of signal 2, and suggests that *in vitro* apoptosis measurements may reveal the induction of apoptosis even when changes in levels of apoptotic cells are blunted by rapid apoptotic cell disposal *in vivo*.

A physiologically important example of the relationship of activation to apoptosis is the Fcγ-receptor "off signal", whereby IgG immune complexes, or their surrogate whole anti-Ig, actively inhibit proliferation and differentiation of B cells. This dominant inhibitory signal requires crosslinking surface Ig to FcγRII on the B cell surface[17]. IL-4 bypasses the off-signal so that whole anti-Ig (WIg) + IL-4 stimulates proliferation similar to $F(ab')_2$ anti-Ig[18]. We have shown that polyclonal WIg causes early acceleration of B cell apoptosis, whereas equimolar $F(ab')_2$ anti-Ig does not (Fig. 8). Not only does IL-4 block this increase (Fig. 8), but so do several other apoptosis inhibitors (Fig. 9). Thus apoptosis acceleration

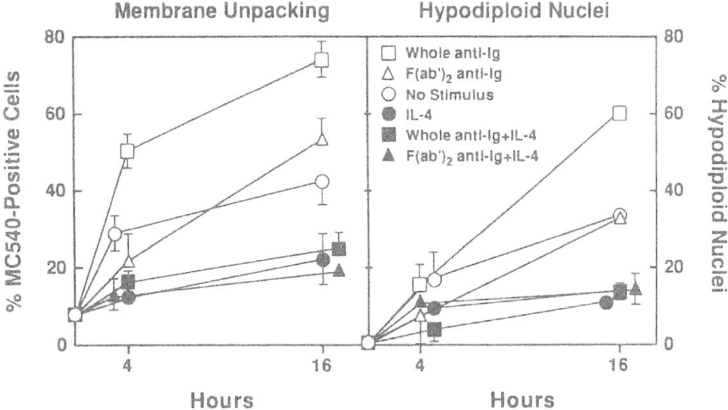

Figure 8. Whole anti-Ig accelerated B cell apoptosis. Small dense B cells were cultured with medium alone (O), rabbit $F(ab')_2$ anti-mouse IgG heavy and light chains at the optimal concentration for cell cycle entry, 25 μg/ml (Δ), an equimolar concentration of whole rabbit anti-mouse IgG (37.5 μg/ml) (□), 25 U/ml IL-4 (●), $F(ab')_2$ anti-Ig + IL-4 (▲), and whole anti-Ig + IL-4 (■). By flow cytometry, membrane unpacking was measured with merocyanine 540 (left panel) and hypodiploid nuclei with propidium iodide (right panel). Means ± SEM of three complete experiments. Brackets indicating 1 SEM are shown where they are larger than the symbol.

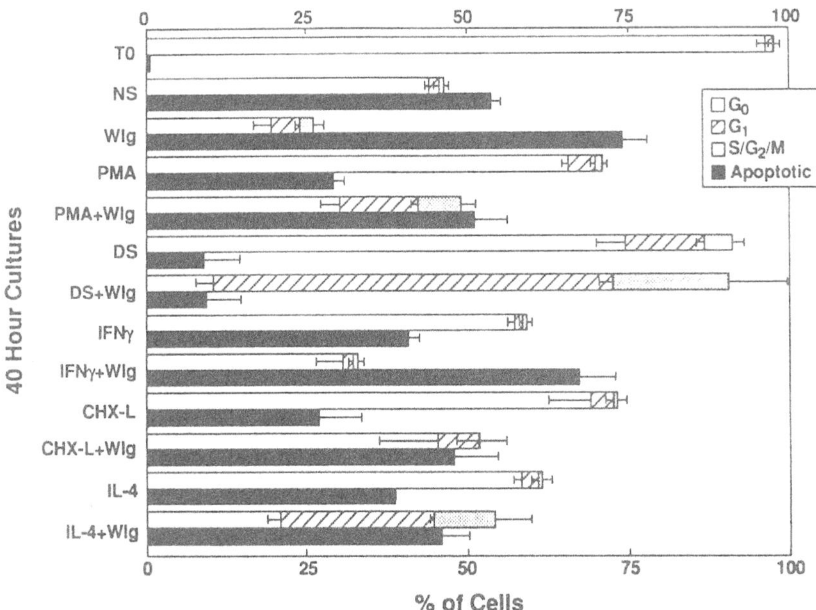

Figure 9. Do reagents which protect B cells from apoptosis acceleration by whole anti-Ig also advance them into cell cycle? Small dense B cells were cultured for 40 hr with medium alone (NS), with whole anti-Ig alone (WIg), with protective agents at the concentrations used in Figs. 1 and 5, or with both WIg and a protective agent. Using flow cytometry with acridine orange, both cell cycle progress and hypodiploid cells (black bars) were assessed on the same samples. Stippled bars show the percent of cells in S/G2/M whereas crosshatched bars show the percent of cells in G1, and open bars show cells in G0. The brackets indicate one SEM from 3 experiments.

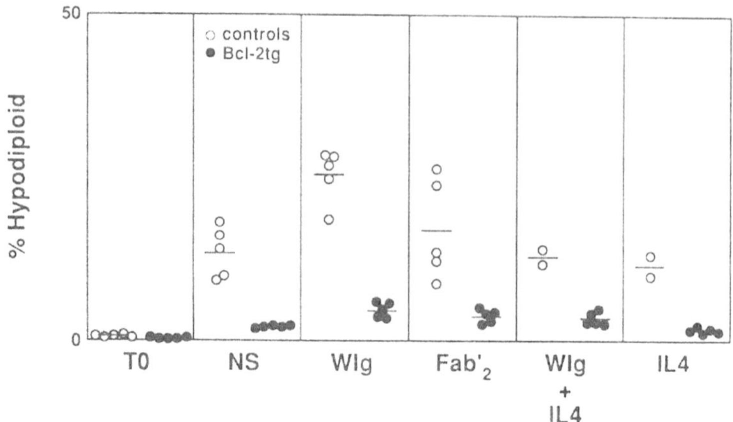

Figure 10. Does bcl-2 overexpression protect B cells from the acceleration of apoptosis by whole anti-Ig? Small dense B cells from the spleens of bcl-2-transgenic mice (each mouse a black symbol) and nontransgenic littermates (each mouse an open symbol) were cultured for 4 hr in the presence of WIg, F(ab')$_2$, anti-Ig, IL-4, or WIg plus IL-4 as in Fig. 8, and assayed for hypodiploid nuclei. Compiled from three experiments. P values for the difference between bcl-2 transgenic and control: $P < .04$ for T0 $< .005$ for NS (no stimulus), $<.0006$ for WIg, and $< .03$ for F(ab')$_2$.

contributes to the inhibition of proliferation under "off signal" conditions. The finding of D'Ambrosio et al.[19] that protein tyrosine phosphatase IC (PTPIC) is essential for WIg induced inhibition of proliferation, together with the role of PTPIC in fas-induced apoptosis[20] and our results, may indicate a broader role for PTPIC in apoptosis regulation than previously supposed.

How do changes in bcl-2 levels relate to the apoptosis regulatory phenomena we have been describing? Acknowledging that bcl-2 itself is only one of several bcl-2 family members worthy of study in this context, we have noted that in bcl-2-IgH-transgenic (TG) mice[21], the rate of apoptosis *in vitro* was slower not only in WIg cultures but across the board, whether apoptosis regulators were present or not (Fig. 10). Nevertheless, the regulatory effect of these

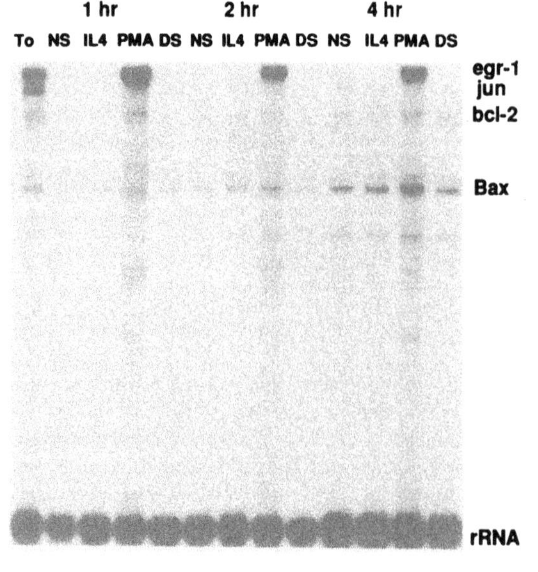

Figure 11. Rnase protection. 2.5 x 10^6 B cells/assay group were treated with medium alone (NS), IL-4, PMA, or DS for 1, 2, or 4 hours, and RNA was harvested. The RNA was hybridized to ^{32}P-labelled RNA probes for egr-1, jun, bcl-2, bax, and rRNA as a load control. The hybridization reactions were treated with Rnase T1 and Rnase A, and after extraction the protected products were run on a denaturing 6% polyacrylamide gel. The gel was dried and exposed to film. All of the RNAs were detectable in freshly isolated (T0) cells. PMA induced egr-1, but no changes in bcl2 or bax mRNAs were seen under conditions of apoptosis protection.

agents was qualitatively preserved in bcl-2-TG mice. No consistent effect of apoptosis inhibitors on B cell bcl-2 or bax mRNA levels was evident in normal mice by RNAase protection (Fig. 11).

We conclude that the apoptosis rate observed *in vitro* appears to be the result of conditions intrinsic to the cell which have already been determined by its developmental stage or previous activation (example: bcl-2 levels, exposure to anti-δ), and extrinsic influences such as lymphokines and polysulfated polyglycans, the various forms of signal 2. Because *in vitro*, the apoptosis behavior of subsets of lymphocytes can be quantitatively examined under circumstances where concentrations of reagents and other variables can be controlled, experiments on apoptosis regulation *in vitro* have much to contribute to our knowledge of the important role of apoptosis in immunology.

ACKNOWLEDGMENTS

The authors wish to thank Claudia Bishop for expert manuscript preparation, and Kristy Miller for superb technical assistance. Without Justin Fishbaugh's wizardry in the UI Flow Cytometry Facility, this data could not have been produced. The antibodies used in Fig. 7 were the generous gift of Dr. Fred Finkelman. Dr. David Lafrenz provided vital assistance with the experimental technique used in Fig. 11. Our work is supported by the American Cancer Society (Dr. Ashman) and the Department of Veteran's Affairs (Dr. Ashman and Dr. Stunz).

REFERENCES

1. MacLennan ICM, Liu Y-J, Johnson G. 1992. Maturation and dispersal of B cell clones during T-dependent antibody responses. Immunol Rev 126:143–161.
2. Grabstein KH, Waldschmidt TJ, Finkelman FD, Hess BW, Alpert AR, Boiani NE, Namen AE, Morrissey PJ. 1993. Inhibition of murine B and T lymphopoiesis in vivo by an anti-interleukin 7 monoclonal antibody. J. of Exper. Med. 178:257–264.
3. Forster I, and Rajewsky K. 1990. The bulk of the peripheral B-cell pool in mice is stable and not rapidly renewed from the bone marrow. Proc. Natl. Acad. Sci. USA 87(12):4781–4789.
4. Illera VA, Perandones CE, Stunz LL, Mower DA, Jr., and Ashman RF. 1993. Apoptosis in splenic B lymphocytes: Regulation by protein kinase C and IL-4. J. of Immunol., 151:2965.
5. Perandones CD, Illera VA, Peckham D, Stunz LL, and Ashman RF. 1993. Regulation of apoptosis in vitro in mature murine spleen T cells. J. of Immunol., 151:3521.
6. Cohen JJ, and Duke RC. 1984. Clucocorticoid activation of a calcium-dependent endonuclease in thymocyte nuclei leads to cell death. J. Immunol., 132:38.
7. Fadok VA, Voelker DR, Campbell PA, Cohen JJ, Bratton DL, and Henson PM. 1992. Exposure of phosphatidylserine on the surface of apoptotic lymphocytes triggers specific recognition and removal by macrophages. J. Immunol. 148:2207–2216.
8. Wetzel GD, Swain SL, Dutton RW, and Kettman JR. 1984. Evidence for two distinct activation states available to B lymphocytes. J. Immunol. 133:2327.
9. Minchin SA, Leitenberg D, Stunz LL, and Feldbush TL. 1990. Polyclonal activation of rat B cells. J. of Immunol. 145:2427.
10. Bradbury MG, and Parish CR. 1991. Characterization of lymphocyte receptors for glycosaminoglycans. Immunol. 72:231.
11. Mower DA, Jr., Peckham DW, Illera VA, Fishbaugh JK, Stunz LL, and Ashman RF. 1994. Decreased membrane phospholipid packing and decreased cell size precede DNA cleavage in mature moust B cell apoptosis. J. Immunol. 152:4832.
12. Bell DA, and Morrison B. 1991. The spontaneous apoptotic cell death of normal human lymphocytes in vitro: the release of, and immunoproliferative response to, nucleosomes in vitro. Clin. Immunol. Immunopath 60:13.

13. Green DR, Mahboubi A, Nishioka W, Oja S, Echeverri F, Shi Y, Glynn J, Yang Y, Ashwell J, and Bissonnette R. 1994. Promotion and inhibition of activation-induced apoptosis in T-cell hybridomas by oncogenes and related signals. Immunol. Rev., 142:321–342.
14. . Shi Y, Bissonnette RP, Glynn JM, Guilbert LJ, Cotter TG, and Green DR. 1992. Role for c-myc in activation-induced apoptotic cell death in T cell hybridomas. Science 257:212.
15. Finkelman FD, Scher I, Mond JJ, Kung JT, and Metcalf ES. 1982. Polyclonal activation of the murine immune system by an antibody to IgD. I. Increase in cell size and DNA synthesis. J. Immunol. 129:629–637.
16. Finkelman FD, Holmes JM, Dulchanina OL, and Morris SC. 1996. Crosslinking of membrane IgD, in the absence of T cell help, kills mature B cells *in vivo.* (ms submitted).
17. Phillips NE, and Parker DC. 1984. Cross-linking of B lymphocyte Fcγ receptors and membrane immunoglobulin inhibits anti-immunoglobulin-induced blastogenesis. J. Immunol. 6:1777.
18. O'Garra A, Rigley KP, Holman M, McGlaughlan JB, and Klaus GGB. 1987. B cell stimulatory factor 1 reverses Fc receptor mediated inhibition of B lymphocyte activation. Proc. Natl. Acad. Sci. USA, 84:6254.
19. D'Ambrosio D, Hippen KL, Minskoff SA, Mellman I, Pani G, Siminovitch KA, and Cambier JC. 1995. Recruitment and activation of PTP1C in negative regulation of antigen receptor signaling by FcγRIIβ1. Science 268:293.
20. Su X, Zhou T, Wang Z, Yang PA, Jope RS, and Mountz JD. 1995. Defective expression of hematopoietic cell protein tyrosine phosphatase (HCP) in lymphoid cells blocks fas-mediated apoptosis. Immunity 2:353–362.
21. McDonnell TJ, Nunez G, Platt FM, Hockenbery D, London L, McKearn JP, and Korsmeyer SJ. 1990. Deregulated bcl-2-immunoglobulin transgene expands a resting but responsive immunoglobulin M and D-expressing B-cell population. Molecular and Cell. Biol. 10:1901–1907.

17

APOPTOTIC CELL DEATH IN THE CHICKEN BURSA OF FABRICIUS

Karen A. Jacobsen,[1] Eustache Paramithiotis,[2] Don L. Ewert,[3] and Michael J. H. Ratcliffe[1]

[1] Department of Microbiology and Immunology, McGill University
3775 University Street, Montreal, Quebec, Canada H3A 2B4.
[2] Division of Developmental and Clinical Immunology, University of Alabama at Birmingham
378 Wallace Tumor Institute, Birmingham, Alabama 35294–3300.
[3] The Wistar Institute
3601 Spruce Street, Philadelphia, Pennsylvania

1. INTRODUCTION

The bursa of Fabricius is the site of primary B cell development in birds, and has been reviewed extensively elsewhere[1–3]. It is productively colonized during embryonic life by a small number (<50,000) of B lineage committed precursors which undergo immunoglobulin gene rearrangements either prior to their migration to the bursa or within the bursal rudiment during a short window of time in embryonic development. By about day 18 of embryonic development (day e18) all Ig gene rearrangements in the chicken have been completed and the adult B cells compartment is derived from this small pool of rearranged B cell precursors[4].

The limited repertoire of B cell specificities generated by Ig gene rearrangement is subsequently diversified by somatic gene conversion events which involve replacement of functionally rearranged V(D)J sequences with homologous sequences from pseudo-V_L or pseudo-VD_H genes. As far as is currently understood, these gene conversion events continue to diversify the chicken V(D)J repertoire within the bursa until the bursa involutes at several months of age[4,5].

B cell precursors which have undergone productive Ig gene rearrangement form, and subsequently proliferate in, epithelial buds of the embryo bursa, each of which ultimately develops into a lymphoid follicle. There are about 10,000 follicles in the normal bursa. After hatch, each follicle comprises a medulla and a cortex, separated by a complex cortico-medullary layer of cells. Cortical lymphocytes are small, densely packed cells whereas medullary cells are larger and more loosely packed.

Each bursal follicle is connected to the bursal lumen by the epithelial tuft containing specialized cells which pinocytose and transport the contents of the bursal lumen into the medulla of the bursal follicle. Since the bursal lumen is connected via the bursal duct to the intestinal lumen, the medullary compartment of the bursal follicle is therefore exposed to antigens derived from the intestinal lumen. Finally since the terminal 1–2cm of the intestine, i.e. the region from the cloaca to the bursal duct, can undergo reverse peristalsis, antigens from the outside environment can be actively transported up the intestine, along the bursal duct, into the bursal lumen and from there into the medulla of lymphoid follicles in the bursa.

2. THE BALANCE BETWEEN LIFE, EMIGRATION AND DEATH IN THE BURSA

2.1 B Cell Production

Following hatch, there is no further immigration of B cell precursors to the bursa. Consequently B cell production in the bursa can be quantitated simply by measuring rates of bursal cell division which can be accomplished in a number of ways. Based on propidium iodide based cell cycle analysis, the bursa of a three week old chicken contains about 12% of cells in S phase of the cell cycle with about the same number of cells in G2/M phases of the cell cycle[6]. Consequently a substantial proportion of bursal cells are rapidly cycling.

The rate at which cells enter the cell cycle can be assessed by metaphase arrest techniques in which the rate of accumulation of cells in metaphase is quantitated following exposure of chickens to substances which block cell cycle progression in metaphase such as vincristine sulfate[7]. In three week old chickens, this approach demonstrated that the rate of entry of cells in the follicular cortex into the cell cycle was substantially different as compared to the cells in the follicular medulla. Rate of accumulation of cortical cells corresponded to a population doubling time of about 12 hours whereas the rate of accumulation of medullary cells corresponded to a doubling time of about 72 hours. In consequence, cortical cells are undergoing rapid cell division, with little cell division being observed in the medulla.

As an alternative to measuring cell cycle progression by metaphase arrest, the rate of entry of cells into S phase of the cell cycle can be estimated by the incorporation of bromodeoxyuridine (BrdUrd) into the DNA of cells during S phase[8, 9]. Short term pulsing of chickens with BrdUrd confirmed that cells in the follicular cortex undergo rapid cell division whereas cells in the follicular medulla are much less rapidly dividing[8]. However, longer pulsing with BrdUrd (~24 hours) demonstrated that virtually all bursal cells become labeled with BrdUrd during this time period[8], including those of the follicular medulla[9]. Thus, following cell division in the cortex, some cells might drop out of cell division and migrate post-mitotically from the cortex to the medulla, thereby reconciling the observed appearance of BrdUrd labeled cells in the follicular medulla with the minimal *in situ* cell division measured by metaphase arrest techniques.

In the three week old chicken the rate of cell division within the cortex of bursal follicles is equivalent to about a doubling in size of the bursa each day. At this age however the bursa increases in cellularity by only about 10% per day[10].

2.2 Emigration to the Periphery

By labeling bursal cells *in situ* with the fluorescent dye fluorescein isothiocyanate (FITC), the emigration rate of B cells from the bursa to the periphery can be determined by the appearance of FITC labeled B cells in peripheral blood or tissues[11,12] and corresponds to an emigration rate of about 1% of the peripheral B cell pool per hour. Taking into account the relative sizes of the bursa and peripheral B cell compartments in three week old chicks, emigration to the periphery therefore accounts for only about 5–10% of the cells produced in the bursa each day.

2.3 Cell Death in the Bursa

The difference between bursal cell production rates and emigration rates to the periphery suggested the possibility that most bursal cells might die *in situ*. Indeed this was confirmed and bursal cell death was shown to occur by apoptosis[13,14]. The death of bursal cells *in vivo* has all the classic hallmarks of apoptotic cell death, including nuclear condensation and the fragmentation of DNA into nucleosome size fragments. From a quantitative standpoint, the amount of low molecular weight DNA in the chicken bursa is almost an order of magnitude higher than is observed in the mammalian or avian thymus, consistent with the large difference between the levels of bursal cell production and the rate of bursal cell emigration to the periphery.

Apoptotic cells can be readily identified in single cell suspensions from bursa tissue. Typically about 5–8% of bursal cells contain the <2N amounts of DNA typical of apoptotic cells[14] again considerably more than is observed in the thymus. Nonetheless, electron microscopy demonstrates that at any given point of time, most of the apoptotic bodies are engulfed within phagocytic cells (e.g. Fig 1). Therefore the quantitation of so called "free apoptotic cells" i.e. those present in single cell suspensions is likely a substantial underestimate of the absolute amount of apoptosis going on in the bursa. At this point it is unclear whether the free apoptotic cells are representative of all apoptotic cells and phagocytosed remnants in the bursa at any given point in time, or represent a discrete subset of apoptotic cells.

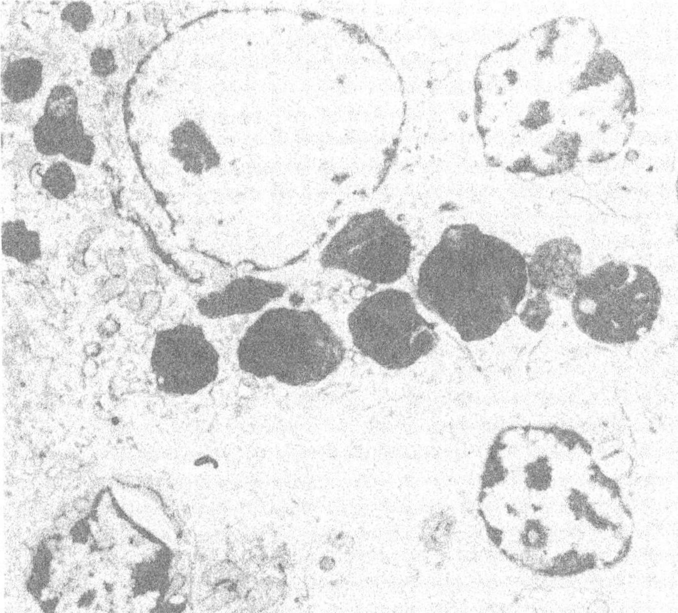

Figure 1. Electron micrograph of a bursal section showing a phagocytic cell containing several condensed nuclei of phagocytosed apoptotic cells.

2.4 Role of sIg Expression in Bursal Cell Apoptosis

We have demonstrated that those cells containing <2N DNA express low to undetectable levels of cell surface immunoglobulin (sIg), in contrast to cycling bursal B cells[14]. The identification of these cells as B lineage followed from their content of Ig heavy and light chain gene rearrangements. In contrast, bursal cells cultured for 12 hours *in vitro* to induce extensive apoptotic cell death, expressed levels of sIg indistinguishable from *ex vivo* sIg^+ bursal cells. Consequently the induction of apoptotic cell death did not in itself induce the loss of bursal cell sIg leading to the conclusion that loss of sIg among bursal cells *in vivo* might induce the apoptotic cell death.

At this point it is unclear why bursal cells might lose the expression of sIg. However, since all developing B cells in the chicken undergo Ig gene rearrangement in the embryo, non-productive gene rearrangement can be eliminated as a mechanism. Non-productive gene conversion events have been observed in a transformed cell line which continues to undergo gene conversion *in vitro*[15]. It seems reasonable to suggest that any bursal cells which have undergone a non-productive gene conversion event leading to a loss of bursal cell sIg expression might be eliminated. Consequently, continued expression of sIg is likely a requirement for continued progression of B cell development.

However, it seems unlikely that gene conversion can be sufficiently inefficient as to result in the elimination of most bursal cells produced on a daily basis. It is therefore possible that limitations in cell-cell contact and/or cytokines required for continued bursal cell survival might also play a role in maintaining the cellularity of the bursa. In this regard, it is therefore possible that those bursal cells observed as phagocytosed apoptotic remnants might have been eliminated for different reasons than those observed as free apoptotic cells, the latter, minor, population including the consequences of non-productive gene conversion events.

2.5 The Interrelationship between Bursal Cell Division and Bursal Cell Death

Apoptotic cells containing <2N DNA are observed to contain BrdUrd within about 4 hours following the incorporation of BrdUrd into the DNA of dividing cells[14]. The proportion of BrdUrd labeled apoptotic cells rapidly increases to about 60–70% of bursal cells within about 12 hours demonstrating that these apoptotic cells are derived from rapidly dividing precursors. This is entirely consistent with the apoptotic cells observed within the cortex of bursal follicles by electron microscopy.

However, following the rapid labeling of two thirds of the apoptotic cells with BrdUrd, labeling of the remaining population takes a further 12 hours. Consequently this suggests that about one third of the bursal cells dying *in situ* do so approximately 12 - 24 hours after their final cell division. This might be a reflection of the observed cell death in the medulla of bursal follicles, which as discussed earlier contains cells which may have migrated in from the cortex of bursal follicles having exited from the cell cycle following division in the cortex.

From a biological standpoint, it seems entirely possible that the reasons for cell death within the bursal cortex might differ from those in the medulla. We have suggested elsewhere that the cortex of bursal follicles is a site for the generation of antibody diversity, whereas the bursal medulla might be a site for the interaction of a diversified pool of specific B cells with antigens derived from the outside environment[16]. Consequently cortical death might reflect a requirement for maintained sIg expression and bursal homeostasis whereas medullary death might reflect selection events imposed by exogenous antigen.

3. Bcl-2 AND ITS EFFECT ON B CELL DEVELOPMENT IN THE BURSA

3.1 Expression of Human *bcl*-2 in the Chicken Bursa

Bcl-2 is a member of a growing family of proteins which regulate programmed cell death, as reviewed elsewhere[17] as well as in this volume. We were interested therefore in determining whether the apoptotic cell death observed in the chicken bursa could be affected by the expression of the *bcl*-2 proto-oncogene. To approach this question, we used a recombinant avian RCAS retrovirus containing the human *bcl*-2 gene[18] (hu-*bcl*-2). The RCAS viruses are replication competent and so infection of chick embryos at about day 10 of embryonic development leads to rapid infection of cells within the developing embryo, production of virus and dissemination throughout a wide range of tissues including both B and T lymphoid cells[19,20].

Typically in RCAS-hu-*bcl*-2 infected chicks, bursal cells expressed substantial levels of the human *bcl*-2 gene product as detected by a monoclonal antibody specific for the human (as opposed to the chicken) *bcl*-2 gene product (Fig 2). *Bcl*-2 expression in such infected chicks was typically biphasic with roughly one half of the bursal cells expressing high levels of hu-*bcl*-2 and one half expressing lower but still significant levels of the transferred gene product. At this point, the reason for the biphasic expression of the transferred gene product is unclear, but may reflect the number of RCAS-*bcl*-2 proviruses in different populations of cells and/or differences in the regulation of viral promoter in subpopulations of bursal cells. Nonetheless, chicks infected as embryos with the RCAS-hu-*bcl*-2 virus contain a high proportion of bursal cells which express high levels of the human *bcl*-2 gene product.

3.2 Human *bcl*-2 Is Functional in Chicken Bursal Cells *in Vitro*

Bcl-2 typically protects cells from death under *in vitro* conditions such as growth factor removal. To demonstrate the functionality of the human *bcl*-2 gene product in chicken cells, bursal cells from RCAS-hu-*bcl*-2 infected chicks were cultured *in vitro*. Under these conditions, bursal cells from normal chickens rapidly undergo apoptotic cell death such that within about 16 hours >90% of cells exhibit the characteristics of apoptotic cell death, containing <2N DNA[14]. We therefore assessed, by two color flow cytometry, cellular DNA

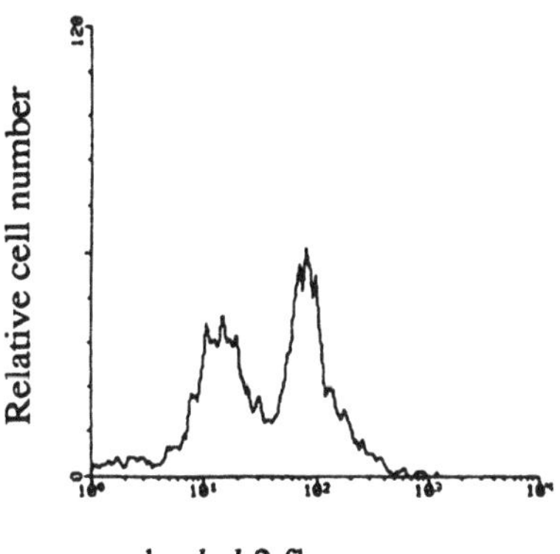

Figure 2. Bursal cells from a three week old chicken infected with RCAS-hu-*bcl*-2 at day e10 were fixed and stained with a monoclonal antibody specific for human (as opposed to chicken) *bcl*-2. The profile from 10,000 events gated on forward and side scatter is shown.

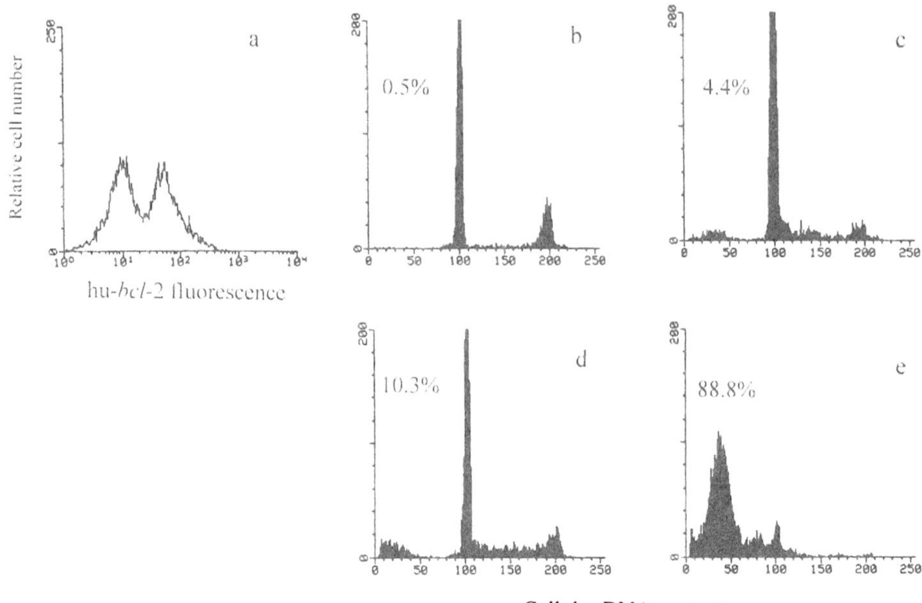

Figure 3. Bursal cells from a three week old chicken infected with RCAS-hu-*bcl*-2 at day e10 were cultured in Iscove's Modified Dulbecco's Medium supplemented as described elsewhere[21], and including 2% normal chicken serum for 4 (b, d) or 12 (c, e) hours. After culture cells were fixed and stained for the expression of human *bcl*-2 and cellular DNA content. Cell cycle analyses from cells expressing high (b, c) or low (d, e) hu-*bcl*-2 expression (e.g. a) are shown from 10,000 gated cells.

content of those bursal cells expressing high levels of hu-*bcl*-2 as compared to those cells containing lower levels of hu-*bcl*-2 (Fig 3). The vast majority of cells expressing high levels of hu-*bcl*-2 were protected from apoptotic cell death within the first 12 hours of culture with fewer than 5% of cells having <2N DNA content. In contrast, those cells expressing lower levels of hu-*bcl*-2 were largely committed to apoptotic cell death with almost 90% cells having <2N DNA content after 12 hours of culture. We can conclude therefore, that the human *bcl*-2 gene product is functional in protecting chicken lymphoid cells from apoptotic cell death induced by tissue culture.

3.3 Human *bcl*-2 Expression Reduces Apoptotic Cell Death in the Chicken Bursa *in Vivo*

Analysis of the frequency of apoptotic cell death from *ex vivo* bursa from chickens infected with the RCAS-hu-*bcl*-2 virus revealed that among those cells expressing high levels of hu-*bcl*-2, the frequency of cells having <2N DNA was <0.5% (in contrast to the 5% - 8% typically observed from the normal bursa[14]) (Fig 4). In contrast, among those cells expressing lower levels of hu-*bcl*-2, about 3% had <2N DNA content. Consequently we can conclude that the expression of high levels of hu-*bcl*-2 gene product prevents the appearance of free apoptotic cells in the bursa. The frequency of apoptotic cells among those bursal cells expressing lower levels of hu-*bcl*-2 was somewhat lower than typically observed in the bursa of either normal chickens or chickens infected with irrelevant RCAS viruses at the same age (about 3 weeks), suggesting that even low level expression of hu-*bcl*-2 may have provided some protection from apoptotic cell death.

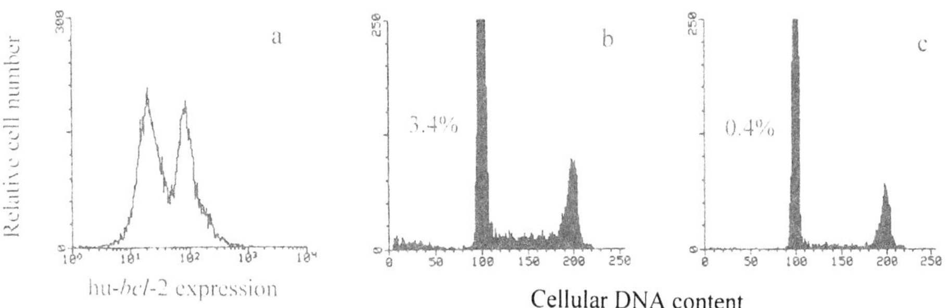

Figure 4. Bursal cells from a three week old chicken infected at day e10 with RCAS-hu-*bcl*-2 were fixed and stained for hu-*bcl*-2 expression (a). The cellular DNA content of those cells expressing low(b) or high (c) levels of hu-*bcl*-2 is shown from 10,000 gated cells.

In addition to the lower rates of apoptotic cell death observed by flow cytometry at the single cell level among those cells expressing high levels of hu-*bcl*-2, electron microscopy revealed considerably fewer overtly apoptotic cells or remnants within the bursa of those chickens infected with the RCAS-hu-*bcl*-2 virus (Fig 5). In sections of the bursa from normal chickens, it is typically difficult to find fields which do not contain apoptotic cells (e.g. Fig 1). In contrast, from the RCAS-hu-*bcl*-2 infected bursae, many fields were free of overtly apoptotic cells. We can conclude, therefore, that the expression of the human *bcl*-2 gene product protects bursal cells from apoptotic cell death *in vivo*.

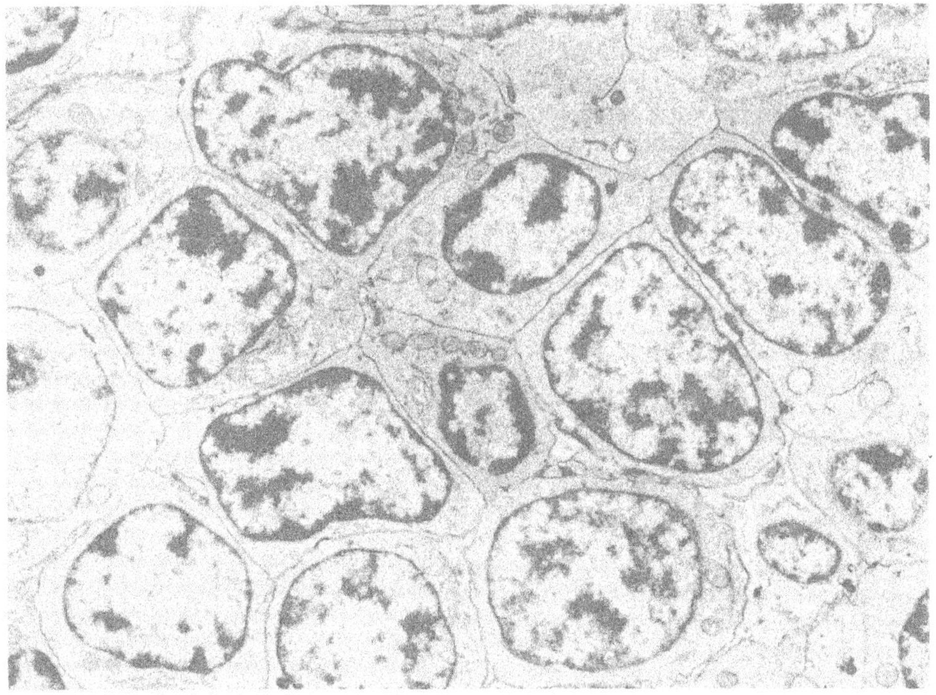

Figure 5. Electron micrograph of a bursal section from a three week old RCAS-hu-*bcl*-2 infected chicken showing no overtly apoptotic cells.

3.4 The Balance between B Cell Production and Emigration from the Bursa in RCAS-hu-*bcl*-2 Infected Chickens

From the distribution of cells in the cell cycle shown in Fig 4, it is also clear that among those cells expressing high levels of hu-*bcl*-2, a smaller proportion of cells is in the S/G2/M phases of the cell cycle. Analysis of the rates of entry of bursal cells into the cell cycle, as judged by BrdUrd incorporation, support this observation (data not shown). There are two possible explanations for this observation. Either the rate of cell division is reduced among those cells expressing high levels of hu-*bcl*-2 i.e. cycling cells take longer to progress through the cell cycle or, alternatively, the increased proportion of cells in G0/G1 reflects those cells which would normally be eliminated by apoptotic cell death, with the actively cycling cells going through the cell cycle at a normal rate. This point is addressed in more detail elsewhere (Ewert, et al., manuscript in preparation). Nonetheless, we can conclude that while the rate of production of bursal cells expressing high levels of hu-*bcl*-2 is reduced as compared to normal bursal cells, this reduction is not likely to be more than a factor of two-fold as compared to those cells expressing lower levels of hu-*bcl*-2. Obviously, the rate of cell division among cells expressing high levels of hu-*bcl*-2 must be high enough to maintain this population of cells within the developing bursa.

The emigration of cells from the bursa was quantitated by intrabursal labeling of cells with FITC[12]. Analysis of the rate of emigration of cells from the bursae of RCAS-hu-*bcl*-2 infected chickens demonstrated that when assessed as a proportion of peripheral blood B cells, the apparent rate of emigration was reduced (Fig 6a). However, the peripheral blood of RCAS-hu-*bcl*-2 infected chickens typically contained about 2 - 4 fold the absolute number of B cells per ml that are contained in the peripheral blood of normal chickens (Fig 6b) and so the absolute emigration rate of cells from the bursae of RCAS-hu-*bcl*-2 infected chickens is slightly increased as compared to normal chickens (Fig 6c).

Overall, therefore, the rate of bursal cell production is decreased by about a factor of two-fold and the rate of emigration from the bursa is increased by less than two-fold in RCAS-hu-*bcl*-2 infected chickens. These are relatively small differences, given the dramatic reduction in apoptotic cell death observed among cells expressing high levels of hu-*bcl*-2 and the observation that the cellularity of RCAS-hu-*bcl*-2 infected bursae is not increased over normal bursal cellularity or the cellularity of the bursae from chickens infected with irrelevant RCAS viruses.

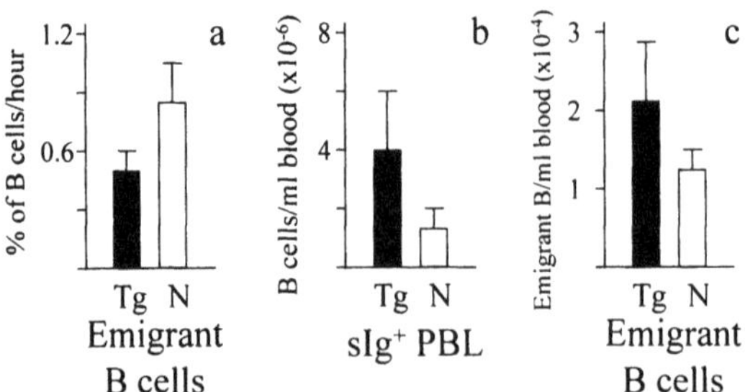

Figure 6. Bursal cells in three week old normal or RCAS-hu-*bcl*-2 infected chickens were labeled *in situ* with FITC and six hours later FITC labeled cells were detected by flow cytometry in the peripheral blood as described elsewhere[12]. The emigration rate determined as a percentage of peripheral blood B cells (a), combined with the absolute number of B cells per ml of blood (b) provides a measure of the absolute rate of bursal cell emigration to the peripheral blood (c).

4. ELIMINATION OF BURSAL CELLS IN RCAS-HU-Bcl-2 INFECTED CHICKENS

The relatively small differences in production and emigration rates in the bursae of RCAS-hu-*bcl*-2 infected chickens were difficult to reconcile with dramatic reductions in the amount of apoptotic cell death observed in single cell suspensions (Fig 4) or in electron microscope sections (Fig 5). However, electron microscopy of sections from RCAS-hu-*bcl*-2 infected bursae revealed the presence of apparently normal nuclei engulfed by phagocytic cells (Fig 7). We have not observed this phenomenon in sections from normal bursae. It seems possible, therefore, that those cells which would normally undergo apoptotic cell death and phagocytosis, leading to the typical pattern of phagocytosed apoptotic remnants seen in Figure 1, are still phagocytosed despite expressing the human *bcl*-2 gene product which protects them from some of the morphological changes normally associated with apoptotic cell death. Under such circumstances, the relatively normal bursal cellularity of RCAS-hu-*bcl*-2 infected chicks would be maintained.

5. CONCLUSIONS

The avian bursa of Fabricius is a site of extensive B cell proliferation as well as B cell death by apoptosis. We have demonstrated that the expression of a transgenic human bcl-2 proto-oncogene protects bursal lymphocytes from apoptotic cell death *in vitro* and dramatically reduces the appearance of cells with the phenotypic characteristics of apoptotic cells *in vivo*. Nonetheless, bursal cellularity is maintained in chickens expressing the human *bcl*-2 gene product, despite relatively small differences in the rates of B cell production and emigration as compared to normal birds. We suggest that this may be, at least in part, due to the phagocytic clearance of morphologically normal cells which would be eliminated by apoptosis and phagocytosis in the absence of the hu-*bcl*-2 transgene.

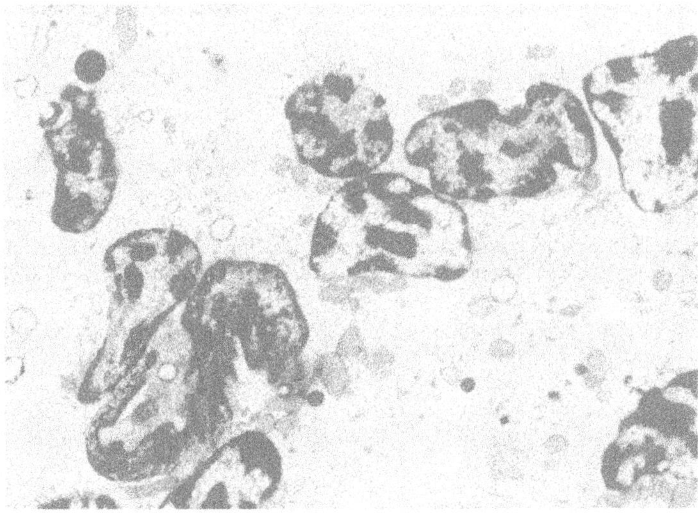

Figure 7. Electron micrograph of a bursal section from a three week old RCAS-hu-*bcl*-2 infected chicken demonstrating numerous apparently normal nuclei contiguous with the mitochondria of a phagocytic cell within the bursa.

Murine neutrophils have a short life-span in the blood and are normally eliminated by phagocytic clearance following apoptotic cell death. Expression of a transgenic *bcl*-2 blocked the induction of apoptosis in neutrophils but did not inhibit their phagocytic uptake *in vitro*[22]. Consequently, the results demonstrated here suggest that the dissociation of apoptotic cell death from the phagocytic clearance of apoptotic cells observed *in vitro* can be extended to the elimination of chicken bursal cells *in vivo*.

6. ACKNOWLEDGMENTS

This work was supported by the National Cancer Institute of Canada, the Medical Research Council of Canada (MT1040) and NIH (CA57516). MJHR is a senior Chercheur-boursier of the Fonds de la Recherche en Santé du Québec, EP is a fellow of the Irvington Institute for Immunological Research and the Natural Sciences and Engineering Council of Canada and KAJ is the recipient of a postdoctoral fellowship from the Medical Research Council of Canada.

7. REFERENCES

1. M.J.H. Ratcliffe, Development of the avian B lymphocyte lineage, *CRC Crit. Rev. Poult. Biol.* 2:207 (1989).
2. J.-C. Weill, and C.-A. Reynaud, The chicken B cell compartment, *Science (Wash. DC)* 238:1094 (1987).
3. M.J.H. Ratcliffe, and E. Paramithiotis, The end can justify the means, *Semin. Immunol.* 2:165 (1990).
4. W.T. McCormack, L.W. Tjoelker, and C.B. Thompson, Avian B-cell development: generation of an immunoglobulin repertoire by gene conversion. *Ann. Rev. Immunol.* 9:219 (1991).
5. M.J.H. Ratcliffe, and K.A. Jacobsen, Rearrangement of chicken immunoglobulin genes in chicken B cell development, *Semin. Immunol.* 6:175 (1994).
6. M.J.H. Ratcliffe, and L. Tkalec, Cross-linking of the surface immunoglobulin on lymphocytes from the bursa of Fabricius results in second messenger generation. *Eur. J. Immunol.* 20:1073 (1990).
7. J.D. Reynolds, Mitotic rate maturation in the Peyer's patches of fetal sheep and in the bursa of Fabricius of the chick embryo, *Eur. J. Immunol.* 17:503 (1987)
8. E. Paramithiotis, and M.J.H. Ratcliffe, B cell emigration directly from the cortex of lymphoid follicles in the bursa of Fabricius, *Eur. J. Immunol.* 24:458 (1994).
9. S. Ekino, Role of environmental antigens in B cell proliferation in the bursa of Fabricius at neonatal stage. *Eur. J. Immunol.* 23:772 (1993).
10. P.M. Lydyard, G.E. Grossi, and M.D. Cooper, Ontogeny of B cells in the chicken I. Sequential development of clonal diversity in the bursa, *J. Exp. Med.* 144:79 (1976).
11. O. Lassila, Emigration of B cells from the chicken bursa of Fabricius, *Eur. J. Immunol.* 19:955 (1989).
12. E. Paramithiotis, and M.J.H. Ratcliffe, Bursa-dependent subpopulations of peripheral blood B lymphocytes in chicken blood, *Eur. J. Immunol.* 23:96 (1993).
13. B. Motyka, and J.D. Reynolds, Apoptosis is associated with the extensive B cell death in the sheep ileal Peyer's patches and the chicken bursa of Fabricius: a possible role in B cell selection, *Eur. J. Immunol.* 21:1951 (1991).
14. E. Paramithiotis, K.A. Jacobsen, and M.J.H. Ratcliffe, Loss of surface immunoglobulin expression precedes B cell death by apoptosis in the bursa of Fabricius. *J. Exp. Med.* 181:105 (1995).
15. J.-M. Buerstedde, C.-A. Reynaud, E.H. Humphries, W. Olsen, D.L. Ewert, and J.-C. Weill, Light chain gene conversion continues at a high rate in an ALV-induced cell line, *EMBO J.* 9:921 (1990).
16. M.J.H. Ratcliffe, Ligation of cell surface immunoglobulin in the chicken bursa of Fabricius, *Res. Immunol.* 144:421 (1993).
17. S. Cory, Regulation of lymphocyte survival by the bcl-2 gene family, Ann. Rev. Immunol. 13:513 (1995).
18. I. Givol, I. Tsarfaty, J. Resau, S. Rulong, P.P. da Silva, G. Nasioulas, J. DuHadaway, S.H. Hughes, and D.L. Ewert. Bcl-2 expressed using a retroviral vector is localized primarily in the nuclear membrane and the endoplasmic reticulum of chicken embryo fibroblasts, *Cell Growth Diff.* 5:419 (1994).
19. C.J. Petropoulos, and S.H. Hughes, Replication-competent retrovirus vectors for the transfer and expression of gene cassettes in avian cells, *J. Virol.* 65:3728 (1991).

20. C.J. Petropoulos, W. Payne, D.W. Salter, and S.H. Hughes, Using avian retroviral vectors for gene transfer, *J. Virol.* 66:3391 (1992).
21. T. Benatar, S. Iacampo, L. Tkalec, and M.J.H. Ratcliffe, Expression of immunoglobulin genes in the avian embryo bone marrow revealed by retroviral transformation. *Eur. J. Immunol.* 21:2529 (1991).
22. E. Lagasse, and I.L. Weissman, bcl-2 inhibits apoptosis of neutrophils but not their engulfment by macrophages. *J. Exp. Med.* 179:1047 (1994).

GENERATION AND REGULATION OF B CELL AUTOREACTIVITY ARISING IN THE PERIPHERY

Philip Kuo, Daniel Michael, Boaz Tadmor, and Betty Diamond

Albert Einstein College of Medicine
1300 Morris Park Avenue
Bronx, New York 10461

Recent work on the role of environmental antigens in activating autoreactivity underscores a potentially confounding complication to current models of peripheral tolerance. Cross-reactive B cells that bind both autoantigen and foreign antigen are capable of mediating tissue damage, yet may be subject to positive selection as a consequence of interaction with antigen and T cell help. This discussion will review our work with cross-reactive antibodies and suggest a hypothesis to explain their generation and subsequent regulation.

1. THE PREIMMUNE B CELL REPERTOIRE

The preimmune B cell repertoire is generated by rearrangement of heavy (V,D,J) and light (V,J) chain variable region gene segments in the bone marrow (1). During immunoglobulin gene rearrangement, three méchanisms operate to delete or otherwise inactivate B cells that express autoreactive membrane immunoglobulin. For the purposes of this discussion, B cell anergy is defined as a condition of functional paralysis in which the cell can no longer be activated by contact with antigen and appropriate T cell help (2) and deletion is the elimination of autoreactive cells by programmed cell death, or apoptosis (3,4). Although anergy has been characterized in transgenic mouse models, it is unclear whether it exists under physiologic conditions and, if so, whether it is merely a stage in the process of deletion. Elegant studies have demonstrated the importance of yet another process, termed "receptor editing," whereby an autoreactive B cell undergoes additional light chain and heavy chain rearrangements until an immunoglobulin that no longer displays autoreactivity is formed (5–7).

The rules that determine whether an autoreactive B cell in the bone marrow is deleted, becomes anergic, or is subject to receptor editing remain unclear. Transgenic mouse studies have demonstrated that soluble antigen may mediate anergy (8), while membrane bound antigen may lead to apoptosis (9,10). Alternatively, other studies suggest that affinity

Mechanisms of Lymphocyte Activation and Immune Regulation VI, Edited by Gupta and Cohen
Plenum Press, New York, 1996

differences may underlie this critical distinction (4,11). While our understanding of these mechanisms is not yet complete, there is a considerable accumulation of data suggesting that they are quite effective at protecting the organism from autoreactivity arising in the bone marrow.

Although a B cell which exits the bone marrow is presumably not reactive against self antigens that are present within the bone marrow, it may display autoreactivity against peripheral tissues. Bretscher and Cohen hypothesized in 1970 that a B cell in the periphery requires two signals-- antigen and T cell help-- in order to be activated, and further predicted that the lack of a second signal would lead to downregulation of the B cell (12,13). This requirement protects against autoimmunity by inactivating B cells which bind autoantigen in the absence of T cell help. In support of this, a number of studies have shown that both anergy (8,14) and deletion (15) take place in the periphery. Receptor editing, however, appears to be restricted to the bone marrow where the recombinases, RAG-1 and RAG-2, are expressed (16).

2. PERIPHERAL DIVERSIFICATION OF THE B CELL REPERTOIRE

After immunoglobulin rearrangement in the bone marrow, there is a second occasion in the life of a B cell when new antigenic specificities may be formed. As B cells become activated by antigen and antigen specific T cells in secondary lymphoid organs they form germinal centers (17,18). Within the germinal center, somatic point mutation of heavy and light chain variable regions occurs at an estimated rate of 1/1000 base pairs per cell per generation (19,20). Murine studies appear to show intrinsic hot spots of mutation, but point mutation occurs throughout the variable region gene (21,22).

It has long been appreciated that somatic mutation is responsible for affinity maturation in the immune response. Some investigators have suggested that somatic mutation can only alter affinity and fine specificity, but cannot generate novel antigenic specificities. It has become clear, however, that expansion of the chicken B cell repertoire is largely mediated by somatic mutation via gene conversion events in the Bursa of Fabricius (23). In sheep, somatic point mutation in Peyers patches appears to diversify the B cell repertoire prior to antigen exposure, suggesting that mutation can create new antigenic specificities in addition to altering affinity (24). We have speculated that novel specificities may be generated even in the mouse, where somatic mutation is initiated only after exposure to antigen (25).

3. ROLE OF SOMATIC MUTATION IN AUTOREACTIVITY

Studies of autoantibodies from our own laboratory and from many others have shown that they are encoded by the same V gene segments that encode presumably protective anti-microbial antibodies (26,27). Furthermore, the vast majority of autoantibodies appear to be somatically mutated (28–33). While it is not always possible to identify with certainty the germline template for each autoantibody encoding V gene, in most instances where a germline gene can be unequivocally identified, the V gene segment encoding the autoantibody has accumulated a number of point mutations. Analysis of the mutations present in these variable region genes suggests the force of antigen selection (33). Autoantibodies display two features which have been interpreted to show antigen selection: 1) a clustering of replacement mutations (which alter the amino acid encoded) in complementarity determining regions (CDRs) rather than framework regions (FRs); and 2) a higher than random ratio of replacement (R) to silent(S) mutations. A clustering of R mutations in CDRs suggests selection for amino acid substitutions within the antigen binding site. A greater than random ratio of R to S mutations also suggests selection for an altered amino acid sequence. These features of autoantibodies suggest they derive from antigen activated responses, but our understanding of the structural features of antigen binding is too limited to permit identification of the selecting antigen by mutational analysis.

4. RELATION OF PROTECTIVE IMMUNITY TO AUTOREACTIVITY

An important observation that has been made about autoantibodies is that many cross-react with foreign antigen (34–38). Unfortunately, the affinity of a given antibody for self antigen and for foreign antigen has rarely been measured. It is, therefore, not possible to speculate on the importance of either specificity in activation of the B cell. Studies of non-autoimmune individuals demonstrate the activation of autoreactivity in the serum following encounter with foreign antigen (39). Although some investigators have suggested this autoreactivity results from polyclonal activation, others have demonstrated idiotypic similarities between the autoantibodies and the antibodies to foreign antigen (40,41), and still others have demonstrated antigenic cross-reactivity (42,43).

Antibodies against the bacterial cell wall component phosphorylcholine (PC) can protect mice from lethal pneumococcal infection. An in vitro study of somatic mutation had previously demonstrated that a single amino acid substitution in the heavy chain of an antibody to PC converts the antigenic specificity to double stranded DNA (dsDNA) (43). For many years this observation has been a paradigm for the potential generation of autoreactivity by somatic mutation, but additional in vitro or in vivo examples have not been readily observable. In genealogies of hybridomas from autoimmune mice, it has been shown that an antibody to single stranded DNA (ssDNA) can mutate to acquire reactivity to dsDNA (31), but such examples do not represent an acquisition of autospecificity arising during a protective B cell response. Gerhard and colleagues have shown that in the course of a response to influenza hemagglutinin, a B cell clone making antibody to one epitope can give rise to daughter cells making antibody to a second distinct epitope (44). Somatic mutation, in this instance, generates a new antigenic specificity, but not an anti-self specificity. The cross-reactive, anti-self anti-foreign antibodies identified to date have not been analyzed to determine whether somatic mutation accounts for their autoreactivity. It has, therefore, remained a question whether somatic mutation might routinely generate autospecificities in the course of a response to foreign antigen.

5. STUDIES OF AUTOANTIBODIES IN THE NON-AUTOIMMUNE HOST

We have tried for several years to immunize non-autoimmune mice with PC and use hybridoma technology to immortalize anti-DNA producing B cells generated during the anti-PC response. While we have obtained one or two antibodies that bind both PC and dsDNA from occasional fusions, we have not routinely obtained hybridomas making such antibodies (42). Consideration of the function of germinal centers led us to speculate that B cells developing autospecificity might be deleted or anergized. A study by Dipasquale and Youle in 1992 demonstrated that somatic cell fusion does not rescue T cells induced to undergo apoptosis (45). We, therefore, reasoned that autoreactive B cells undergoing apoptosis might not form viable antibody secreting hybridomas with a conventional fusion partner.

B cell survival appears to depend in part on bcl-2 expression. Mice with a targeted disruption of the bcl-2 gene begin to develop a lymphoid cell repertoire, but by 5 to 6 weeks of age they demonstrate an absence of all mature B and T cells (46). In contrast, mice transgenic for constitutive expression of bcl-2 in B cells show marked lymphoid hyperplasia and, in some genetic backgrounds, elevated serum titers of autoantibodies (47–49).

B cells within germinal centers express high levels of bcl-2 mRNA, but protein expression is limited to a subset of B cells in the apical light zone (50). Production of bcl-2 protein is correlated with exposure to both antigen and T cell help (51). This observation has been interpreted to signify an important role for bcl-2 in permitting B cells to exit the germinal center and move into a plasma cell or memory cell compartment. Recent data have also suggested a role for bcl-x in B cell survival in germinal centers (52–54).

Based on the above mentioned observations, we hypothesized that autoreactive B cells were destined to undergo apoptosis and, furthermore, that B cells lacking bcl-2 protein production might undergo apoptosis even after fusion to a hybridoma partner. We, therefore, transfected NSO cells, a standard B cell fusion partner, with a construct leading to constitutive expression of bcl-2 (55). NSO^{bcl-2} cells were shown to have a greater fusion efficiency and an enhanced ability to form viable antibody secreting hybridomas with autoreactive B cells (55). NSO^{bcl-2} was, therefore, used to test whether autospecificities routinely arise by somatic mutation and whether regulation of bcl-2 expression is critical in determining the fate of these autoreactive B cells.

To determine whether DNA binding antibodies were being produced by clonal progeny of anti-PC producing B cells, we performed fusions of spleen cells from PC immunized mice at various times following immunization. BALB/c mice were immunized intraperitoneally with 100 ug keyhole limpet hemocyanin (KLH) emulsified in complete Freund's adjuvant in order to prime the carrier-specific T cell response. After 2 weeks, the mice were given an intrasplenic injection of 50ug PC-KLH in saline. Because somatic mutation occurs only after day 4 or 5 of the germinal center response, we performed fusions on days 2 and 5 following PC immunization and then later on days 10 and 11 following immunization. PC binding hybridomas were subcloned on soft agarose or by limiting dilution cloning. Culture supernatants from hybridoma clones were normalized for immunoglobulin concentration and then assayed for binding to PC-BSA and to dsDNA by ELISA.

As seen in Table 1, anti-PC antibodies were produced by hybridomas derived from fusions of NSO^{wt} and NSO^{bcl-2} splenocytes on days 2 and 5 following PC immunization, but none demonstrated cross-reactivity with dsDNA (56). In contrast, approximately 40% of anti-PC antibodies obtained from day 10 and 11 hybridomas cross-reacted with dsDNA. These cross-reactive, anti-foreign anti-self hybridomas could only be obtained by fusion with NSO^{bcl-2}. Fusions with NSO^{wt} yielded a single hybridoma making a cross-reactive, anti-PC anti-dsDNA antibody. This discrepancy strongly suggests that bcl-2 expression is critical to the survival of these autoreactive B cells. Presumably, they undergo apoptosis in the mouse *in vivo* and also proceed to apoptosis following fusion with NSO^{wt}. The presence of bcl-2 in the NSO^{bcl-2} cell line is necessary to rescue them from this suicidal pathway. Only a very few (total of 3) clones were recovered which bound to dsDNA but not PC. If this specificity arises frequently from anti-PC progenitor cells, then this subset of autoreactive B cells is not rescued by the presence of bcl-2 in the fusion partner, but it is currently unclear whether this specificity arises infrequently or whether it is subject to different regulatory forces than cross-reactive B cells.

The observation that cross-reactive B cells could only be recovered on day 10 or 11 following PC immunization suggested that the specificity for dsDNA might be generated by

Table 1. Analysis of anti-PC hybridomas derived from fusions with NSO^{bcl-2}*

Time points	anti-PC only	anti-PC, anti-DNA
Day 2 and 5	100% (20)	0%**(0)M
Day 10 and 11	62% (21)	38%**(13)

*NSO^{wt} fusions for two mice were performed and demonstrated that only one of 18 anti-PC antibodies bound dsDNA.
**percentage of total anti-PC-carrier reactive clones that are cross-reactive with dsDNA.

somatic mutation. To test this hypothesis, the heavy chains of all the cross-reactive antibodies were sequenced (56). Two observations were made by analyzing heavy chain sequences. First, a variety of different V_H gene families were utilized to encode these anti-PC antibodies. While the serum response to PC in BALB/c mice is dominated by antibodies expressing the S107 V_1 heavy chain gene segment (57), anti-PC hybridomas obtained by fusion to NSO^{bcl-2} express multiple genes. The presence of B cells in the spleen that contribute only slightly to the serum response suggests that *in vivo* selection plays a role in the dominance of S107 like antibodies. Secondly, all heavy chains displayed evidence of somatic mutation. Since autoreactivity is apparent only in hybridomas derived from day 10 and 11 fusions, autoreactivity may be generated by somatic mutation. It is important to note, however, that only in one case is it proven that autospecificity is generated by somatic mutation. Most of the antibodies have not been reverted to their germ line encoded sequences to prove the role of mutation in self-specificity.

6. CHARACTERIZATION OF CROSS-REACTIVE ANTIBODIES

The kinetics of the appearance of cross-reactive antibodies and the demonstration that they are encoded by somatically mutated immunoglobulin genes strongly suggest that they are produced by cells that have matured in germinal centers. We are currently deriving hybridomas from isolated germinal center cells to confirm this hypothesis. Selection in germinal centers presumably depends on competition for antigen (58). B cells making higher affinity antibodies or with increased density of membrane immunoglobulin will more avidly bind antigen, a necessary signal for survival and differentiation. It is possible that the acquisition of DNA binding results in a loss of affinity for PC and that B cells making cross-reactive antibodies might fare poorly in the competition for antigen within germinal centers. We, therefore, tested the cross-reactive antibodies for relative binding to PC-KLH. Several displayed binding equivalent to that of PC2μ, a canonical anti-PC antibody, while one, B9–5, showed slightly increased binding. Although some of the antibodies made by the hybridomas displayed binding with a substantially lower affinity than the binding of PC2μ, a number of the B cells captured as NSO^{bcl-2} hybridomas might be expected to compete effectively for antigen. It is important to note that we assayed binding on PC-carrier which was the immunogen used *in vivo*. Binding to soluble PC was also assayed by inhibition studies and none of the cross-reactive antibodies bind PC as well as the canonical anti-PC antibody.

To test whether these cross-reactive antibodies have pathogenic potential, we injected hybridoma cells into SCID mice that had been pretested for absence of serum IgM (42,56). Because the amount of antibody secreted by these cells can be quite variable, we assayed all mice for concentration of serum IgM prior to sacrifice and only analyzed kidneys of mice with comparable titers of serum antibody. The cross-reactive antibodies deposited in

Table 2. Hybridomas from PC-immunized bcl-2 Tg mice

	Primary response	Secondary response
anti-PC	58% (7)	100% (13)
Cross-reactive anti-PC, anti-dsDNA	42% (5)	0% (0)

glomeruli at serum concentrations of 10–50 μg/ml. These cross-reactive antibodies that arise in response to antigen can, therefore, cause renal damage if expressed in the serum.

7. AUTOREACTIVITY IN bcl-2 TRANSGENIC MICE

We are also studying the PC response in transgenic mice which express bcl-2 constitutively in the B cell compartment. Constitutive expression of bcl-2 might be expected to rescue B cells at all checkpoints for elimination of autoreactive cells, including the bone marrow and germinal centers. BALB/c mice expressing a bcl-2 transgene show marked lymphoid hyperplasia, but do not spontaneously secrete anti-DNA antibodies. Mice were immunized as previously described, with KLH in adjuvant, followed by PC-KLH in saline. Preliminary studies suggest that the serum anti-PC response is slightly delayed, but otherwise normal in terms of class switching, antibody titer, and lack of anti-dsDNA reactivity. Interestingly, however, we were able to recover cross-reactive anti-PC, anti-DNA clones by hybridoma technology with NSO^{wt} after primary immunization (Table 2). This implies that constitutive expression of bcl-2 may prolong survival of pre-germinal center or germinal center cells expressing autoreactivity, but cannot overcome anergy or deletion as these cells do not contribute to the expressed serum antibody repertoire. We obtained significantly fewer cross-reactive antibodies from the secondary response, again suggesting that expression of the bcl-2 transgene is not sufficient for survival of these clones through the memory compartment (59).

8. HYPOTHESIS REGARDING THE REGULATION OF CROSS-REACTIVE B CELLS

Studies of B cell genealogies have demonstrated that positive selection occurs in germinal centers. Both antigen and T cell help appear to be necessary for positive selection, just as they are both needed for B cell activation. Presumably those B cells binding antigen most avidly are the most likely to undergo positive selection. It is clear that two classes of B cell will not undergo positive selection: those that lose reactivity with the eliciting antigen, and those expressing a novel antigenic specificity. In the latter case, even if antigen is fortuitously present, it is unlikely that specific T cell help would also be present to provide the second signal needed for positive selection.

Cross-reactive anti-foreign, anti-self B cells present another problem. Since both foreign antigen and T cell help are present, it seems likely that positive selection would occur. Studies by Goodnow and Nossal showing negative selection by apoptosis in germinal centers (60,61) in the presence of soluble hapten suggest that deletion of antigen specific cells can also occur in germinal centers when B cells encounter antigen in the absence of T cell help.

Previous studies have suggested that there is a vulnerable period for a B cell following positive selection within the germinal center when the B cell can be rendered anergic or deleted by encounter with antigen alone (62,63). Our results also lead us to speculate that cross-reactive B cells may undergo positive selection in germinal centers, but are vulnerable either in the germinal center or subsequently in the memory cell compartment to negative regulation, anergy or deletion, following exposure to self antigen in the absence of T cell help.

9. CONCLUSION

Pathogenic autoantibodies mediate tissue damage in a number of autoimmune diseases. Studies of patients with autoimmune disease have not revealed the origin of these autoantibodies. Microbial infection remains one hypothesis for the inciting antigenic trigger in autoimmune disease. Evidence for this hypothesis has not been easy to obtain, although there are some observations that are consistent with this theory. Some naturally arising (or transgenic rodent strains that develop autoimmune disease have an attenuated clinical presentation if raised under germ free conditions (64,65). Lupus-prone mice bearing the *xid* mutation display an inability to mount an antibody response to bacterial polysaccharides and also develop an attenuated form of disease (66). In addition, there is a high degree of non-concordance among identical twins for autoimmune disease (67), suggesting an environmental trigger might contribute to the etiology of the disease.

Studies of the germinal center response have recently begun to elucidate some aspects of the selection of B cells within germinal centers. It is clear that large numbers of B cells do not survive germinal center selection and, instead, undergo apoptosis within the germinal center. In order to examine those B cells that were not selected for survival, and that constitute part of the graveyard of germinal center B cells, we generated a B cell fusion partner with constitutive expression of bcl-2. Using this fusion partner it is possible to routinely obtain hybridomas making antibodies that bind foreign antigen PC, yet also cross react with self antigen dsDNA and display pathogenic potential. These autoantibodies arise with kinetics that suggest a critical role for somatic mutation in the generation of anti-self specificity. We have more recently asked whether other foreign antigens also elicit the generation of cross-reactive, anti-foreign anti-self antibodies with pathogenic potential. We opted to immunize mice with the hapten azophenylarsonate (Ars) coupled to a protein carrier. Previous studies have shown that the variable region gene segments that encode an anti-Ars response can encode antibodies to ssDNA (68). Manser and colleagues have recently demonstrated that *in vitro* mutations in CDR2 of the heavy chain of an anti-Ars antibody can generate antibodies with increased affinity for Ars that also bind ssDNA (69). We have preliminary data demonstrating that approximately one third of the anti-Ars hybridomas obtained by fusion of spleen cells from an immunized mouse with NSO^{bcl-2} bind both Ars-carrier and dsDNA. These antibodies have not yet been assayed for pathogenic potential. It appears, however, that cross-reactivity between PC-carrier and dsDNA does not represent a unique case of structural homology, and that structural homology between self antigens and foreign antigens is likely to be frequent. Autoreactivity may routinely be just one or two amino acid substitutions away from protection.

In addition, studies by Craig Thompson and his colleagues have suggested that bcl-2 is not the only anti-apoptosis gene that is important in survival of B cells (52). Bcl-x has been identified as a gene that is activated in germinal centers but not in naive B cells, suggesting a possible role in survival during the germinal center response (54). We have recently generated an NSO line expressing bcl-x, which like NSO^{bcl-2} yields a greater number of hybridomas than NSO^{wt}. We are currently examining whether this line rescues a different population of B cells than NSO^{bcl-2}.

There is increasing evidence that negative as well as positive selection occurs in germinal centers. The interaction of Fas on germinal center B cells with Fas ligand expressed on some T helper cells can mediate B cell apoptosis (70,71). Furthermore, soluble hapten has been demonstrated to cause apoptosis in B cells in germinal centers responding to hapten-carrier immunization (60,61). It is not yet clear whether there are additional pathways mediating apoptosis in this B cell population. The observation that autoreactivity is presumably acquired by somatic mutation in response to foreign antigen suggests that defects in the regulation of cross-reactive B cells could lead to autoimmune disease. An actual link, however, between dysregulation of B cell function in germinal centers and autoimmune disease remains to be shown.

REFERENCES

1. Blackwell, T.K. and Alt, F.W. Mechanism and developmental program of immunoglobulin gene rearrangement in mammals. Ann. Rev. Genetics 23:605, 1989.
2. Goodnow, C.C. Transgenic mice and analysis of B cell tolerance. Annu. Rev. Immunol. 10:489, 1992.
3. MacLennan, I.C. Autoimmunity. Deletion of autoreactive B cells. Curr. Biol. 5(2):103–6, 1995.
4. Nemazee, D., Russell, D., Arnold, B., Haemmerling, G., Allison, J., Miller, J.F., Morahan, G. and Buerki, K. Clonal deletion of autospecific B lymphocytes. Immunol Rev. 122:117–32, 1991.
5. Radic, M., Erikson, J., Litwin, S. and Weigert, M. B lymphocytes may escape tolerance by revising their Ag receptors. J. Exp. Med. 177:1165, 1993.
6. Tiegs, S., Russell, D. and Nemazee, D. Receptor editing in self reactive bone marrow B cells. J. Exp. Med. 177:1009, 1993.
7. Gay, D., Saunders, T., Camper, S. and Weigert, M., Receptor editing: an approach by autoreactive B cells to escape tolerance. J. Exp. Med. 177:999, 1993.
8. Goodnow, C.C., Crosbie, T., Adelstein, S., Lavoie, T.B., Smith-Gill, S. J., Brink, R.A. Pritchard-Briscoe, H., Witherspoon, J.S., Lobley, R.H., Raphael K., Trent, R.J. and Basten, A. Altered immunoglobulin expression and functional silencing of self reactive B lymphocytes in transgenic mice. Nature 334: 676, 1988.
9. Nemazee, D.A. and Burki, K. Clonal deletion of B lymphocytes in a transgenic mouse bearing anti-MHC class I antibody genes. Nature. 337:562, 1989.
10. Hartley, S.B., Crosbie, J., Brink, R.A., Kantor, A.B., Basten, A. and Goodnow, C.C. Elimination from peripheral lymph tissue of self reactive B lymphocytes recognizing membrane bound Ag. Nature 353:765, 1991.
11. Erikson, J., Radic, M.Z., Camper, S.A., Hardy, R.R., Carmack, C. and Weigert, M. Expression of anti-DNA immunoglobulin transgene in non-autoimmune mice. Nature. 349:331, 1991.
12. Bretscher, P. and Cohen M. A theory of self-nonself discrimination. Science 169:1042, 1970.
13. Bretscher, P. The two signal model of lymphocyte activation twenty-one years later. Immunol. Today 13:74, 1992.
14. Offen, D., Spatz, L., Escowitz, H., Factor, S. and Diamond, B. Induction of tolerance to an IgG autoantibody. Proc. Natl. Acad. Sci. 89:8332, 1992.
15. Russell, D.M., Dembic, Z., Morahan, G., Miller, J.F., Burki, K. and Nemazee, D. Peripheral deletion of self-reactive B cells. Nature. 354:308, 1991.
16. Oettinger, M.A., Schatz, D.G., Gorka, C. and Baltimore, D. RAG-1 and RAG-2, adjacent genes that synergistically activate V(D)J recombination. Science. 248:1517, 1990.
17. Gray, D. Regulation of immunological memory. Curr Opin Immunol. 6:425, 1994.
18. Thorbecke, G.J., Amin, A.R. and Tsiagbe, V.K. Biology of germinal centers in lymphoid tissue. FASEB J. 8:832, 1994.
19. Apel, M. and Berek, C. Somatic mutations in antibodies expressed by germinal center B cells early after primary immunization. Int. Immunol. 2:813, 1990.
20. Jacob, J., Kelsoe, Rajewsky, K. and Weiss, U. Intraclonal generation of antibody mutants in germinal centers. Nature 354:389, 1991.
21. Betz, A.G., Neuberger, M.S. and Milstein, C. Discriminating intrinsic and antigen-selected mutational hotspots in immunoglobulin V genes. Immunol. Today 14:405, 1993.
22. Betz, A.G., Rada, C., Pannel, R., Milstein, C. and Neuberger, M.S. Passenger transgenes reveal intrinsic specificity of the antibody hypermutation mechanism: clustering, polarity, and specific hot spots. Proc. Nat. Acad. Sci. USA 90:2385, 1993.
23. Reynaud, C.A., Anquez, V., Duhan, A. and Weill, J.L. A single rearrangement event generates most of the chicken immunoglobulin light chain diversity. Cell 40:283, 1985
24. Reynaud, L.A., Garcia, C., Hein, W.R. and Weill, J.L. Hypermutation generating the sheep immunoglobulin repertoire is an antigen-independent process. Cell 80:115, 1995.

25. Diamond, B., Katz, J.B., Paul, E., Aranow, C., Lustgarten, D. and Scharff, M.D. The role of somatic mutation in the pathogenic anti-DNA response. Ann. Rev. Immunol. 10:731, 1992.
26. Manheimer-Lory, A., Monhian, R., Splaver, A., Gaynor, B. and Diamond, B. Analysis of the V_kI family: Germline genes and expressed antibodies. Autoimmunity 20:259, 1995.
27. Manheimer-Lory, A., Irigoyen, M., Gaynor, B., Monhian, R., Splaver, A. and Diamond, B. Analysis of V_KI and V_KII light chain genes in the expressed B cell repertoire. Annals of the N.Y. Acad. Sci. 764:301, 1995.
28. Manheimer-Lory, A., Katz, J.B., Pillinger, M., Ghossein, C., Smith, A. and Diamond, B. Molecular characteristics of antibodies bearing an anti-DNA associated idiotype. J. Exp. Med. 174:1639, 1991.
29. Paul, E. and Diamond, B. Characterization of two human anti-DNA antibodies bearing the pathogenic idiotype 8.12. Autoimmunity 16:13, 1993.
30. Radic, M.Z. and Weigert, M. Genetic and structural evidence for Ag selection of anti-DNA antibodies. Ann. Rev. Immunol. 12:487. 1994.
31. Shlomchik, M., Mascelli, M., Shan, H., Radic, M.F., Pisetsky, D., Marshak-Rothstein, A. and Weigert, M. Anti-DNA antibodies from autoimmune mice arise by clonal expansion and somatic mutation. J. Exp. Med. 171:765, 1990.
32. Marion, T.N., Tillman, D.M., Jou, N.T. and Hill, R.J. Selection of immunoglobulin variable regions in autoimmunity to DNA. Immunol Rev. 128:123, 1992.
33. Shlomchik, M.J., Aucoin, A.H., Pisetsky, D.S. and Weigert, M.G. Structure and function of anti-DNA autoantibodies derived from a single autoimmune mouse. Proc. Natl. Acad. Sci. USA. 84: 9150, 1987.
34. Shoenfeld, Y., Vilner, Y., Coates, A.R.M, Rauch, J., Lavie, G., Shaul, D. and Pinkhas, J. Monoclonal anti-tuberculosis antibodies react with DNA and monoclonal anti-DNA antibodies react with Mycobacterium tuberculosis. Clin Exp Immunol. 66:255, 1986.
35. Carroll, P., Stafford, D., Schwartz, A. and Stollar, B.D. Murine monoclonal anti-DNA autoantibodies bind to endogenous bacteria. J Immunol. 135:1086, 1985.
36. Vashishtha, A. and Fischetti, V.A. Surface-exposed conserved region of the streptococcal M protein induces antibodies cross-reactive with denatured forms of myosin. J. Immunol. 150:4693, 1993.
37. Cunningham, M.W., Antone, S.M . Gulizia, J.M., McManus, B.M., Fischetti, V.A. and Gauntt, C.J. Cytotoxic and viral neutralizing antibodies cross-react with streptococcal M protein, enteroviruses and human cardiac myosin. Immunology 89:1320, 1992.
38. Adderson, E.E., Shackelford, P.G., Quinn, A., Wilson, P.M., Cunningham, M.W., Insel, R.A. and Carroll, W.L. Restricted immunoglobulin V_H usage and VDJ combinations in the human response to Haemophilus influenzae type b capsular polysaccharide. Nucleotide sequences of monospecific anti-Haemophilus antibodies and polyspecific antibodies cross-reacting with self antigens. J. Clin. Invest. 91:3734, 1993.
39. El-Roiey, A., Sela, O., Isenberg, D.A., Feldman, R., Collaco, B.C., Kennedy, R. and Shoenfeld, Y. The sera of patients with Klebsiella infections contain a common anti-DNA (16/6Id) and polynucleotide activity. Clin. Exp. Immunol. 67:507, 1987.
40. Grayzel, A., Solomon, A., Aranow, C. and Diamond, B. Antibodies elicited by pneumococcal antigens bear an anti-DNA associated idiotype. J Clin. Invest. 87:842, 1991.
41. Monestier, M., Bonin, B., Migliorini, P., Dang, H., Datta, S., Kuppers, R., Rose, N., Maurer, P., Talal, N. and Bona, C. Autoantibodies of various specificities encoded by genes from the VHJ558 family bind to foreign antigens and share idiotypes of antibodies specific for self and foreign antigens. J Exp Med. 166:1109, 1987.
42. Limpanasithikul, W., Ray, S. and Diamond, B. Cross reactive antibodies have both protective and pathogenic potential. J. Immunol. 155:967, 1995.
43. Diamond, B. and Scharff, M.D. Somatic mutation of the T15 heavy chain gives rise to an antibody with autoantibody specificity. Proc. Natl. Acad. Sci. USA 81:5841, 1984.
44. Clarke, S.H., Staudt, L.M., Kavaler, J., Schwartz, D., Gerhard, W. and Weigert, M. Inter- and intraclonal diversity in the antibody response to influenza hemagglutinin. J. Exp. Med. 161:687, 1985.
45. Dipasquale, B. and Youle, R.J. Programmed cell death in heterokaryons. Am. J. Pathology 141:1471, 1992.
46. Nakayama, K., Negishi, I., Kuida, K., Shinkai, Y., Louie, M.D., Fields, L.E., Lucas, P.J., Stewart, V., Alt, F.W. and Loh, D.Y. Disappearance of the lymphoid system in bcl-2 homozygous mutant chimeric mice. Science 261:1584, 1993.
47. McDonnell, T.J. and Korsmeyer, S.J. Progression from lymphoid hyperplasia to high-grade malignant lymphoma in mice transgenic for the t(14; 18). Nature 349:254, 1991
48. Strasser, A., Whittingham, S., Vaux, D.L., Bath, M.L., Adamas, J.M. and Cory, S., and Harris, A.W. Enforced BCL2 expression in B-lymphoid cells prolongs antibody responses and elicits autoimmune disease. Proc. Natl. Acad. Sci. USA 88:8661, 1991.

49. Strasser, A., Harris, A.W. and Cory, S. The role of bcl-2 in lymphoid differentiation and neoplastic transformation. Curr Top in Micro and Imm. 182:299, 1992.
50. Hockenbery, D.M., Zutter, M., Hickey, W., Nahm, M. and Korsmeyer, S.J. Bcl-2 protein is topographically restricted in tissues characterized by apoptotic cell death. Proc. Natl. Acad. Sci. USA 88:6961, 1991.
51. Liu, Y.J., Mason, D.Y., Johnson, G.D., Abbott, S., Gregory, C.D., Hardie, D.L., Gordon, H.J. and MacLennan, I.C. Germinal center cells express bcl-2 protein after activation by signals which prevent their entry into apoptosis. Eur. J. Immunol. 21:1905, 1991.
52. Boise, L.H., Gonzalez-Garcia, M., Postema, C.E., Ding, L., Lindstein, T., Turka, L.A., Mao, X., Nunez, G. and Thompson, C.B. bcl-x, a bcl-2-related gene that functions as a dominant regulator of apoptotic cell death. Cell 74:597, 1993.
53. Nunez, G., Merino, R., Grillot, D. and Gonzalez-Garcia, M. Bcl-2 and Bcl-x: regulatory switches for lymphoid death and survival. Immunol. Today 15:582, 1994.
54. Ohta K., Iwai K., Kasahara Y., Taniguchi N., Krajewski S., Reed J.C. and Miyawaki T. Immunoblot analysis of cellular expression of Bcl-2 family proteins, Bcl-2, Bax, Bcl-X and Mcl-1 in human peripheral blood and lymphoid tissues. Int. Immunol. 7:1817, 1995.
55. Ray, S. and Diamond, B. Generation of a novel fusion partner to sample the repertoire of splenic B cells destined for apoptosis. Proc. Natl. Acad. Sci. USA 91:5548, 1994.
56. Ray S., Putterman C. and Diamond B. Pathogenic autoantibodies are routinely generated during the response to foreign antigen: a paradigm for autoimmune disease. Proc. Natl. Acad. Sci. USA (In press).
57. Crews, S., Griffin, J., Huang, H., Calame, K. and Hood, L. A single V_H gene encodes the immunological response to PC. Cell 25:59, 1981.
58. George, J., Penner, S.J., Weber, J., Berry, J. and Claflin, J.L. Influence of membrane Ig receptor density and affinity on B cell signaling by antigen. Implications for affinity maturation. J. Immunol. 151:5955, 1993.
59. Hartley, S.B., Cooke, M.P., Fulcher, D.A., Harris, A.W., Cory, S., Basten, A. and Goodnow, C.C. Elimination of self-reactive B lymphocytes proceeds in two stages: arrested development and cell death. Cell 72:325, 1993.
60. Pulendran, B., Klannourakis, G., Nouri, S., Smith, K.G.C. and Nossal, G.J.V. Soluble antigen can cause enhanced apoptosis of germinal-center B cells. Nature 375:331, 1995.
61. Shokat, K.M. and Goodnow, C.C. Antigen-induced B-cell death and elimination during germinal-center immune responses. Nature 375:334, 1995.
62. Linton, P., Rudie, A. and Klinman, N.R. Tolerance susceptibility of newly generating memory B cells. J. Immunol. 146:4099, 1991.
63. Johnson, J.G. and Jemmerson, J.G. Tolerance induction in resting memory B cells specific for a protein antigen. J. Immunol. 148:2682, 1992.
64. Unni, K.K., Holley, K.E., McDuffie, F.C. and Titus, J.L. Comparative study of NZB mice under germ free and conventional conditions. J. Rheumatol. 2:36, 1975.
65. Taurog, J.D., Richardson, J.A., Croft, J.T., Simmons, W.A., Zhou, M., Fernandez-Sueiro, J.L., Balish, E. and Hammer, R.E. The germfree state prevents development of gut and joint inflammatory disease in HLA-B27 transgenic rats. J. Exp. Med. 180:2359, 1994.
66. Fieser, T.M., Gershwin, M.E., Steinberg, A.D., Dixon, F.J. and Theofilopoulos, A.N. Abrogation of murine lupus by the xid gene is associated with reduced responsiveness of B cells to T-cell-helper signals. Cellular Immunol. 87:708, 1984.
67. Arnett, F.C. The genetic basis of lupus erythematosus. In Dubois' Lupus Erythematosus, 4e. Wallace, D.J. and Hahn, B.H. eds. pp. 13–36. Philadelphia, Lea and Febiger, 1993.
68. Naparstek, Y., Andre-Schwartz, J., Manser, T., Wysocki, L.J., Breitman, L., Stollar, B.D., Gefter, M. and Schwartz, R.S. A single germline V_H gene segment of normal A/J mice encodes autoantibodies characteristics of systemic lupus erythematosus. J. Exp. Med. 164:614, 1986.
69. Casson, L.P. and Manser, T. Random mutagenesis of two complementarity determining region amino acids yields an unexpectedly high frequency of antibodies with increased affinity for both cognate antigen and autoantigen J. Exp. Med. 182:743, 1995.
70. Lagresie, C., Bella C., Daniel, P.T., Kramer, P.H. and Defrance, T. Regulation of germinal center B cell differentiation. Role of the human APO-1/FAS (CD95) molecule. J. Immunol. 154:5746, 1995.
71. Watanabe D., Suda T. and Nagata S. Expression of Fas in B cells of the mouse germinal center and Fas-dependent killing of activated B cells. Int Immunol. 12:1949, 1995.

INDUCIBLE RESISTANCE TO Fas-MEDIATED APOPTOSIS IN PRIMARY B LYMPHOCYTES

Thomas L. Rothstein,[1,2*] Linda C. Foote,[2] Thomas J. Schneider,[2] Gavin M. Fischer,[2] Bruce A. Jacobson,[2] David H. Lynch,[3] Shry-Te Ju,[1,4] and Ann Marshak-Rothstein[2,4]

[1] The Department of Medicine and
The Evans Memorial Department of Clinical Research
[2] The Department of Microbiology
Boston University Medical Center
Boston, Massachusetts, 02118
[3] Department of Immunobiology
Immunex Research and Development Corporation
Seattle, Washington 98101
[4] The Department of Pathology
Boston University Medical Center
Boston, Massachusetts, 02118

INTRODUCTION

The regulated initiation of programs leading to cell death is responsible, at least in part, for maintaining homeostasis within the immune system. Two general mechanisms for the production of cytotoxicity in susceptible targets by T effector cells have been described, one involving Ca^{++} dependent, perforin/granzyme exocytosis, and the other involving Fas antigen (CD95) engagement[1]. Fas functions in target cells as a receptor that signals for apoptosis, and Fas has been implicated in the process of reducing lymphocyte numbers following acute immune responses, termed activation induced cell death[2–8]. Although classical $CD8^+$ cytotoxic T lymphocytes have recently been shown to express both types of cytotoxic activity, a distinct class of T cells, with the phenotype of $CD4^+$ Th1 cells, mediates target cell death primarily in a Fas-dependent fashion[9–14].

Primary B cells express little Fas and are poor targets for Th1-induced cytotoxicity; after activation, however, Fas expression is induced and B cells acquire susceptibility to Th1 effector cells[13,15–18]. Early experiments demonstrating activation-induced B cell susceptibil-

* Address correspondence to: Dr. T.L. Rothstein, Room E-556, Boston University Medical Center Hospital, 88 East Newton Street, Boston, MA 02118, Tel: 617–638–7028, FAX: 617–638–7530, e-mail: trothstein@med-med1.bu.edu.

Mechanisms of Lymphocyte Activation and Immune Regulation VI, Edited by Gupta and Cohen
Plenum Press, New York, 1996

ity to Fas killing in the murine system relied on one form of stimulation, that provided by the polyclonal mitogen, lipopolysaccharide (LPS). However, different forms of B cell stimulation that engage separate receptors may trigger distinct intracellular signaling pathways and downstream events[19–21]. This has raised the question as to whether all forms of activation produce the same outcome in terms of B cell sensitivity to Fas-mediated apoptosis, or whether this might be regulated in a receptor-specific fashion.

Two receptors are of particular interest: surface immunoglobulin, the B cell receptor for antigen, and CD40, the B cell receptor for T-dependent stimulation. Engagement of either of these receptors is fully capable of driving primary B cells to exit the resting state and enter S phase, and thus each provides complete mitogenic stimulation. However, sIgM and CD40 appear to employ somewhat different intracellular signaling pathways. Signaling via sIgM relies heavily on *src* kinase activity that eventuates in the phosphorylation and activation of PLC (among other substrates), producing mediators that trigger PKC and increase intracellular Ca^{++}[22]. In contrast, CD40 engagement produces little tyrosine kinase activity, and PKC activation has not been observed[23,24]. In keeping with these results, many outcomes of sIgM crosslinking, both on a molecular scale such as induction of nuclear NF-κB expression and on a cellular scale such as proliferation, are eliminated by PKC depletion or inhibition, whereas the same outcomes for CD40 engagement are PKC-independent[19,25,26].

In recent work engagement of these mitogenic receptors was found to produce completely different results in terms of subsequent B cell sensitivity to Th1-induced, Fas-mediated cell death[18]. Antigen receptor crosslinking failed to induce susceptibility to Th1-induced cytotoxicity, whereas triggering through CD40 led to marked susceptibility to Fas-dependent killing that was, in fact, greater than that produced by LPS. Importantly, B cells stimulated through both sIgM and CD40 were not lysed by Th1 effectors, indicating that sIgM-derived mediators not only fail to induce Th1-sensitivity but, in fact, produce protection against Fas-dependent killing, or Fas-resistance. Fas-resistance was induced equally well by specific antigen in immunoglobulin transgenic mice and by IgM-specific $F(ab')_2$ antibody fragments, indicating that protection against Fas-mediated apoptosis in B cells is an outcome of physiological, as well as pharmacological, forms of antigen receptor engagement. Thus, sIgM-derived signals, induced either by anti-IgM or by specific antigen, induce a state of Fas-resistance that must be viewed as an active, dominant process that suppresses Fas signaling for cell death in otherwise sensitive targets.

The mechanisms responsible for inducible Fas-resistance in B cells, and the consequences for normal and aberrant B cell function, remain poorly understood. In the studies discussed below we sought to further elucidate the signals and mechanisms that impart protection against Fas-dependent Th1 effector cell activity in mature, primary B cells.

RESULTS

1. sIgM-Mediated Fas-Resistance Depends on Both PKC and Ca^{++} Signals

PLC is one of several mediators that are tyrosine phosphorylated following sIgM engagement and *src* kinase activation[22]. The role of the PLC pathway in fostering B cell Fas-resistance was initially explored by examining the need for PKC, using prolonged incubation with the agonist phorbol myristate acetate (PMA) to deplete the enzyme, as previously described[19–21,27,28]. Primary (murine splenic) B cells were prepared as described[18] and cultured for a total of 48 hours; B cells were incubated for the first 24 hours with CD40L plus PMA (or CD40L plus the relatively inactive phorbol ester analog, 4αPMA) after which

Table 1. Stimulation of B cell targets in various ways dictates susceptibility to Th1-induced cytotoxicity

B cell treatment		
-48 hrs	-24 hrs	Lysis
CD40L	–	+++
CD40L	anti-IgM	0
CD40L + 4αPMA	–	+++
CD40L + PMA	–	+++
CD40L + 4αPMA	anti-IgM	0
CD40L + PMA	anti-IgM	++
CD40L	HA1004 + anti-IgM	0
CD40L	H7 + anti-IgM	+++
CD40L	DMSO	+++
CD40L	PMA	+
CD40L	Ionomycin	++
CD40L	PMA + Ionomycin	0

Purified splenic B cells from BALB/cByJ mice were stimulated for 2 days and then tested as targets for Th1-mediated, Fas-dependent cytotoxicity in standard 4 hr lectin-dependent ^{51}Cr release assays using AE7 effectors, as previously described[18]. CD40L was always present during the 2 day culture period. Additional reagents were added at the same time as CD40L (-48 hrs), or 24 hours before the end of the culture (-24 hrs). The kinase inhibitors HA1004 and H7 were added at -25 hrs, one hour before anti-IgM was added (at -24 hrs). Per cent specific lysis was calculated on the basis of ^{51}Cr release in experimental and control (no effectors) samples, as described[18], and was generally 40-50% for B cells stimulated by CD40L alone at effector:target ratios of 3:1 to 9:1. To assemble the table, in each experiment per cent specific lysis for each treatment was related to the corresponding per cent specific lysis of sensitive B cells stimulated by CD40L alone, to calculate lysis for each treatment as a proportion of maximal lysis. Two or more experiments for each condition were averaged; values were assigned as follows: 0, 0-10%; +/-, 10-25%; +, 25-50%; ++, 50-75%; +++, 75-100%. Reagents were as follows. CD40L, supernatant from J558L cells that secrete a CD40L-CD8α fusion protein at 1:10, plus supernatant from the 53-6-72 hybridoma containing anti-CD8α antibody at 1:80, as previously described[18]; anti-Ig, F(ab')$_2$ goat anti-mouse IgM antibody at 10 μg/ml; PMA, phorbol myristate acetate at 100 ng/ml; 4αPMA, 4α–phorbol myristate acetate at 100 ng/ml; HA1004, 25 μM; H7, 25 μM; ionomycin, 400 ng/ml. B cells rendered sensitive to Th1-induced lysis by CD40L treatment, as assessed by ^{51}Cr release assay, were also found to display increased fluorescence intensity on TUNEL assay in which nicked DNA ends are labeled; further, B cells rendered resistant to Th1-induced lysis by combined treatment with CD40L plus anti-IgM were found to display essentially background levels of fluorescence intensity by TUNEL. These results indicate that the ^{51}Cr release assay as used here with Th1 effectors accurately reflects apoptosis as judged by DNA cleavage.

anti-IgM was added for the final 24 hours. In this way PKC depletion was produced prior to the initiation of sIgM engagement. B cells were then tested in lectin-dependent ^{51}Cr release assays as targets for cytotoxicity produced by AE7 Th1 effector cells, that have previously been shown to kill in a Fas-dependent fashion[13,18]. B cells cultured with CD40L alone for 48 hours were susceptible to Th1-induced cytotoxicity, as expected, and CD40-mediated induction of Fas-sensitivity was not affected by early exposure to PMA or 4αPMA. However, prior PKC depletion was found to markedly interfere with the induction of Fas-resistance by subsequently added anti-IgM, in contrast to B cells cultured with CD40L plus 4αPMA, in which subsequently added anti-IgM produced typical levels of protection against Th1-induced cytotoxicity. These results suggest that Fas-resistance mediated via sIgM requires PKC.

This result was confirmed by the use of two isoquinolinesulfonamide derivatives that are protein kinase inhibitors expressing more (H7) or less (HA1004) activity toward PKC[29].

B cells were cultured for a total of 48 hours with CD40L; one of the isoquinolinesulfonamides was added at 25 hours prior to the end of culture, after which anti-IgM was added for the final 24 hours; B cells were then tested for susceptibility to Th1 effectors. Inhibition of PKC with H7 was found to block induction of Fas-resistance by anti-IgM, in that B cells cultured with CD40L, H7 and anti-IgM were not protected against Th1-induced cytotoxicity but were instead as sensitive as B cells cultured only with CD40L. In contrast, B cells cultured with CD40L, HA1004, and anti-IgM were Fas-resistant. Thus, results obtained with both PKC-depletion and PKC-inhibition indicate that sIgM-mediated Fas-resistance requires functional PKC.

Many, but not all, PKC-dependent responses require a Ca^{++} signal as well; for example, B cell proliferation and NF-AT induction require Ca^{++} in addition to PKC activity, whereas NF-κB induction and CREB phosphorylation do not[28,30–34]. For this reason we tested the capacity of PKC alone and in combination with Ca^{++} to mimic sIgM engagement and produce protection against Th1-mediated apoptosis. B cells were stimulated by CD40L for 48 hours; the PKC agonist PMA, and the calcium ionophore ionomycin, were added either individually or together 24 hours before the end of culture, after which B cells were tested as targets in cytotoxicity assays with Th1 effectors. PMA or ionomycin each produced some protection against Th1-induced apoptosis, but the extent of Fas-resistance was limited. The combination of PMA plus ionomycin, however, produced virtually complete resistance to Fas-mediated apoptosis induced by Th1 cells, paralleling the effect of anti-IgM added to a separate group of CD40L-stimulated B cells. Thus, only the combination (but not individual application) of 2 pharmacologic reagents that mimic the effects of PLC-derived second messengers fully reproduced sIgM-induced Fas-resistance. Together with the inhibitor studies, these results suggest that sIgM mediates protection against Th1 effectors via activation of the PLC pathway; this work also identifies a set of pharmacological reagents that can be used in place of anti-Ig to bring about Fas-resistance.

2. Fas-Resistance Results from an Intracellular Change in the Target Cell

The mechanism underlying sIgM-mediated Fas-resistance could involve either extracellular or intracellular events. In previous work we have shown that anti-IgM treatment does not diminish CD40L-stimulated Fas expression, as detected at the level of mRNA and at the level of surface protein[18]. However, anti-IgM signaling could alter the conformation or accessibility of Fas such that it is no longer engaged by T cell-bound FasL. Alternatively, Fas-resistance may depend on the inducible surface expression of a protective receptor, or on the loss of a co-apoptotic surface receptor, that is engaged by a T cell ligand. Excluding these possibilities, the effects of sIgM could be expressed through intracellular mechanisms that oppose Fas signaling for target cell death. Experiments were undertaken to evaluate these possibilities.

The possible alteration or sequestration of Fas was evaluated by making use of an unlabeled (cold) target inhibition strategy, in which graded numbers of unlabeled B cells were added to ^{51}Cr-labeled, CD40L-stimulated (Th1-sensitive) targets just prior to cytotoxicity assay with Th1 effectors. Unlabeled CD40L-stimulated (Th1-sensitive) B cells successfully competed, and thus reduced, the lysis of Th1-sensitive labeled B cell targets in a dose-dependent fashion, as expected. Importantly, comparable dose-dependent inhibition of Th1-induced cytotoxicity was observed when unlabeled competitors were B cells that had been treated with CD40L plus anti-IgM and were thus Fas-resistant. Separate experiments using Fas-deficient MRL/*lpr*[35] B cell competitors and MRL/+ B cell targets demonstrated that successful competition depends on Fas expression. Thus, Fas expressed by dual (CD40L

Table 2. Th1-Resistant B cells are protected against rFASL but succumb to Fas-specific antibody

B cell treatment		Lysis by different mediators		
-48 hrs	-24 hrs	Th1	rFasL	Jo-2
CD40L	–	+++	++	+++
CD40L	anti-IgM	+/–	0	++
CD40L	IL-4	+	0	+++

Primary B cells were treated for 2 days and tested in ^{51}Cr release assays as described in the legend to Table 1, with the exception that B cells were additionally treated with rIL-4 at 500 units/ml, and that apoptosis was mediated by AE7 effector cells, supernatant from SF21 cells transfected with recombinant FasL, and Fas-specific antibody Jo-2.

plus anti-IgM) stimulated B cells (that are resistant to Th1-induced cytotoxicity) engages Th1 effector cells as readily as Fas expressed by B cell targets stimulated by CD40L alone (that are Fas-sensitive) and for this reason Fas-resistance does not depend on interference with functional recognition of Fas by Th1 effector cells. These results also eliminate potential mechanisms for Fas-resistance that involve induction of a surface protein that interferes with, or deletion of a protein that is essential for, intercellular contact (adhesion) between T cells and B cells.

The possible induction of a protective receptor, or the loss of a co-apoptotic receptor, that is activated by a T cell ligand was evaluated by determining the susceptibility of Th1-sensitive (CD40L-stimulated for 48 hours) and Th1-resistant (CD40L-stimulated for 48 hours with anti-IgM during the last 24 hours) B cell targets to soluble, recombinant FasL. B cells stimulated by CD40L alone were efficiently lysed by rFasL to the same extent as that produced by Th1 effector cells, strongly suggesting that Th1-induced B cell cytotoxicity does not require co-apoptotic T cell-B cell interactions beyond FasL-Fas, as suggested before[13,18]. Importantly, B cells stimulated by CD40L plus anti-IgM that were protected against Th1-induced cytotoxicity similarly resisted cytotoxicity induced by rFasL. Thus, Fas-resistance, like Fas-sensitivity, does not require T cell-B cell interactions beyond FasL-Fas, and protection against Th1 effector cell activity does not depend on the FasL trigger being presented in the context of an activated T cell.

Taken together, the results of cold competition and rFasL studies strongly suggest sIgM engagement produces Fas-resistance through an intracellular change that opposes Fas signaling for cell death rather than through an alteration in Fas or other surface receptor expression. This is supported by our inability to detect any soluble activity in the supernatants of dual stimulated (Fas-resistant) B cells that induces Fas-resistance in fresh CD40L-stimulated B cells (unpublished observations).

In studying cell-free means of triggering Fas, the monoclonal (Fas-specific) antibody, Jo-2[36], was tested along with rFasL. However, here the results were strikingly different. Dual (CD40L plus anti-IgM) stimulated B cells that resist lysis by Th1 effector cells and rFasL were not well protected against cytotoxicity induced by Jo-2. Thus, "Fas-resistant" B cells succumbed to Jo-2 Fas engagement. This implies that the intracellular change(s) triggered by sIgM suppresses, but does not completely eliminate, the capacity of Fas to signal for cell death, inasmuch as critical elements of the Fas death pathway must remain intact at some level that can function following Jo-2 crosslinking. In other words, Fas-resistance appears to be relative, rather than absolute. Further, induction of cell death by Jo-2 in B cells that are protected against cytotoxicity mediated by FasL as found on Th1 effectors and as represented by a soluble recombinant protein implies that Jo-2 triggers Fas in a manner that does not recapitulate physiological engagement of the receptor.

In these experiments Jo-2 was found to induce B cell cytotoxicity at very low concentrations (eg, 0.5 ng/ml), much below that required for killing T cell targets (unpublished observations). This suggested that hypercrosslinking of Fas-bound antibody by Fc receptors might play a role in bringing about Jo-2 cytotoxicity. To test this, the FcR blocking antibody, 2.4G2[37], was included along with Jo-2 in cytotoxicity assays. FcR-specific antibody was found to inhibit Jo-2 killing in a dose-dependent fashion, with virtually complete inhibition at 0.3 μg/ml, suggesting that Jo-2 overcomes Fas-resistance through a particularly strong interaction with target cell Fas that depends on FcR binding.

3. Fas-Resistance Occurs Slowly, Requires Protein Synthesis, and May Relate to Enhanced bcl-x

The intracellular change that mediates sIgM-induced Fas-resistance may involve rapid post-translational modification of pre-existing proteins, or may require transcriptional activation and new protein synthesis. To address this issue the time course of inducible Fas-resistance was evaluated by varying the time of addition of anti-IgM to B cells that were stimulated by CD40L for 48 hours prior to cytotoxicity assay with Th1 effectors. Protection against Th1-induced cytotoxicity was found to develop progressively over a period of hours, reaching a maximum at about 12 hours of exposure to anti-IgM, with a T1/2 for maximal Fas-resistance of 2–6 hours. The several hour delay in achieving half-maximal protection against Th1-induced cytotoxicity suggests that Fas-resistance follows a complex series of events, possibly including new protein synthesis, although the detection of modest Fas-resistance early on raises the possibility that more than one mechanism may be at play.

The need for new protein synthesis for induction of Fas-resistance was tested more directly by taking advantage of the early onset of partial protection. Protein synthesis inhibitors become toxic for primary B cells in a time-dependent fashion at exposure times greater than 6–8 hours. Within this period, however, anti-IgM produces a greater than 50% decline in Th1-induced cytotoxicity. When cycloheximide was added to CD40L-stimulated B cells just prior to the addition of anti-IgM at 6 hours, Fas-resistance was completely reversed, in that subsequent Th1-induced cytotoxicity was indistinguishable from that observed with B cells stimulated by CD40L alone. Thus it appears both by time course and sensitivity to cycloheximide that sIgM-mediated Fas-resistance depends on one or more delayed events that include new protein synthesis.

Bcl-2 and its homologs and interacting proteins have been shown to influence cell fate in lymphocytes[38–40]. In the course of evaluating these proteins, it was found that anti-Ig treatment stimulates *bcl-x*[39–42] expression in CD40L-stimulated B cells as judged by RT-PCR,

Table 3. Protein synthesis inhibition blocks induction of Fas-resistance

B cell treatment		
-48 hrs	-6 hrs	Lysis
CD40L	–	+++
CD40L	anti-IgM	+
CD40L	CHX + anti-IgM	+++

Primary B cells were treated for 2 days and tested as targets for Th1-mediated, Fas-dependent cytotoxicity in ^{51}Cr release assays as described in the legend to Table 1, with the exception that anti-IgM was added 6 hours before the end of the culture. CHX, cycloheximide at 2μg/ml, added 1 hour before anti-IgM.

and that Bcl-x protein is also substantially induced, as judged by Western blotting. Dual treated B cells expressed levels of *bcl-x*/Bcl-x that were more than 5-fold greater than that observed with B cells stimulated by CD40L alone. These results suggest the possibility that regulation of Bcl-x expression may be related to induction of Fas-resistance in primary B cells, but this conclusion awaits more definitive functional data that would demonstrate reliance on Bcl-x as a causative factor. Some indirect support for this notion is provided by the observation that dual treated, Fas-resistant B cells are less sensitive to apoptosis induced by C2 ceramide than CD40L stimulated B cells, suggesting that the mechanism responsible for Fas-resistance acts distally, downstream of ceramide[39,40,43,44].

4. Il-4 Induces Fas-Resistance

Although not a complete mitogen, IL-4 activates B cells and in so doing enhances the proliferative responses to complete mitogens such as anti-Ig[45]. Because of the numerous stimulatory and differentiative effects of IL-4, this lymphokine was tested for the capacity to regulate B cell susceptibility to Fas-mediated apoptosis. Experiments were carried out as described above with B cells stimulated by CD40L for 48 hours; IL-4 was added 24 hours before the end of culture, after which B cells were tested as targets in cytotoxicity assays using Th1 effector cells. IL-4 induced protection against Th1-induced cytotoxicity that was readily apparent but not quite as complete as that produced by anti-Ig; that is, even at optimal doses, IL-4-treated (CD40L-stimulated) B cells were lysed to a small degree at high effector:target cell ratios (eg, 9:1) and to a greater extent than anti-Ig-treated (CD40L-stimulated) B cells. This difference in the degree of protection afforded by IL-4 and anti-Ig was observed in numerous experiments. In broad terms, IL-4 produced about 10-fold protection against Th1 effectors, meaning about 10-fold more Th1 cells were needed to produce equivalent levels of cytotoxicity in resistant as opposed to sensitive B cells; by way of comparison, protection produced by anti-Ig amounted to more than 20-fold. The protective effect of IL-4 was reversed by first adding 11B11 neutralizing anti-IL-4 antibody[46], demonstrating the specificity of this lymphokine-induced Fas-resistance.

IL-4 is not known to produce phosphorylation or activation of PLC, and, in keeping with this, PKC inhibition does not affect IL-4-induced B cell responses[47–49] although PKC is required for sIgM-mediated Fas-resistance. To address the role of PKC, IL-4-induced protection against Th1 killing was evaluated in PKC-depleted B cells, produced by adding PMA (or the relatively inactive analog, 4αPMA) with CD40L at the start of 48 hour B cell cultures, and then adding IL-4 24 hours before the end of culture and subsequent cytotoxicity

Table 4. Il-4 influences susceptibility of B cell targets to Th1-induced cytotoxicity

B cell treatment		
-48 hrs	-24 hrs	Lysis
CD40L	–	+++
CD40L	IL-4	+
CD40L + 4αPMA	–	+++
CD40L + PMA	–	+++
CD40L + 4αPMA	IL-4	+
CD40L + PMA	IL-4	+

Primary B cells were treated for 2 days and tested as targets for Th1-mediated, Fas-dependent cytotoxicity in ^{51}Cr release assays as described in the legend to Tables 1 and 3.

assay. There was little diminution in the level of Fas-resistance produced by IL-4 when B cells treated with PMA (PKC-depleted) were compared with B cells treated with (inactive) 4αPMA. In contrast, and as expected (see above), protection against Th1 effectors produced by anti-Ig was reversed as a result of PKC-depletion. These results indicate that the signaling pathway that couples IL-4R with Fas-resistance does not require PKC, unlike the pathway coupling sIgM and Fas-resistance, and further suggest that more than one intermediate pathway leads to the phenotype of protection against Th1-induced, Fas-mediated apoptosis.

The time course for IL-4 induction of protection against Th1 killing lends further support to the hypothesis that different signaling pathways couple sIgM and IL-4R to Fas-resistance. To evaluate the duration of treatment needed to oppose Fas signaling, IL-4 was added at various times to B cells that were stimulated by CD40L for 48 hours prior to cytotoxicity assay with Th1 effectors. Protection against Th1-induced cytotoxicity was found to develop progressively over a period of hours, but the time required for induction of Fas-resistance was greater for IL-4 than for anti-Ig. Thus, at 6 hours of exposure, there was little protection against Th1 effector cells with IL-4 (about 1/4 of maximum protection), whereas, in direct contrast, at this time point substantial protection was produced by anti-IgM (about 2/3 of maximal protection). (For this reason it was impossible to assess the role of new protein synthesis in contributing to IL-4-induced Fas-resistance.) However, both IL-4 and anti-IgM induced maximal Fas-resistance within 24 hours of B cell treatment. These results suggest that different pathways account for anti-Ig- and IL-4-induced Fas-resistance, although it may be that these receptors share downstream mediators in common.

The mechanism underlying IL-4-induced Fas-resistance could involve either extracellular or intracellular events, as discussed earlier with respect to anti-Ig. Flow cytometric analysis of fluorescently stained B cells demonstrated that CD40L-stimulated Fas expression was little affected by the presence of IL-4, and thus an alteration in Fas expression cannot explain Fas-resistance in this case, although cold (unlabeled) target inhibition studies have not been carried out to eliminate the possibility that the functional form or accessibility of Fas is affected by IL-4.

IL-4-treated (CD40L-stimulated) B cells that were resistant to Th1 effectors were also observed to be resistant to soluble, recombinant FasL, and thus T cell:B cell interactions beyond FasL:Fas do not appear to be involved in the induction of Fas-resistance. At the same time, IL-4 treatment does not entirely eliminate the Fas death machinery because, as noted above with regard to anti-Ig-treated B cells, Jo-2 effects lysis of dual (CD40L plus IL-4) stimulated, Th1-resistant B cells to the same extent as that observed with B cells stimulated by CD40L alone. These results strongly suggest that engagement of IL-4R, like sIgM, produces Fas-resistance through an intracellular change that produces relative, but not absolute, opposition to Fas signaling for cell death, although this takes place in a PKC-independent fashion for IL-4R and a PKC-dependent fashion for sIgM.

The differences in IL-4R and sIgM signaling suggested that these receptors might act in concert to produce Fas-resistance. To test this, CD40L-stimulated (for 48 hours) B cells were treated with suboptimal doses of IL-4 and anti-IgM, individually or together, for the last 24 hours of culture, after which cytotoxicity assay was carried out. At these suboptimal doses, specific target cell lysis produced by treatment (of CD40L-stimulated B cells) with either IL-4 or anti-IgM amounted to 15–30% (in comparison to about 50% specific lysis of B cells stimulated by CD40L alone), whereas there was no specific lysis of CD40L-stimulated B cells treated with the combination of IL-4 and anti-IgM. These results suggest that antigen and IL-4 receptors act in synergy to produce resistance against Th1-induced, Fas-dependent, cytotoxicity, such that effective Fas-resistance is induced by lower doses of the corresponding ligands when applied together.

In the course of these studies additional soluble T cell products were evaluated for the capacity to regulate B cell susceptibility to Fas-mediated apoptosis. Th1 lymphokines, including IL-2, IFNγ, and TNFα, had little effect. The same was true for additional Th2 lymphokines, including IL-5 and IL-6. Thus, the effect of IL-4 appears at this point to be relatively distinct among soluble T cell derived mediators.

DISCUSSION

The most important conclusion to be drawn from this work is that B cells are not simply passive targets for Th1-induced cytotoxicity, but rather, susceptibility to Fas-directed cell death is modulated by specific receptor interactions that detect environmental cues in the form of antigen and lymphokine. Fas-resistance appears to result from an intracellular change in B cell targets that leads to relative, but not absolute, suppression of Fas signaling that counteracts physiological ligands mediating apoptosis. Surface immunoglobulin signals for Fas-resistance in a PKC-dependent fashion, and Fas-resistance induced by sIg is paralleled by the combination of two pharmacologic agents that activate PKC and increase Ca^{++} and thus mimic PLC-derived second messengers. In contrast, CD40 signals for Fas-resistance in a largely PKC-independent fashion. Together, these results suggest that distinct intracellular signaling pathways may be responsible for induction of Fas-resistance, in a receptor-specific fashion.

CD40, sIgM, and IL-4R express varied effects on B cell fate in other systems. For example, CD40 and sIgM engagement similarly rescue germinal center B cells from spontaneous apoptosis[50], but express completely different activities with respect to Fas-sensitivity. Along the same lines, CD40 and IL-4R individually and together rescue primary B cells from apoptosis induced by hypercrosslinking sIgM[51], yet oppose each other in regulating Fas killing. Thus, the regulation of Fas-mediated apoptosis differs from that of other forms of apoptosis and the roles of CD40, sIgM and IL-4R cannot be predicted from previous studies in other systems.

The proximate mechanism by which Fas-resistance is established remains unknown. The time course for induction of Fas-resistance suggests a progressive response consistent with new protein synthesis, which has been confirmed for anti-IgM by experiments with cycloheximide. In the case of sIgM, preliminary results implicate Bcl-x in modulating Fas-sensitivity, although the evidence to date is only associative, and causation has not been demonstrated. Of some interest, *bcl-2*, a related molecule, contains a CRE site that responds to CREB[52], which is phosphorylated and activated by sIgM, but not by CD40, signaling[20]. Although CD40 (which mediates Fas-sensitivity) and sIg (which reverses susceptibility to Fas killing) trigger induction of a wide variety of similar outcomes on a molecular scale, differential transcription factor regulation as in the case of CREB may play a role in inducing gene expression that leads to Fas-resistance.

At this point it is not clear that IL-4 mediates Fas-resistance in the same way as anti-Ig, particularly in view of the more prolonged time course required for the IL-4-induced protective response and its distinct PKC-independence. If IL-4 and sIgM signaling do rely on induction of a similar protective protein, or proteins, however, a consideration of signaling properties that they express in common might provide insight into one or more aspects of the mechanism(s) underlying Fas-resistance, although IL-4 and sIgM are known to share little in the way of downstream signaling events.

The role that inducible Fas-resistance plays in B cell physiology is undefined and remains speculative at present. Fas-deficient mice are characterized by substantial autoantibody production[35] and autoreactive B cells are found within the T cell areas of the periarteriolar lymphoid sheath (PALS)[53], suggesting that the failure of T cell mediated cytotoxicity at this stage, prior to germinal center formation, leads to aberrant survival and subsequent differentiation of autoantibody producing B cells. Further, Fas has been shown to be critically involved in the disposition of anergic B cells, whose lifespan is foreshortened as a result of CD4+ T cell mediated, Fas-dependent cytotoxicity[54]. Taken together these results suggest that Fas-mediated apoptosis represents a principal mechanism for eliminating autoreactive B cells that have been anergized.

A potential role for inducible Fas-resistance may then lie in the regulation of anergic, autoreactive B cell fate. Autoreactive antibodies represent forbidden specificities, and B cells with (relatively low avidity) self-recognizing antigen receptors are anergized[55], resulting in a brief state of suspended animation with respect to specific antigen, which is followed by Fas-mediated apoptosis. Importantly, the anergic pathway results in relatively more prolonged survival of autoreactive B cells than deletional tolerance (which rapidly eliminates B cells expressing receptors that recognize autologous antigen with higher avidity). During anergy, B cell surface IgM is diminished and sIg-mediated signaling is modulated; however, responses to CD40L and IL-4 remain intact[56,57].

Anergic, autoreactive B cells are thought to be present within peripheral lymphoid tissue of normal individuals on the basis of phenotypic parallels with B cells from double transgenic mice[58]. Presumably there is a continual flux of B cells into (as a result of specific receptor expression) and out of (as a result of Fas-mediated apoptosis) the anergic pool. We suggest that anergic B cells serve as a potential resource, even as they represent a potential risk, as follows. Anergic B cells constitute a sequestered or cloaked version of the autoreactive elements missing from the serological immune repertoire. Viewed this way, anergic B cells may be thought of as an inverse image of the expressed repertoire. Interference with the ultimate deletion of such anergic B cells would likely predispose to autoreactivity. However, in the event of extreme immunological stress, perhaps produced by invasive pathogenic organisms never previously experienced in evolutionary time, it may be advantageous to prevent Fas-mediated deletion and thereby recover anergic B cells whose receptor specificities include foreign determinants that fail to be recognized by the active B cell pool because they cross-react with self. This would obviously carry the risk of producing autoimmunity along with protective immunity, but from the organism and species standpoint,

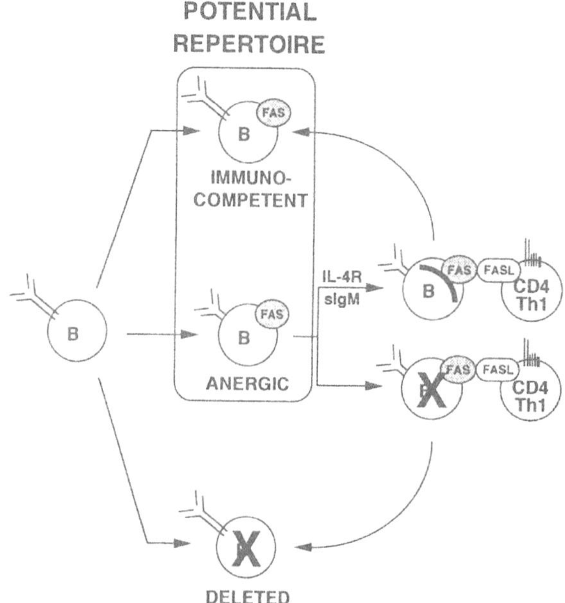

Figure 1. Inducible resistance to Fas-mediated apoptosis may contribute to autoantibody formation. A speculative model relating Fas-resistance, anergy and autoreactive antibody formation is presented. In this model, initial antigen receptor interaction leads to tolerance in the form of deletion (denoted "X"), or anergy, or to signals that produce immunocompetence. Anergic B cells that are ultimately deleted through a Fas-dependent mechanism are returned to the immunocompetent B cell pool if Fas-resistance (shown as an arc) is induced via the receptor for IL-4, alone or in conjunction with the antigen receptor. Thus, anergic B cells may represent part of a potential repertoire, that escapes deletion and becomes activated if appropriate signals trigger Fas-resistance. See text for additional details.

autoreactivity may be a better outcome than termination of existence. Thus, anergic B cells should perhaps be considered a potential pool of immunoglobulin specificities.

Immunological stress capable of reversing Fas-sensitivity might reasonably be expected to express itself through IL-4 production (as Th2 cells drive the serological response[59]), perhaps augmented by sIg engagement (by microbial antigen). Thus, we predict that IL-4R engagement, alone or in combination with sIg triggering, will be found to induce resistance against Fas-mediated apoptosis in anergic B cells, allowing these B cells to persist, to become activated, and subsequently to undergo further stimulatory and differentiative responses resulting, intentionally or unintentionally, in expansion of the serological immune repertoire manifested by the addition of autoreactive specificities and the risk of frank autoimmune disease. In this model the recognized increase in the frequency of autoantibodies with increasing age could be interpreted as reflecting the cumulative effect of many episodes of immunological stress that have interfered with the Fas-dependent disposition of anergic B cells. This highly speculative model is now being tested.

ACKNOWLEDGMENTS

This work was supported by United States Public Health Service Grants AI29690, T32-AI07309, and T32-CA64070.

REFERENCES

1. G. Berke. The CTL's kiss of death. *Cell* 81:9 (1995).
2. S. Yonehara, A. Ishii, and M. Yonehara. A cell-killing monoclonal antibody (anti-Fas) to a cell surface antigen co-downregulated with the receptor of tumor necrosis factor. *J. Exp. Med.* 169:1747 (1989).
3. B. C. Trauth, C. Klas, A. M. J. Peters, S. Matzku, P. Moller, W. Falk, K.-M. Debatin, and P. H. Krammer. Monoclonal antibody-mediated tumor regression by induction of apoptosis. *Science* 245:301. (1989).
4. C. Watanabe-Fukunaga, I. Brannan, N. G. Copeland, N. A. Jenkins, and S. Nagata. Lymphoproliferation disorder in mice explained by defects in Fas antigen that mediates apoptosis. *Nature* 356:314. (1992).
5. S.-T. Ju, D. J. Panka, H. Cui, R. Ettinger, M. El-Khatib, D. H. Sherr, B. Z. Stanger, and A. Marshak-Rothstein. Fas(CD95)/FasL interactions required for programmed cell death after T-cell activation. *Nature* 373:444. (1995).
6. J. Dhein, H. Walczak, C. Baumler, K.-M. Debatin, and P. H. Krammer. Autocrine T-cell suicide mediated by APO-1/(Fas/CD95). *Nature* 373:438. (1995).
7. T. Brunner, R. J. Mogil, D. LaFace, N. J. Yoo, A. Mahboubi, F. Echeverri, S. J. Martin, W. R. Force, D. H. Lynch, C. F. Ware, and D. R. Green. Cell-autonomous Fas (CD95)/Fas-ligand interaction mediates activation-induced apoptosis in T-cell hybridomas. *Nature* 373:441. (1995).
8. S. Nagata, and P. Golstein. The Fas death factor. *Science* 267:1449. (1995).
9. D. Kägi, F. Vignaux, B. Lederemann, K. Bürki, V. Depraetere, S. Nagata, H. Hengartner, and P. Golstein. Fas and perforin pathways as major mechanisms of T cell-mediated cytotoxicity. *Science* 265:528 (1994).
10. B. Lowin, M. Hahne, C. Mattmann, and J. Tschopp. Cytolytic T-cll cytotoxicity is mediated through perforin and Fas lytic pathways. *Nature* 370:650 (1994).
11. E. Rouvier, M.-F. Luciani, and P. Golstein. Fas involvement in Ca^{2+}-independent T cell-mediated cytotoxicity. *J. Exp. Med.* 177:195 (1993).
12. T. Stalder, S. Hahn, and P. Erb. Fas antigen is the major target molecule for $CD4^+$ T cell-mediated cytotoxicity. *J. Immunol.* 152:1127 (1994).
13. S.-T. Ju, H. Cui, D. J. Panka, R. Ettinger, and A. Marshak-Rothstein. Participation of target Fas protein in apoptosis pathway induced by $CD4^+$ Th1 and $CD8^+$ cytotoxic T cells. *Proc. Natl. Acad. Sci. USA* 91:4185 (1994).
14. S. Hanabuchi, M. Koyanagi, A. Kawasaki, N. Shinohara, A. Matsuzawa, Y. Nishimura, Y. Kobayashi, S. Yonehara, H. Yagita, and K. Okumura. Fas and its ligand in a general mechanism of T-cell-mediated cytotoxicity. *Proc. Natl. Acad. Sci. USA* 91:4930 (1994).

15. P. Erb, D. Grogg, M. Troxler, M. Kennedy, and M. Fluri. CD4+ T cell-mediated killing of MHC class II-positive antigen-presenting cells. I. Characterization of target cell recognition by in vivo or in vitro activated CD4+ killer T cells. J. Immunol. 144:790 (1990).
16. L. B. Owen-Schaub, S. Yonehara, W. L. Crump III, and E. A. Grimm. DNA fragmentation and cell death is selectively triggered in activated human lymphocytes by Fas antigen engagement. Cell. Immunol. 140:197 (1992).
17. Daniel, P. T., and P. H. Krammer. Activation induces sensitivity toward APO-1 (CD95)-mediated apoptosis in human B cells. *J. Immunol.* 152:5624. (1994).
18. Rothstein, T. L., J. K. M. Wang, D. J. Panka, L. C. Foote, Z. Wang, B. Stanger, H. Cui, S.-T. Ju, and A. Marshak-Rothstein. Protection against Fas-dependent Th1-mediated apoptosis by antigen receptor engagement in B cells. *Nature* 374:163 (1995).
19. K. Kawakami, and D. C. Parker. Antigen and helper T lymphocytes activate B lymphocytes by distinct signaling pathways. Eur. J. Immunol. 23:77 (1993).
20. H. Xie, and T. L. Rothstein. Protein kinase C mediates activation of nuclear cAMP response element-binding protein (CREB) in B lymphocytes stimulated through surface Ig. *J. Immunol.* 154:1717 (1995).
21. L. Huo, and T. L. Rothstein. Receptor-specific induction of individual AP-1 components in B lymphocytes. *J. Immunol.* 154:3300 (1995).
22. J. C. Cambier, C. M. Pleiman, and M. R. Clark. Signal transduction by the B cell antigen receptor and its coreceptors. *Ann. Rev. Immunol.* 12:457 (1994).
23. T. Kato, T. Kokuho, T. Tamura, and H. Nariuchi. Mechanisms of T cell contact-dependent B cell activation. *J. Immunol.* 152:2130 (1994).
24. L. S. Marshall, D. M. Shepherd, J. A. Ledbetter, a. Aruffo, and R. J. Noelle. Signaling events during helper T cell-dependent B cell activation. I. Analysis of the signal transduction pathways triggered by activated helper T cells in resting B cells. *J. Immunol.* 152:4816 (1994).
25. A.-C. Lalmanach-Girard, T. C. Chiles, D. C. Parker, and T. L. Rothstein. T cell-dependent induction of NF-κB in B cells. *J. Exp. Med.* 177:1215 (1993).
26. D. A. Francis, J. G. Karras, X. Ke, R. Sen, and T. L. Rothstein. Induction of the transcription factors NF-κB, AP-1 and NF-AT during B cell stimulation through the CD40 receptor. *Intl. Immunol.* 7:151 (1995).
27. J. J. Mond, N. Feuerstein, F. D. Finkelman, F. Huang, K.-P. Huang, and G. Dennis. B-lymphocyte activation mediated by anti-immunoglobulin antibody in the absence of protein kinase C. *Proc. Natl. Acad. Sci. USA* 84:8588 (1987).
28. J. Liu, T. C. Chiles, R. Sen, and T. L. Rothstein. Inducible nuclear expression of NF-κB in primary B cells stimulated through the surface Ig receptor. *J. Immunol.* 146:1685 (1991).
29. H. Hidaka, M. Inagaki, S. Kawamoto, and Y. Sasaki. Isoquinolinesulfonamides, novel and potent inhibitors of cyclic nucleotide dependent protein kinase and protein kinase C. *Biochem.* 23:5036 (1984).
30. G. G. B. Klaus, A. O'Garra, M. K. Bijsterbosch, and M. Holman. Activation and proliferation signals in mouse B cells. VIII. Induction of DNA synthesis in B cells by a combination of calcium ionophores and phorbol myristate acetate. *Eur. J. Immunol.* 16:92 (1986).
31. T. L. Rothstein, T. R. Baeker, R. A. Miller, and D. L. Kolber. Stimulation of murine B cells by the combination of calcium ionophore plus phorbol ester. *Cell. Immunol.* 102:364 (1986).
32. W. M. Flanagan, B. Corthesy, R. J. Bram, and G. R. Crabtree. Nuclear association of a T cell transcription factor blocked by FK506 and cyclosporin A. Nature 352:803 (1991).
33. L. Venkataraman, D. A. Francis, Z. Wang, J. Liu, T. L. Rothstein, and R. Sen. Cyclosporin-A sensitive induction of NF-AT in murine B cells. Immunity 1:189 (1994).
34. H. Xie, Z. Wang, and T. L. Rothstein. Signaling pathways for antigen receptor-mediated induction of transcription factor CREB in B lymphocytes. *Cell. Immunol.*, in press (1996).
35. S. Nagata, and T. Suda. Fas and Fas ligand: *lpr* and *gld* mutations. *Immunol. Today* 16:39 (1995).
36. J. Ogasawara, R. Watanabe-Fukunaga, M. Adachi, A. Matsuzawa, T. Kasugai, Y. Kitamura, N. Itoh, T. Suda, and S. Nagata. Lethal effect of the anti-Fas antibody in mice. *Nature* 364:806 (1993).
37. J. C. Unkeless. Characterization of a monoclonal antibody directed against mouse macrophage and lymphocyte Fc receptors. *J. Exp. Med.* 150:580 (1979).
38. T. Sato, M. Hanada, S. bodrug, S. Irie, N. Iwama, L. H. Boise, C. B. Thompson, E. Golemis, L. Fong, H.-G. Wang, and J. C. Reed. Interactions among members of the Bcl-2 protein family analyzed with a yeast two-hybrid system. *Proc. Natl. Acad. Sci. USA* 91:9238 (1994).
39. G. Núñez, R. Merino, D. Grillot, and M. González-García. Bcl-2 and Bcl-x: regulatory switches for lymphoid death and survival. *Immunol. Today* 15:582 (1994).
40. S. Cory. Regulation of lymphocyte survival by the Bcl-2 gene family. *Ann. Rev. Immunol.* 13:513 (1995).

41. L. H. Boise, M. González-García, C. e. Postema, L. Ding, T. Lindsten, L. A. Turka, X. Mao, G. Núñez, and C. B. Thompson. *bcl-x*, a *bcl-2*-related gene that functions as a dominant regulator of apoptotic cell death. *Cell* 74:597 (1993).
42. D. T. Chao, G. P. Linette, L. H. Boise, L. S. White, C. B. Thompson, and S. J. Korsmeyer. Bcl-xL and Bcl-2 repress a common pathway of cell death. *J. Exp. Med.* 182:821 (1995).
43. E. Gulbins, R. Bissonnette, A. Mahboubi, S. Martin, W. Nishioka, T. Brunner, G. Baier, G. Baier-Bitterlich, C. Byrd, F. Lang, R. Kolesnick, A. Altman, and D. Green. FAS-induced apoptosis is mediated via a ceramide-initiated RAS signaling pathway. *Immunity* 2:341 (1995).
44. C. G. Tepper, S. Jayadev, B. Liu, A. Bielawska, R. Wolff, S. Yonehara, Y. A. Hannun, and M. F. Seldin. Role for ceramide as an endogenous mediator of Fas-induced cytotoxicity. Proc. Natl. Acad. Sci. USA 92:8443 (1995).
45. W. E. Paul, and J. Ohara. B-cell stimulatory factor-1/interleukin 4. *Ann. Rev. Immunol.* 5:429 (1987).
46. J. Ohara, and W. E. Paul. B cell stimulatory factor BSF-1: Production of a monoclonal antibody and molecular characterization. *Nature* 315:333 (1985).
47. A. D. Keegan, K. Nelms, L.-M. Wang, J. H. Pierece, and W. E. Paul. Interleukin 4 receptor: signaling mechanisms. *Immunol. Today* 15:423 (1994).
48. M. Dancescu, C. Wu, M. Rubio, G. Delespesse, and M. Sarfati. IL-4 induces conformational change of CD20 antigen via a protein kinase C-independent pathway. Antagonistic effect of anti-CD40 monoclonal antibody. *J. Immunol.* 148:2411 (1992).
49. C. E. Lee, S. R. Yoon, and K. H. Pyun. Interleukin-4 signals regulating CD23 gene expression in human B cells: protein kinase C-independent signaling pathways. *Cell. Immunol.* 146:171 (1993).
50. I. A. C. MacLennan. Germinal centers. *Ann. Rev. Immunol.* 12:117 (1994).
51. S. L. Parry, J. Hasbold, M. Holman, and G. G. B. Klaus. Hypercross-linking surface IgM or IgD receptors on mature B cells induces apoptosis that is reversed by costimulation with IL-4 and anti-CD40. *J. Immunol.* 152:2821 (1994).
52. B. E. Wilson, E. Mochon, and L. M. Boxer. Induction of *bcl-2* expression by phosphorylated CREB proteins during B cell activation. *Blood* 86:327a (1995).
53. B. A. Jacobson, D. J. Panka, K.-A. Nguyen, J. Erikson, A. K. Abbas, and A. Marshak-Rothstein. Anatomy of autoantibody production: dominant localization of antibody-producing cells to T cell zones in Fas-deficient mice. *Immunity* 3:509 (1995).
54. J. C. Rathmell, M. P. Cooke, W. Y. Ho, J. Grein, S. E. Townsend, M. M. Davis, and C. C. Goodnow. CD95 (Fas)-dependent elimination of self-reactive B cells upon interaction with $CD4^+$ T cells. *Nature* 376:181 (1995).
55. G. J. V. Nossal. Negative selection of lymphocytes. *Cell* 76:229 (1994).
56. M. P. Cooke, A. W. Heath, K. M. Shokat, Y. Zeng, F. D. Finkelman, P. s. Linsley, M. Howard, and C. C. Goodnow. Immunoglobulin signal transduction guides the specificity of B cell-T cell interactions and is blocked in tolerant self-reactive B cells. *J. Exp. Med.* 179:425 (1994).
57. J. M. Eris, A. Basten, R. Brink, K. Doherety, M. R. Kehry, and P. D. Hodgkin. Anergic self-reactive B cells present self antigen and respond normally to CD40-dependent T-cell signals but are defective in antigen-receptor-mediated functions. *Proc. Natl. Acad. Sci. USA* 92:4392 (1994).
58. C. C. Goodnow, J. Crosbie, H. Jorgensen, R. A. Brink, and A. Basten. Induction of self-tolerance in mature peripheral B lymphocytes. *Nature* 342:385 (1989).
59. W. E. Paul, and R. A. Seder. Lymphocyte responses and cytokines. *Cell* 76:241 (1994).

20

THE THYMUS AND T CELL DEATH

Jonathan Sprent,[1] Hidehiro Kishimoto,[1] Zeling Cai,[1] Charles D. Surh,[1]
Anders Brunmark,[2] Michael R. Jackson,[2] and Per A. Peterson[2]

[1] The Scripps Research Institute
Department of Immunology, IMM4
0666 N. Torrey Pines Road, La Jolla, California 92037.
[2] R.W. Johnson Pharmaceutical Research
3535 General Atomics Ct., Suite 100, San Diego, California 92121

1. INTRODUCTION

Through the combined effects of positive and negative selection, T cell differentiation in the thymus generates a repertoire of mature T cells that is tailored to tolerate self antigens but mount strong responses to foreign antigens[1,2]. Thymic selection is associated with extensive cell death, and only a very small proportion of thymocytes are selected for survival and export to the extrathymic environment. This article provides an overview of thymic selection and the fate of thymocytes.

1.1. Positive Selection

Typical αβ T cells arise in the thymus by a process of positive selection directed to self peptides bound to class I and II major histocompatibility complex (MHC) molecules expressed on cortical epithelium[1–7]. Positive selection occurs at the level of double-positive (DP) $CD4^+8^+$ cells expressing a low density of αβ T cell receptor (TCR) and CD3 molecules and is presumed to involve the delivery of a low-level protective signal; this signal prevents the cells from succumbing to apoptosis ("death by neglect") and thus promotes cell survival. The signalling involved in positive selection is mediated via TCR molecules and CD8 or CD4 coreceptors (which bind class I and class II molecules, respectively) and causes upregulation of CD69 and differentiation into single-positive (SP) $CD4^+8^-$ and $CD4^-8^+$ cells. These cells migrate from the cortex to the medulla and, after negative selection, are then exported to the periphery.

As the result of positive selection, the T cells leaving the thymus retain covert reactivity to the selecting MHC molecules. Thus, the cells ignore self MHC molecules plus self peptides but are strongly reactive to self MHC molecules complexed with foreign peptides. In general, the precursor frequency of mature T cells for foreign peptides is higher when these peptides are presented by self rather than nonself (allo) MHC molecules, the phenomenon of MHC restriction[8]. In light of this specificity, positive selection is thought to be restricted to a subset of thymocytes displaying significant but low-level ("physiological")

Mechanisms of Lymphocyte Activation and Immune Regulation VI, Edited by Gupta and Cohen
Plenum Press, New York, 1996

TCR affinity for the self MHC molecules expressed in the thymus plus the various self peptides bound to these molecules; the range of self peptides that control positive selection is still unclear, although weak (antagonist) peptides seem to be particularly important[2].

By selecting for physiological reactivity to self MHC molecules, positive selection generates a T cell repertoire that is designed to function optimally in the extrathymic environment. Because of MHC polymorphism, positive selection thus generates a unique T cell repertoire for each individual. Since the range of MHC alleles is enormous, the MHC specificity of the reservoir of early DP thymocytes has to be very extensive and collectively embrace reactivity to all of the different MHC alleles expressed in the species as a whole. Thus, in any one individual, the proportion of DP cells with specificity for the particular "self" MHC molecules expressed in the thymus is likely to be very low. It follows therefore that positive selection affects only a very small proportion of total thymocytes and that most DP cells are superfluous and need to be destroyed. This topic is discussed below.

1.2. Death from Lack of Positive Selection

Despite the large size of the thymus, the number of T cells exported from the thymus is very small, i.e. 1–2 x 10^6 cells/day in young mice, and is limited to SP cells[9]. Since the young thymus contains in the order of 1 x 10^8 cells (85% DP cells), it follows that the vast majority (98%) of thymocytes, particularly DP cells, are destined to die in situ, presumably via apoptosis. This has long been a paradox, however, because conventional histological sections of thymus show little or no evidence of cell death. Nevertheless, applying the TUNEL technique to frozen sections of thymus has revealed appreciable numbers of apoptotic cells scattered throughout the cortex[10]. These cells are as prominent in the thymus of combined MHC class I^-II^- mice as in the normal thymus, which indicates that apoptosis in the cortex is primarily a reflection of lack of positive selection (death by neglect) rather than negative selection.

Interestingly, double staining for macrophage markers shows that nearly all of the apoptotic cells in the cortex are situated inside F4/80^+ macrophages[10]. This finding suggests that dying thymocytes are phagocytosed at a very early stage of apoptosis, i.e. before the cells develop DNA strand breaks (which is the basis of the TUNEL staining method). Early phagocytosis of apoptotic cells by macrophages is presumed to be stimulated by alterations in the cell surface of apoptotic cells, especially the appearance of phosphatidylserine residues[11]. These surface changes are readily detected by macrophages and lead to rapid ingestion of apoptotic cells.

The prodigious capacity of cortical macrophages to ingest apoptotic cells is revealed by the effects of treating the thymus with irradiation or injecting anti-CD3 antibodies or hydrocortisone intravenously[10]. These treatments cause an abrupt onset of apoptosis in the cortex and rapid ingestion of the apoptotic cells by adjacent macrophages. Within 1–2 hours of irradiation, the macrophages in the cortex become massively engorged with apoptotic cells and debris and show very strong TUNEL staining. DNA breakdown ensues rapidly, and within 12–18 hours the macrophages regain their normal size and show a marked reduction in TUNEL staining. In view of this highly efficient capacity of macrophages to clear apoptotic cells, it is not surprising that histological evidence of apoptosis in conventional sections of normal thymus is very limited.

1.3. Cell Death by Negative Selection

As discussed earlier, positive selection is thought to be limited to T cells displaying low-level specificity for self peptide/MHC complexes. Some T cells will inevitably have strong specificity for these self antigens, and would pose a danger of autoimmune disease if allowed to exit to the post-thymic environment. It is now well established that these cells are destroyed in situ[10,12–15]. This process of negative selection (central tolerance) is restricted to peptides that are expressed constitutively in the thymus or reach the thymus via the blood stream. Tissue-specific self antigens fail to induce central tolerance, and the failure of the immune system to react against these antigens is probably largely a reflection of "ignorance": tissue-specific antigens remain sequestered whereas T cells are confined to the recirculating lymphocyte pool[15].

Since the range of self peptides expressed on thymic epithelium is much the same as on spleen cells[16], the peptides involved in positive and negative selection are probably generated from the same or similar pools of intracellular proteins. However, the selecting peptides may be quite different in terms of their TCR and MHC affinity. Thus, in contrast to positive selection, negative selection could be largely limited to peptides with high TCR/MHC affinity. In addition, the peptides controlling negative selection could be displayed at a higher concentration on the cell surface than positively-selecting peptides. In this respect there are two reports that strong (agonist) peptides can induce positive selection at low concentration but negative selection at high concentration[5,6]. This finding supports an affinity/avidity model for thymic selection[3,7] and suggests that the main difference between positive and negative selection is the strength of TCR signalling: weak signals induce positive selection whereas strong signals cause negative selection.

1.4. Sites of Negative Selection

The issue of whether negative selection is induced primarily in the cortex or the medulla is highly controversial[17]. Studies on tolerance to alloantigens in bone marrow (BM) chimeras and thymic organ cultures indicate that BM-derived cells, especially dendritic cells, play a key role in negative selection[8,13,18]. Since these cells reside almost exclusively in the medulla and corticomedullary junction, it has long been argued that the medulla is the principal site of negative selection[8]. This notion is in line with the evidence that circulating self proteins have ready access to the medulla but are largely excluded from the cortex (except in peptide form)[17]. Direct evidence that negative selection can occur in the medulla has come from TUNEL staining of TCR transgenic mice specific for endogenous superantigens[10].

Since most self proteins, e.g. mitochondrial enzymes, are not tissue specific, it is highly likely that the majority of peptides involved in negative selection are expressed constitutively in the cortex on epithelial cells, e.g. peptides derived from Krebs cycle enzymes etc. A priori, one could therefore argue that most T cells destined for negative selection are deleted in the cortex rather than the medulla. Direct evidence that the cortex is conducive to negative selection is provided by the finding that injecting specific peptides into TCR transgenic mice causes massive apoptosis in the cortex followed by marked thymic atrophy[19]. Although these and other data point to the cortex as a prime site for negative selection, it should be remembered that TCR expression in transgenic mice is generally higher than in normal mice and occurs at a very early stage of thymocyte differentiation. Hence demonstrating negative selection in the cortex in TCR transgenic mice does not necessarily prove that the same applies in normal mice. We will return to this issue later.

1.5. Second Signals for Negative Selection

For mature T cells, it is well accepted that T cell stimulation requires two types of signals, one resulting from TCR contact with peptide/MHC complexes (Signal 1) and the other elicited through CD28/B7 interaction (Signal 2 or costimulation)[20,21]. Delivery of these two signals generally requires joint recognition of peptide/MHC and B7 (B7–1 or B7–2) on viable antigen-presenting cells (APC)[20–22]. In addition to B7, a number of other molecules on APC, including ICAM-1 and HSA, express costimulatory function for T cells[23,24]. Hence Signal 2 may well involve multiple signals.

The role of second signals in negative selection is controversial[17]. This issue is difficult to address under normal physiological conditions, and most of the evidence on the requirements for inducing negative selection has come from studies on dispersed thymocytes cultured in vitro. With this approach it has been found that cross linking TCR/CD3 molecules on purified thymocytes fails to cause apoptosis unless the cells receive second signals, e.g. through contact with other cell types[25–27]. Interestingly, studies with peptide-specific TCR transgenic mice indicate that a wide variety of cell types, including thymic epithelial lines, are capable of causing peptide-induced cell death in culture[26,28].

Although these data indicate that negative selection does require costimulation, which particular cell-surface molecules are required for costimulation is still unclear. Particular attention has been focused on the fact that negative selection can occur without CD28/B7 interaction. Thus $B7^-$ cell lines can cause peptide-induced apoptosis in culture[28], and negative selection is apparent in CD28-deficient mice[29]. However, it does not necessarily follow that CD28/B7 interaction plays no role in negative selection. Because of "redundancy", it is possible that a multiplicity of different cell-surface molecules contribute to negative selection. Thus, the contribution of individual molecules such as B7 could be masked by the presence of other costimulatory molecules.

One approach for determining which particular costimulatory molecules participate in negative selection is to examine the function of individual molecules expressed in cell lines. With this approach we have attempted to define the minimal requirements for inducing negative selection of TCR transgenic thymocytes in vitro.

1.6. Minimal Requirements for Negative Selection

The class I-restricted 2C line of TCR transgenic mice undergoes positive selection to $H\text{-}2^b$ [30,31] and displays strong alloreactivity to $H\text{-}2^d$ (L^d) molecules complexed with an 8-mer peptide, p2Ca, derived from a Krebs cycle enzyme[32–34]. This peptide has intermediate affinity for soluble L^d molecules and, when complexed to L^d, high affinity for the 2C TCR (Table 1). A closely-related 9-mer peptide, QL9, has 1000-fold higher affinity for L^d and 5-fold higher affinity for the 2C TCR. Using p2Ca and QL9 peptides in synthetic form, we have examined the requirements for inducing negative selection of 2C thymocytes in vitro. As APC for these experiments, it was considered imperative to use a *neutral* cell line for transfection, i.e. a cell line with a minimal background level of costimulatory molecules. Since most cell lines are not totally negative for costimulatory molecules, we chose a Drosophila cell line for transfection. Being only distantly related to mammalian cells, Drosophila cells are most unlikely to express molecules with costimulatory function for mammalian T cells. In addition, Drosophila cells have the advantage of lacking TAP-1,2 peptide transporters[35]. This means that, after transfection with MHC class I genes, the class

Table 1. Features of the interaction of the 2C TCR with L^d and self peptides

		Binding affinities (M^{-1}) (K_a)	
Peptides	Sequence	Affinity for L^d	Affinity of 2C TCR for peptide-L^d
p2Ca	–LSPFPFDL	4×10^6	2×10^6
QL9	QLSPFPFDL	2×10^8	2×10^7

Data from refs. 33, 34.

I molecules reaching the cell surface are empty and can therefore be loaded with high concentrations of exogenous peptides.

The results of incubating purified DP 2C thymocytes overnight with p2Ca vs QL9 peptides presented by L^d transfected Drosophila cells are summarized in Table 2; apoptosis was measured by TUNEL staining and FACS analysis. With Drosophila cells expressing L^d alone, our expectation was that these APC would fail to cause apoptosis, reflecting a lack of costimulation. For the p2Ca peptide, this was indeed the case. For the strong QL9 peptide, however, Ld-Drosophila cells reproducibly cause a low but significant level of apoptosis (though only at a high concentration of peptide). As expected, coexpression of B7 (B7–1) and L^d on Drosophila cells enhances apoptosis induced by QL9 peptide and reveals low-level apoptosis with p2Ca. Interestingly, quite different results are observed with Drosophila cells expressing ICAM-1 and L^d. With these APC, the surprising finding is that apoptosis of 2C thymocytes is virtually undetectable, either to p2Ca or QL9. When ICAM-1 is coexpressed with B7–1 (and L^d), however, apoptosis is high to both peptides.

These findings imply that B7 and ICAM-1 have radically different roles in negative selection. When coexpressed on the same cells, these two molecules act synergistically and cause strong negative selection. When expressed separately, however, B7 potentiates negative selection whereas ICAM-1 has the opposite effect and protects against negative selection. In other experiments we have found that the protective function of ICAM-1 can be prevented by adding anti-ICAM-1 antibody to the cultures; in this situation apoptosis increases to the level seen with APC expressing L^d alone (L^d APC). Conversely, the level of apoptosis induced by L^d APC can be reduced to background by signalling T cells via the counter receptor for ICAM-1, e.g. by adding soluble ICAM-1 to the cultures.

These data on negative selection of DP thymocytes contrast with the role of B7 and ICAM-1 for mature T cells. Thus, when LN T cells ($CD8^+$ cells) from 2C mice are cultured with L^d-transfected Drosophila cells, coexpression of either B7–1 or ICAM-1 is sufficient to cause low but significant proliferative responses to QL9 peptide (although much higher responses occur when both molecules are coexpressed) (unpublished data of the authors). These findings imply that, in contrast to their roles in negative selection, B7 and ICAM-1 behave quite similarly in terms of stimulating mature T cells.

1.7. Concluding Comments

The system described above has the potential to yield useful insights on the molecular interactions involved in negative selection. The results obtained to date are based on just two

Table 2. Different roles of B7–1 and ICAM-1 in negative selection

Molecules expressed on Drosophila cells	Extent of apoptosis after addition of peptides to transfected Drosophila cells	
	p2Ca	QL9
L^d	–	+
L^d + B7-1	+	++
L^d + ICAM-1	–	–
L^d + B7-1 + ICAM-1	+++	++++

The data summarize the effects of culturing purified $CD4^+8^+$ 2C thymocytes with transfected Drosophila cells plus peptides overnight and then measuring apoptosis by TUNEL staining and FACS analysis. Background apoptosis induced in the absence of peptide (15-25%) has been subtracted. Scale: –, 0-5%; +, 15-25%; ++, 25-35%, +++, 35-45%; ++++, 45-55%.

molecules, B7–1 and ICAM-1, and are therefore incomplete. However, several points can be made.

In the case of the weaker p2Ca peptide, the lack of apoptosis seen with APC expressing L^d alone is in line with the notion that negative selection requires costimulation. However, it is surprising that these APC induce significant apoptosis with the stronger QL9 peptide. This could reflect that the requirement for costimulation can be partly overcome by very strong signalling through the TCR.

The finding that B7 expression on Drosophila cells greatly accentuates apoptosis is in agreement with data on the effects of cross-linked anti-CD28 and anti-TCR/CD3 antibodies[25] and suggests that B7 is an important mediator of negative selection. Our expectation is that several other cell-surface molecules share this property. Studies with other molecules are in progress.

The curious reduction in apoptosis found with APC expressing ICAM-1 (without B7) has yet to be explained. At face value, this finding suggests that signalling via the T cell counter-receptors for ICAM-1 (LFA-1 and/or CD43) induces a protective signal which somehow nullifies the negative signal delivered via the TCR. However, the protective signal delivered via ICAM-1 is reversed and becomes a negative signal when APC express both B7 and ICAM-1. The key question is whether these odd findings are relevant to negative selection occurring in the normal thymus.

In considering this question, it is notable that the distribution of B7 and ICAM-1 in the thymus is quite different. ICAM-1 has a wide tissue distribution and is prominent throughout the thymus, including cortical epithelium and the APCs of the medulla. By contrast, B7 expression is prominent only in the medulla and is largely absent in the cortex. The implication therefore is that positive selection in the cortex occurs under conditions where T cells confront peptide/MHC complexes on cells (cortical epithelium) that express only ICAM-1 and not B7. Based on the data discussed above, one can argue that these conditions are inhibitory for negative selection and thus favor positive selection. When positively-selected cells move from the cortex to the medulla, however, the combined expression of B7 and ICAM-1 on APC ensures strong negative selection. Thymic selection is thus compartmentalized, with the cortex inducing positive selection and the medulla controlling negative selection.

An obvious problem with this model is that it does not fit with the evidence that negative selection can occur in the cortex in TCR transgenic models. However, it is not entirely clear that these findings with TCR transgenic mice are relevant to normal negative selection. As discussed earlier, there is the concern that regulation of TCR expression in TCR transgenic mice is often abnormal. It is also possible that the cortical apoptosis induced by peptides in TCR transgenic mice is mediated by steroids or toxic cytokines released through stimulation of mature T cells. Hence the evidence that negative selection occurs in the cortex under normal physiological conditions is far from conclusive.

2. REFERENCES

1. Sprent, J. T lymphocytes and the thymus. *in:* "Fundamental Immunology," W. E. Paul, ed., Raven Press, New York (1993).
2. Jameson, S.C., K.A. Hogquist, and M.J. Bevan. Positive selection of thymocytes. *Annu. Rev. Immunol.* 13:93 (1995).
3. Sprent, J., D. Lo, E.K. Gao, and Y. Ron. T cell selection in the thymus. *Immunol. Rev.* 101:173 (1988).
4. Hogquist, K.A., S.C. Jameson, W.R. Heath, J.L. Howard, M.J. Bevan, and F.R. Carbone. T cell receptor antagonist peptides induce positive selection. *Cell* 76:17 (1994).
5. Ashton-Rickardt, P.G., L. Van Kaer, T.N. Schumacher, H.L. Ploegh, and S. Tonegawa. Peptide contributes to the specificity of positive selection of $CD8^+$ T cells in the thymus. *Cell* 73:1041 (1993).

6. Sebzda, E., V.A. Wallace, J. Mayer, R.S. Yeung, T.W. Mak, and P.S. Ohashi. Positive and negative thymocyte selection induced by different concentrations of a single peptide. *Science* 263:1615 (1994).
7. Ashton-Rickardt, P.G. and S. Tonegawa. A differential-avidity model for T-cell selection. *Immunol. Today* 15:362 (1994).
8. Sprent, J. and S.R. Webb. Function and specificity of T cell subsets in the mouse. *Adv. Immunol.* 41:39 (1987).
9. Scollay, R. and K. Shortman. Cell traffic in the adult thymus: Cell entry and exit, cell birth and death. *in:* "Recognition and Regulation in Cell Mediated Immunity," J. D. Watson and J. Marbrook, eds., Dekker, New York (1985).
10. Surh, C.D. and J. Sprent. T-cell apoptosis detected in situ during positive and negative selection in the thymus. *Nature* 372:100 (1994).
11. Savill, J., V. Fadok, P. Henson, and C. Haslett. Phagocyte recognition of cells undergoing apoptosis. *Immunol. Today* 14:131 (1993).
12. Kappler, J.W., N. Roehm, and P. Marrack. T cell tolerance by clonal elimination in the thymus. *Cell* 49:273 (1987).
13. Fowlkes, B.J. and F. Ramsdell. T-cell tolerance. *Curr. Opin. Immunol.* 5:873 (1993).
14. Nossal, G.J.V. Negative selection of lymphocytes. *Cell* 76:229 (1994).
15. Sprent, J. Central tolerance of T cells. *Internat. Rev. Immunol.* 13:95 (1995).
16. Marrack, P., L. Ignatowicz, J.W. Kappler, J. Boymel, and J.H. Freed. Comparison of peptides bound to spleen and thymus class II. *J. Exp. Med.* 178:2173 (1994).
17. Sprent, J. and S.R. Webb. Intrathymic and extrathymic clonal deletion of T cells. *Curr. Opin. Immunol.* 7:196 (1995).
18. Matzinger, P. and S. Guerder. Does T-cell tolerance require a dedicated antigen-presenting cell? *Nature* 338:74 (1989).
19. Murphy, K.M., A.B. Heimberger, and D.H. Loh. Induction by antigen of intrathymic apoptosis of $CD4^+$ $CD8^+$ TCR^{lo} thymocytes in vivo. *Science* 250:1720 (1990).
20. Schwartz, R.H. Costimulation of T lymphocytes: the role of CD28, CTLA-4, and B7/BB1 in interleukin-2 production and immunotherapy. *Cell* 71:1065 (1992).
21. Janeway, C.A. and K. Bottomly. Signals and signs for lymphocyte responses. *Cell* 76:275 (1994).
22. Linsley, P.S. and J.A. Ledbetter. The role of the CD28 receptor during T cell responses to antigen. *Annu. Rev. Immunol.* 11:191 (1993).
23. Liu, Y., B. Jones, W. Brady, and C.A. Janeway,Jr. Co-stimulation of murine CD4 T cell growth: cooperation between B7 and heat-stable antigen. *Eur. J. Immunol.* 22:2855 (1992).
24. Dubey, C., M. Croft, and S.L. Swain. Costimulatory requirements of naive $CD4^+$ T cells: ICAM-1 or B7–1 can stimulate naive CD4 T cell activation but both are required for optimal responses. *J. Immunol.* 155:45 (1995).
25. Punt, J.A., B.A. Osborne, Y. Takahama, S.O. Sharrow, and A. Singer. Negative selection of $CD4^+CD8^+$ thymocytes by T cell receptor- induced apoptosis requires a costimulatory signal that can be provided by CD28. *J. Exp. Med.* 179:709 (1994).
26. Page, D.M., L.P. Kane, J.P. Allison, and S.M. Hedrick. Two signals are required for negative selection of $CD4^+CD8^+$ thymocytes. *J. Immunol.* 151:1868 (1993).
27. Aiba, Y., O. Mazda, M.M. Davis, S. Muramatsu, and Y. Katsura. Requirement of a second signal from antigen presenting cells in the clonal deletion of immature T cells. *Internat. Immunol.* 6:1475 (1994).
28. Iwawabuchi, K., K.I. Nakayama, R.L. McCoy, F. Wang, T. Nishimura, S. Habu, K.W. Murphy, and D.Y. Loh. Cellular and peptide requirements for in vitro clonal deletion of immature thymocytes. *Proc. Natl. Acad. Sci. USA* 89:9000 (1992).
29. Shahinian, A., K. Pfeffer, K.P. Lee, T.M. Kundig, K. Kishihara, A. Wakeham, K. Kawai, P.S. Ohashi, C.B. Thompson, and T.W. Mak. Differential T cell costimulatory requirements in CD28-deficient mice. *Science* 261:609 (1993).
30. Sha, W.C., C.A. Nelson, R.D. Newberry, D.M. Kranz, J.H. Russell, and D.Y. Loh. Selective expression of an antigen receptor on CD8-bearing T lymphocytes in transgenic mice. *Nature* 335:271 (1988).
31. Sha, W.C., C.A. Nelson, R.D. Newberry, J.K. Pullen, L.R. Pease, J.H. Russell, and D.Y. Loh. Positive selection of transgenic receptor-bearing thymocytes by K^b antigen is altered by K^b mutations that involve peptide binding. *Proc. Natl. Acad. Sci. USA* 87:6186 (1990).
32. Udaka, K., T.J. Tsomides, and H.N. Eisen. A naturally occurring peptide recognized by alloreactive $CD8^+$ cytotoxic T lymphocytes in association with a class I MHC protein. *Cell* 69:989 (1992).
33. Sykulev, Y., A. Brunmark, M. Jackson, R.J. Cohen, P.A. Peterson, and H.N. Eisen. Kinetics and affinity of reactions between an antigen-specific T cell receptor and peptide-MHC complexes. *Immunity* 1:15 (1994).

34. Sykulev, Y., A. Brunmark, T.J. Tsomides, S. Kageyama, M. Jackson, P.A. Peterson, and H.N. Eisen. High-affinity reactions between antigen-specific T-cell receptors and peptides associated with allogeneic and syngeneic major histocompatibility complex class I proteins. *Proc. Natl. Acad. Sci. USA* 91:11487 (1994).
35. Yang, Y., K. Fruh, J. Chambers, J.B. Waters, L. Wu, T. Spies, and P.A. Peterson. Major histocompatibility complex (MHC)-encoded HAM2 is necessary for antigenic peptide loading onto class I MHC molecules. *J. Biol. Chem.* 267:11669 (1992).

GENETIC REGULATION OF APOPTOSIS IN THE MOUSE THYMUS

Barbara A. Osborne,[1,2*] Sallie W. Smith,[1] Kelly A. McLaughlin,[2]
Lisa Grimm,[2] Grant Morgan,[1] and Richard A. Goldsby[3]

[1] Department of Veterinary and Animal Sciences
[2] Program in Molecular & Cellular Biology
University of Massachusetts
Amherst, Massachusetts 01003 and
[3] Department of Biology
Amherst College
Amherst, Massachusetts 01002

1. INTRODUCTION

Recently, it has become clear that apoptosis, a form of programmed cell death, plays an active role in the maintenance of homeostasis of cells as well in the morphological sculpting of organisms during development[1,2]. Some of the best characterized examples of apoptosis may be found in the immune system. For example, apoptosis is the mechanism used during negative selection in the thymus to remove self-reactive T cells[3–5]. Thymic T cells also are quite susceptible to induction of apoptosis by either glucocorticoids or ionizing radiation[6,7]. Additionally, more recent data indicate that peripheral lymphocytes undergo apoptosis following a variety of different stimuli and, in many instances, cell death in activated peripheral T cells may be traced to Fas/FasL interactions[8–10].

1.1 Cell Death in the Thymus

The first clear indication that apoptosis may be physiologically relevant to thymocytes, comes from the early work of Wyllie who showed that injection of glucocorticoid induced massive apoptosis in the thymus[6]. While these experiments involved the introduction of synthetic glucocorticoids into animals, more recent evidence from Ashwell and colleagues demonstrate that naturally occurring glucocorticoids induce apoptosis *in vivo*[11,12]. Another stimulus known to induce apoptosis in the thymus is ionizing radiation[7].

* Address all correspondence to: Barbara A. Osborne, Department of Veterinary & Animal Sciences, 304 Paige Laboratory, University of Massachusetts, Amherst, MA 01003. 413 545–4882; 413 545–6326 (Fax); osborne@vasci.umass.edu

Mechanisms of Lymphocyte Activation and Immune Regulation VI, Edited by Gupta and Cohen
Plenum Press, New York, 1996

It has been known for many years that immature thymocytes are susceptible to death and data from Kappler and Marrack demonstrate that superantigens induce death in a large proportion of thymocytes carrying the particular Vβ molecule that binds the given superantigen[13]. Other investigators, using TCR transgenic mice, have shown that encounter with the cognate antigen in the thymus induces deletion of the immature $CD4^+$,$CD8^+$ thymocytes[14–16]. The conclusions that negative selection involves the induction of apoptosis in this immature population of thymocytes has been extended using thymic organ culture[3,17,18].

2. GENES THAT MEDIATE APOPTOSIS

There are many indications from several independent systems that new gene synthesis is required for the induction of some forms of cell death[19]. In particular, we and others have found that cell death in T cells requires transcription[20–22]. In an attempt to elucidate the molecules responsible for the initiation of apoptosis in T cells, we created a cDNA library from mRNA isolated from dying thymocytes. By subtractive screening strategies, we have isolated several genes whose expression is either up-regulated or repressed during apoptosis. We focused our search on genes expressed during cell death induced by negative selection. In order to closely mimic the conditions of negative selection, we chose to use a T cell receptor (TCR) transgenic mouse previously demonstrated by Murphy and colleagues to be susceptible to the induction of apoptosis by intraperitoneal injection of antigen[5]. In this system, mice are injected with the ovalbumin peptide OVA 323–336 and initiation of apoptosis proceeds within a few hours following injection of peptide. We used mRNA isolated from the thymus of such animals 3, 4 and 5 hours following injection of antigen. This mRNA was used to construct a cDNA library that was screened using a plus/minus screening strategy.

The method chosen for screening allowed the isolation of several genes whose expression pattern is up-regulated following cell death, as well as two genes, *apt-1* and *apt-3*, that are repressed during apoptosis. In addition to repressed genes, we found several genes induced during apoptosis in T cells. One gene, originally called *apt-2* but later identified as the immediate-early gene *nur77*, is induced within the first 45 minutes following crosslinking of the TCR This gene is expressed at high levels following induction of cell death and is expressed throughout the process of death. *Nur77* is a member of the steroid-thyroid hormone receptor superfamily, a gene family characterized by its activity as a transcription factor. The expression of *nur77* was first observed in serum stimulated 3T3 cells[23]. Subsequently, others have found the expression of *nur77* in PC12 cells induced to differentiate into neuronal cells[24]. Therefore the expression of this gene has been associated with proliferation, differentiation and, as a consequence of our observations, with apoptosis. This is less surprising in light of data from Green and co-workers who have shown that up-regulation of *c-myc*, a gene traditionally associated with proliferation, is required for TCR-mediated death in T cell hybridomas[21].

The induction of *nur77* occurs shortly following crosslinking of the TCR and is inhibited by actinomycin D, a transcriptional inhibitor that also inhibits cell death in DO11.10 cells. Up-regulation of *nur77* also is seen in the thymus of animals treated with peptide to induce death and in thymic organ cultures treated with peptide. Interestingly, treatment of animals or the DO11.10 cell line with glucocorticoids such as dexamethasone, induces cell death but does not result in up-regulation of *nur77*. These data indicate to us that the expression of *nur77* is specific to signals through the TCR.

To determine if the expression of *nur77* is essential for cell death, we made constructs that express the gene in either a truncated sense or antisense orientation. When DO11.10 cells are transfected with a truncated sense construct in conjunction with CD20, a reporter gene used to mark the transfected cells, and induced to die, only 2% CD20+ cells are observed in the live population of cells, indicating that the truncated sense construct does not protect against apoptosis. However, when the antisense construct is used, a significant number (12.9%) of CD20+ cells are observed in the live population of cells. These data provided convincing evidence, on a cell-by cell basis, that inhibition of *nur77* by antisense prevents cell death[25]. Thus *nur77* is required for cell death in T cell hybrids. Data from Woronicz et al.[26] indicating a *nur77* dominant negative mutant also acts to inhibit TCR-mediated death in T cell hybrids strengthens and extends our conclusions. Most recently, Calnan et al.[27] have provided convincing evidence that aline of transgenic mice carrying a dominant negative *nur77* targeted to the thymus, have defects in negative selection providing further evidence for a requirement of *nur77* function in thymic cell death.

In addition to *nur77*, we have identified another gene, *apt-4*, that is expressed 4–5 hours following induction of cell death. While we have not established the role *apt-4* in cell death, we have data indicating antisense inhibition of this gene prevents apoptosis in DO11.10 suggesting that like *nur77*, *apt-4* plays an essential role in this process. The *apt-4* transcript is over 6.5 kb in length. We have not obtained a complete sequence and, at present, available sequence data does not indicate any homology with known genes. In searching for a full-length *apt-4* transcript, we have isolated a gene with homology to *apt-4*. This transcript, called *82,* is approximately 4–5 kb in size and has a unique 3' end with respect to *apt-4*. It is possible that *82* is a splice variant of *apt-4* or perhaps is encoded by a separate but related gene. *82* is expressed in living cells and is repressed when cells are induced to die which is in contrast to *apt-4*, that is expressed only during cell death. Although there is insufficient evidence to determine the role of either gene, it is tempting to speculate that there may be an *apt-4* family and different family members may have varying roles in acting out the cellular decision to live or die. Such interpretations, however, await the support of data.

3. MUTANT CELL LINES AS AN TOOL TO STUDY APOPTOSIS

In order to further our understanding of the events that mediate cell death in T cells, we created a series of mutant cell lines from the parental line, DO11.10, by selection for cells that survive PMA + A23187 induced death (Figure 1). One such mutant, mutant 51 was

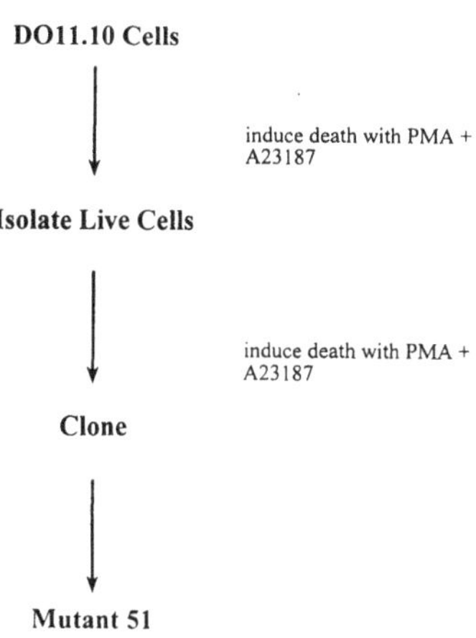

Figure 1. Derivation of mutant 51.

Table 1. Comparison of DO11.10 Versus Mutant 5

	DO11.10 clone 2	Mutant 51
Ability to die in response to PMA + A23187	+	–
Ability to die in response to TCR crosslinking	+	–
Ability to die in response to dexamethansone	+	+
Ability to die in response to g-irradiation	+	+
IL-2 production following TCR crosslinking	+	–
Induction of *nur77* mRNA following TCR crosslinking	+	+
Phosphorylation of Nur77 protein following TCR crosslinking	+	–
Release of intracellular [Ca++] following TCR crosslinking	+	+/–

isolated and this line was shown to be resistant to the induction of cell death via TCR crosslinking or by PMA +A23187. Mutant 51 however is susceptible to death induced via ionizing radiation or dexamethasone (Table 1). While mutant 51 does produce *nur77* mRNA in response to TCR signals, very little Nur77 protein is made and the protein is not phosphorylated. In normal situations, TCR signaling induces *nur77* mRNA and Nur 77 protein that is heavily phosphorylated. Whether the lack of phosphorylated Nur77 protein in mutant 51 is directly associated with the inability of this cell line to die in response to TCR signals remains to be determined.

4. THE ROLE OF OXYGEN IN T CELL DEATH

Despite the differences in the pathways by which different stimuli initiate the apoptosis program in thymocytes, there are similarities in their responses to pharmacological agents that affect the levels of reactive oxygen intermediates (ROI). Treatment of thymocytes with the thiol, N-(2-mercaptoethyl)-1,3 propanediamine (WR-1065), blocks the induction of death in thymocytes by gamma radiation, glucocorticoids or treatment with calcium ionophores [28]. A number of other compounds with a potential for antioxidant activity, including N-acetyl cysteine[29], retinoic acid [30], and vitamin E [31], have been reported to display anti-apoptotic activity in some systems. Also, it has been observed that increasing the level of enzymes that protect against oxidative stress inhibits the induction of apoptosis. Specifically, resistance to growth factor withdrawal-induced apoptosis was conferred on an IL-3 dependent cell line by transfection with a glutathione peroxidase expressing construct[32].

The many observations of the protective effects of ROI scavengers on thymocyte apoptosis induced by signals from the TCR (or its pharmacological equivalents), glucocorticoids or γ-radiation, suggested that this might be a common feature shared by these otherwise different inductive pathways. Despite the correlation between ROI scavenging and inhibition of apoptosis, it was not clear how the pharmacological effects of antioxidants on these pathways might to be interpreted. Although it seemed likely that their action could be traced to a direct lowering of the level of ROIs, other mechanisms could not be ruled out. We initiated a series of experiments to address the question of whether the induction of any or all these T cell deaths are dependent upon oxygen derived species. This was done directly by determining whether or not the capacity of a given agent to induce apoptosis is affected by the withdrawal of molecular oxygen. If T cell apoptosis induced by glucocorticoids or TCR-like signals is unaffected by anaerobiosis, it must be concluded that ROIs do not have an essential role in the induction of death by these stimuli.

4.1 The Induction of Death in a T Cell Hybridoma Requires Oxygen

To examine the role of oxygen in apoptosis in thymocytes, an anaerobic chamber was constructed and thymocytes were induced to die via various stimulii in the absence of oxygen. In all cases, thymocytes were resistent to apoptosis in the absence of oxygen. To determine the extent to which a T cell hybridoma might model the response of a normal cell population to death-inducing stimuli under aerobic and anaerobic conditions, the hybridoma line, DO11.10, was examined. For each of the stimuli, one set of DO11.10 cultures was maintained under an air/CO_2 atmosphere and the other was kept in N_2/CO_2 during treatment. Figure 2 shows that treatment of DO11.10 with concentrations of apoptosis-inducing concentrations of PMA/A23187 produces 65% death in air/CO_2 but yields only 1% under a N_2/CO_2 atmosphere. Figure 2 also shows that treatment of DO11.10 with 1000 cGy of γ-radiation induced 59% death under air/CO_2. However, the same dose produced only 8% death in these cells when administered under a N_2/CO_2 atmosphere. Killing by PMA/A23187 and ionizing radiation show similar oxygen dependencies in this T cell hybridoma and thymocytes. A departure from this pattern was seen in the effects of dexamethasone. Treatment with dexamethasone showed no dependence on the presence of molecular oxygen but thymocytes induced to die via dexamethasone require molecular oxygen (data not shown).

DO11.10 cells stimulated to die through the TCR or the addition of phorbol ester plus calcium ionophore were also protected when 20 mM n-acetyl- cysteine (NAC), a potent antioxidant, was added to cultures prior to stimulation. NAC increased the percentage of DO11.10 cells killed by dexamethasone (data not shown) paralleling what was observed in the absence of molecular oxygen. In this regard the T cell hybridoma shows a marked departure from what was observed with normal T cells where NAC effectively protected against dexamethasone induced death.

5. THE ROLE OF p53 IN T CELL APOPTOSIS

In a collaboration with Tyler Jacks, we and others also have described a role for p53 in apoptosis in T cells[33,34]. Data from Yonish-Rouach and colleagues suggested that p53 may

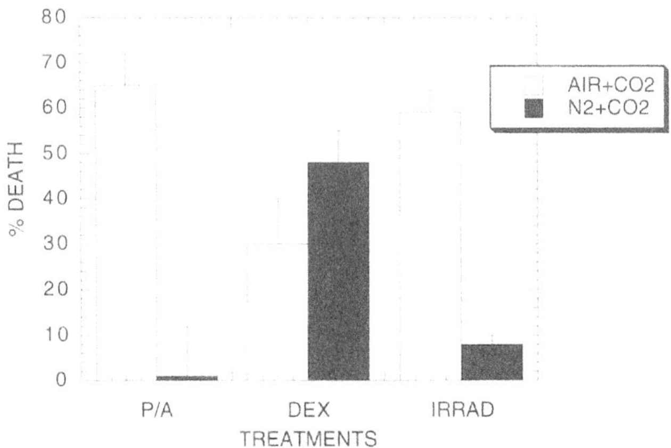

Figure 2. Induction of death by gamma-irradiation, PMA/A23187, or dexamethasone in the presence or absence of molecular oxygen. Treatment of DO11.10 with either 10 nM phorbol myristate (PMA) and 250 nM calcium ionophore A23187, 2.5 uM dexamethasone or 1000 cGys of g-irradiation. Cells were maintained in either Air/CO_2 or N_2/CO_2 as indicated for 14–18 hours before FACScan analysis. Results are expressed as a ± SD of triplicate determination.

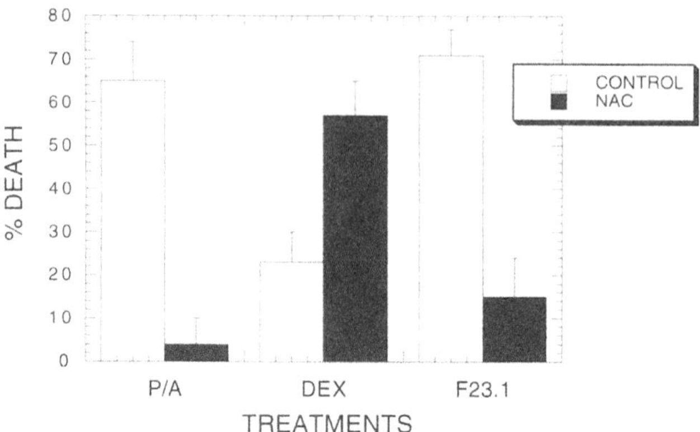

Figure 3. Effects of N-acetylcysteine (NAC) on DO11.10 cell death induced by PMA/A23187 or dexamethasone. Cells were treated with 10 nM PMA plus 250 nM A23187, 10 ug/ml F23.1 (anti-TCR) or 2.5 uM dexamethasone in the presence or absence of 60 mM NAC, incubated under standard conditions and analyzed for imdiction of apoptosis. Results are expressed as a ± SD of triplicate determination.

play a role in the induction of apoptosis[35]. These investigators showed that a mutant form of p53, expressed in a temperature dependent fashion, when transfected into the M1 myeloid leukemia cell line, induced apoptosis at the permissive temperature. Cells grown at the non-permissive temperature did not undergo apoptosis and did not express p53, suggesting that p53 may function as a cell death gene. To test the role of p53 in normal, non-transformed cells, we took advantage of the availability of a p53 null mouse[36]. Using wt +/+, heterozygous +/- and mutant -/- mice, we asked whether p53 is required for apoptosis in thymocytes. The answer was quite clear; in -/- mice, p53 is required for radiation-induced apoptosis but not for glucocorticoid nor TCR-meditated apoptosis. These data suggested that cell death induced by radiation differs from that induced by either TCR or glucocorticoid. In fact, from data described above, we already knew that *nur77* is expressed in TCR-mediated but not glucocorticoid induced death. Taken together, these data indicate there are at least 3 pathways that thymocytes may employ to undergo apoptosis; one pathway induced by gamma radiation and requires p53, another is induced by glucocorticoids and a third, induced by TCR ligation, involves *nur77, egr-1* and *apt-4*. Furthermore, *nur77* is required for the latter pathway.

6. CONCLUSIONS

The decision to die is an active process in many cells. In particular, T lymphocytes appear to require new gene synthesis to die. What signals initiate this process and what genes are required? Some of the signals include T cell receptor engagement, glucocorticoids and ionizing radiation. While the details of glucocorticoid induced apoptosis are not well-understood, it is known that the glucocorticoid receptor (GR) is required and deletion of the DNA-binding domain of the GR interferes with the induction of death[37]. These data imply that GR acts by initiating transcription of genes that are required for the subsequent death of the cells. The identity of these molecules, as well as their mechanism of action, have yet to be elucidated. While Cohen and colleagues have isolated genes induced during glucocorticoid death, their role in this process has not been determined[38].

The mechanism of radiation-induced death requires p53 as described above. The p53 protein can act as a transcription factor suggesting that its role in radiation-induced apoptosis

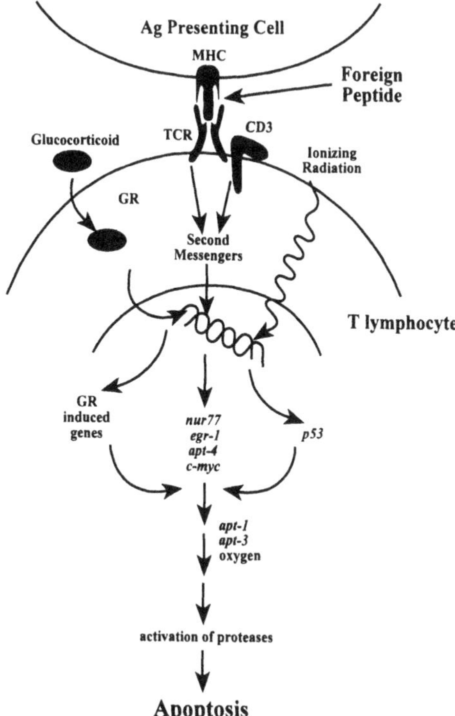

Figure 4. Multiple pathways lead to apoptosis in T cells.

is the initiation of transcription of down-stream gene(s). In fact, a candidate down-stream gene is the recently described p21 gene. The protein encoded by p21 forms a multimeric complex with PCNA resulting in cell cycle arrest[39]. However very recent data indicate that p53 can act to initiate apoptosis in immortalized pituitary cells in the absence of both RNA and protein synthesis[40]. How this relates to p53-mediated death in thymocytes is unknown at present.

Cell death initiated through TCR interactions requires the expression of *c-myc* and *nur77*. Data from Green and co-workers, as well as our own data, demonstrate convincingly that these two genes are required for TCR-mediated death[21,25,26]. As well, cell death through this pathway involves the action of *apt-4*. Preliminary data from our lab suggest this gene is critical, although definitive experiments precisely defining its role in apoptosis is not yet available. As seen in other pathways described above, death through TCR interaction requires the expression of genes encoding transcription factors, indicating a role for genes down-stream of *c-myc* and *nur77*. While these genes appear to be induced by specific apoptotic stimuli, it appears as though all apoptotic pathways in thymocytes and most in T cell lines require molecular oxygen. Supporting this conclusion is the observation that an antioxidant, such as NAC, blocks these cell deaths.

In conclusion, we now have a better, albeit incomplete, understanding about the events that lead to apoptosis in T cells. We know that genes such as *c-myc* and *nur77* are induced by signals through the TCR, that these genes are required and the expression of these genes is unique to TCR-mediated signals. On the other hand, p53 is required for radiation-induced death but does not appear to play a role in either glucocorticoid or TCR-mediated death. These data indicate the existence of multiple signaling pathways leading to apoptosis in T cells. We propose that these pathways converge and form a final common pathway that results in the ultimate demise of the cell (Figure 4). To understand how T cells die by apoptosis, it will be necessary to design experimental protocols that allow us to identify and ultimately dissect the common components of this cell death pathway.

REFERENCES

1. Ellis, R., J. Yuan, and H. R. Horvitz, Mechanisms and functions of cell death. *Annu. Rev. Cell Biol.* 7:663 (1991).
2. Schwartz, L. M. and B. A. Osborne, Programmed cell death, apoptosis and killer genes. *Immunol. Today* 14:582 (1993).
3. Swat, W., L. Ignatowicz, H. von Boehmer, and P. Kisielow, Clonal deletion of immature $CD4^{+}$ $CD8^{+}$ thymocytes in suspension culture by extrathymic antigen-presenting cells. *Nature* 351:150 (1991).
4. Smith, C. A., G. Williams, R. Kingston, E. J. Jenkinson, and J. T. Owen, Antibodies to CD3/T-cell receptor complex induces death by apoptosis in immature T cells in thymic culture. *Nature* 337:181 (1989).
5. Murphy, K. M., A. B. Heimberger, and D. Y. Loh, Induction by antigen of intrathymic apoptosis of $CD4^{+}$ $CD8^{+}$ TCR^{lo} thymocytes *in vivo*. *Science* 250:1720 (1990).
6. Wyllie, A. H., Glucocorticoid induced thymocyte apoptosis is associated with endogenous endonuclease activation. *Nature* 284:555 (1980).
7. Sellins, K. and J.J. Cohen, Gene induction by γ-irradiation leads to DNA fragmentation in lymphocytes. *J. Immunol.* 139:3199 (1987).
8. Dhein, J., H. Walczak, C. Baumler, K-M. Debatin, and P. H. Krammer, Autocrine T-cell suicide mediated by APO-1/(Fas/CD95). *Nature* 373:438 (1995).
9. Brunner, T., R. J. Mogil, D. LaFace, N.J. Yoo, A. Mahoubi, F. Echeverri, S.J. Martin, W.R. Force, D. H. Lynch, C.F. Ware and D.R. Green, Cell-autonomous Fas (CD95)/Fas-ligand interaction mediates activation-induced apoptosis in T-cell hybridomas. *Nature* 373:441 (1995).
10. Ju, S-T, D. J. Panka, H. Cul, R. Ettinger, M. El-Khatib, D. H. Sherr, B. Z. Stanger, and A. Marshak-Rothstein, Fas (CD95)/FasL interactions required for programmed cell death after T-cell activation. *Nature* 373:444 (1995).
11. Vacchio, M., V. Papadopoulus, and J. D. Ashwell, Steroid production in the thymus: implications for thymocyte selection. *J. Exp. Med.* 179:1835 (1994) .
12. King, L.B., M.S. Vacchio, R. Hunziker, D. H. Margulies, and J. D. Ashwell, A targeted glucocorticoid receptor antisense transgene increases thymocyte apoptosis and alters thymocyte development. *Immunity* 5:647 (1995).
13. White, J., A. Herman, A. M. Pullen, R. Kubo, J. Kappler, and P. Marrack, The Vβ-specific superantigen staphylococcal enterotoxin B: Stimulation of mature T cells and clonal deletion in neonatal mice. *Cell* 56:27 (1989).
14. Kisielow, P., H. Bluthmann, U. D. Staerz, M. Steinmetz, and H. von Boehmer, Tolerance in T-cell-receptor transgenic mice involves deletion of nonmature $CD4^{+}$, 8^{+} thymocytes. *Nature* 333:742 (1988).
15. Sha, W., C. Nelson, R. Newberry, D. Kranz, J. Russell, and D. Y. Loh, Positive and negative selection of an antigen receptor on T cells in transgenic mice. *Nature* 336:73 (1988).
16. Berg, L., G. Frank and M.M. Davis, The effects of MHC gene dosage and allelic variation on T cell receptor selection. *Cell* 60:1043 (1990).
17. MacDonald, H. R. and R. K. Lees, Programmed death of autoreactive thymocytes. *Nature* 343:624 (1990).
18. Winslow, G. M., M. T. Scherer, J. W. Kappler, and P. Marrack, Detection and biochemical characterization of the mouse mammary tumor virus 7 superantigen (Mls-1a). *Cell* 71:719 (1992).
19. Schwartz, L. M., The role of cell death genes during development. *BioEssays* 13:389 (1991).
20. Cohen, J. J. and R. Duke, Glucocorticoid activation of a calcium-dependent endonuclease in thymocyte nuclei leads to cell death. *J. Immunol.* 132:38 (1984).
21. Shi, Y., M. Szaly, L. Paskar, M. Boyer, B. Singh, and D. R. Green, Activation-induced cell death in T cell hybridomas is due to apoptosis. *J. Immunol.* 144:3326 (1990).
23. Lau, L. and D. Nathans, Expression of a set of growth-related immediate early genes in BALB/c 3T3 cells: Coordinate regulation with c-fos or c-myc. *Proc. Natl. Acad. Sci. (USA)* 84:1182 (1987).

24. Watson, M. A. and J. Milbrandt, The NGFI-B gene, a transcriptionally inducible member of the steroid receptor gene superfamily: genomic structure and expression in rat brain after seizure induction. *Mol. Cell. Biol.* 9:4213 (1989).
25. Liu, Z.-G., S. W. Smith, K. A. McLauglin, L. M. Schwartz, and B. A. Osborne, Apoptotic signals through the T-cell receptor of a T-cell hybrid require the immediate-early gene *nur77*. *Nature* 36:281 (1994).
26. Woronicz, J. D., B. Calnan, V. Ngo, and A. Winoto, Requirement for the orphan steroid receptor Nur77 in apoptosis of T-cell hybridomas. *Nature* 367:277 (1994).
27. Calnan, B.J., S. Szychowski, F. K-M. Chan, D. Cado, and A. Winoto, A role for the orphan steroid receptor Nur77 in apoptosis accompanying antigen-induced negative selection. *Immunity* 3:273 (1995).
28. Ramakrishnan, N. and G. Catravas, N-(2-mercaptoethyl)-1,3-propanediamine (WR-1065) protects thymocytes from programmed cell death. *J. Immunol.* 148:1817 (1992).
29. Mayer, M. and M. Nobel, N-acetyl-L-cysteine is a pluripotent protector against cell death and enhancer of trophic factor-mediated cell survival in vitro. *Proc. Natl. Acad. Sci. USA* 91:7496 (1994).
30. Iwata, M., M. Mukai, Y. Nakai, and R. Iseki, Retinoic acid inhibits activation-induced apoptosis inT cell-hybridoms and thymocytes. *J. Immunol.* 149:3302 (1992).
31. Sandstrom, P. A., M. Mannie, and T. M. Buttke, Inhibition of activation-induced death in T cell hybridomas by thiol antioxidative stress as a mediator of apoptosis. *J. Leukoc. Biol.* 55:221 (1994).
32. Hockenbery, D., Z. Oltvai, X. Yin, C. Milliman, and S. Korsmeyer, Bcl-2 functions in an antioxidant pathway to prevent apoptosis. *Cell* 75:241 (1993).
33. Lowe, S, E. M. Schmitt, S. W. Smith, B. A. Osborne, and T. Jacks, p53 is required for radiation-induced apoptosis in mouse thymocytes. *Nature* 362:847 (1993).
34. Clarke, A.R., C.A. Purdie, D.J. Harrison, R.G. Morris, C.C. Bird, M.L. Hooper, and A. Wyllie, Thymocyte apoptosis induced by p53- dependent pathways. *Nature* 362:849 (1993).
35. Yonish-Rouach, E., D. Resnitzky, J. Lotem, L. Sachs, A. Kimchi, aand M. Oren, Wild-type p53 induces apoptosis of myeloid leukemic cells that is inhibited by interleukin-6. *Nature* 352:345 (1991).
36. Jacks, T., L. Remington, B. O. Williams, E. Schmidt, S. Halacimi, R. T. Bronson, and R. A. Weinberg, Tumor spectrum analysis in p53 mutant mice. *Curr. Biol.* 4:1 (1994) .
37. Dieken, E. S. and R. L.Miesfeld, Transcriptional transactivation functions localized to the glucocorticoid receptor N terminus are necessary for steroid induction of lymphocyte apoptosis. *Mol. Cell. Biol.* 92:589 (1992).
38. Owens, G. P., W. Hahn, and J. J. Cohen, Identification of mRNAs associated with programmed cell death in immature thymocytes. *Mol. Cell. Biol.* 11:4177 (1991).
39. Waga, S., G. J. Hannon, D. Beach, and B. Stillman, The p21 inhibitor of cyclin-dependent kinases controls DNA replication by interaction with PCNA. *Nature* 369:520 (1994).
40. Caelles, C., A. Helmberg, and M. Karin, p53-dependent apoptosis in the absence of transcriptional activation of p53 target genes. *Nature* 370:220 (1994).
41. Shi, Y., J. M. Glynn, L. J. Guilbert, T. G. Cotter, R. Bissonette, and D. R. Green, Role for c-myc in activation-induced apoptotic death in T cell hybridomas. *Science* 257:212 (1992).

REGULATION OF T CELL ACTIVATION BY CD28 AND CTLA4

Patricia J. Noel,[1] Lawrence H. Boise,[1] and Craig B. Thompson[1,2]

[1] Department of Medicine
Gwen Knapp Center for Lupus and Immunology Research
[2] Howard Hughes Medical Institute
Department of Molecular Genetics and Cell Biology
The University of Chicago
Chicago, Illinois 60637

1. INTRODUCTION

T cell activation is initiated by engagement of the T cell receptor (TCR)-CD3 complex by antigen displayed by major histocompatibility complex (MHC) molecules expressed on the surface of an antigen presenting cell (APC). However, under most circumstances, this initial signal is insufficient to induce a proliferative response. TCR engagement in the absence of a second or costimulatory signal can lead to a state of anergy or cell death[1]. CD28 is a T cell surface receptor capable of providing this critical second signal following ligation of the B7 family of counter-receptors that are expressed on APCs[2,3]. This interaction provides not only proliferative signals, but also crucial survival signals that are important for both initiation and maintenance of an immune response[4,5].

Lymphocytes are an extremely dynamic population of cells. In addition to the constant production of new cells and antigen-specific clonal expansion that can occur, there is a balancing amount of cell death that is necessary to maintain immune homeostasis. In the past, much attention has been focused on T cell proliferation, while relatively little is known about the regulation of cell survival during an immune response and the maintenance of lymphocyte cell number. In this article, we will focus on recent data concerning the role of CD28 and the structurally related receptor CTLA4, and their respective positive and negative roles in T cell activation.

2. THE ROLE OF CD28 IN T CELL SURVIVAL

Costimulation through CD28/B7 has multiple affects on the developing immune response including the upregulation of lymphokines such as IL-2 and the survival gene *bcl-x*$_L$ [6–8]. It has been demonstrated that T cells activated through the TCR alone are more

Mechanisms of Lymphocyte Activation and Immune Regulation VI, Edited by Gupta and Cohen
Plenum Press, New York, 1996

susceptible to apoptosis [9,10]. Thus one potential function of costimulation through CD28 is the induction of survival factors which increase cell viability following primary activation. In order to investigate the role of CD28 costimulation in the survival of activated T lymphocytes, human peripheral blood T cells were activated with anti-CD3 in the presence and absence of anti-CD28 prior to γ-irradiation. Viability was then assessed over a 4 day culture period. Cells activated with anti-CD3 showed enhanced survival as compared to resting cells following irradiation. In cultures where costimulation was provided by anti-CD28 treatment there was a slight but reproducible enhancement of cell survival over that observed in cultures treated with anti-CD3 alone. Since it has been previously demonstrated that growth factors play a role in the extrinsic regulation of cell survival, it seemed possible that IL-2 was functioning in an autocrine fashion to regulate T cell survival. Although cells activated with anti-CD3 survive well in culture following irradiation, washing cells free of conditioned medium, 12 hours after activation and prior to γ-irradiation, leads to a significant reduction in survival over the course of the culture period. This reduction in viability could by completely reversed by the addition of exogenous IL-2 to the culture, demonstrating that, in the absence of costimulation, IL-2 functions as an extrinsic regulator of cell survival.

Interestingly, when costimulation was provided by the addition of anti-CD28 to anti-CD3 stimulated cultures, washing the cells free of lymphokines did not result in decreased viability. Furthermore, the addition of exogenous IL-2 to washed cells did not increase survival beyond that of washed or unwashed cells stimulated with anti-CD3 and anti-CD28. Therefore it seemed likely that, in addition to increasing production of the extrinsic survival factor IL-2, costimulation through CD28 is also capable of regulating the induction of intrinsic factor(s) which are able to prevent cells from undergoing apoptosis.

3 EXPRESSION OF Bcl-2 AND Bcl-x_L DURING T CELL ACTIVATION

Two members of the *bcl-2* family, *bcl-2* and *bcl-x_L*, have been directly implicated in the control of lymphocyte survival [8,11,12]. In order to determine whether CD28/B7 signaling plays a role in the induction of these intrinsic regulators of cell survival, *bcl-2* and *bcl-x_L* mRNA levels were examined in human resting T cells and in cells activated with anti-CD3 and anti-CD3 + anti-CD28. No expression of either *bcl-2* or *bcl-x_L* mRNAs was detected in resting cells, while both messages were detected by 6 hours following activation with anti-CD3 and anti-CD3 + anti-CD28. Interestingly, in cells stimulated with anti-CD3 + anti-CD28 there was significant enhancement of *bcl-x_L* mRNA levels as compared to that seen in cells stimulated with anti-CD3 alone. In contrast, no further induction of *bcl-2* was seen under these conditions. However, previous reports have suggested that levels of *bcl-2* message do not correlate precisely with levels of Bcl-2 protein[13], thus Western blot analysis of both Bcl-2 and Bcl-x_L was performed under the conditions described above. These analyses revealed high levels of Bcl-2 in resting cells, and protein levels remained unchanged in cells stimulated with either anti-CD3 alone or anti-CD3 + anti-CD28. It is likely that the increase in *bcl-2* message observed in activated cells is the result of increased transcriptional rate and/or message stability in order to maintain constant protein levels in a proliferating cell population. This is consistent with the idea that Bcl-2 is critical for maintaining the survival of resting lymphocytes in the periphery, but is unlikely to be required for the survival of activated T cells during an immune response[11,12].

Conversely, expression of Bcl-x_L was not detected in resting cells, but was induced by 6 hours with peak expression occurring by 24 hours following stimulation with anti-CD3 (Figure 1). Although anti-CD28 was unable to induce expression of Bcl-x_L alone, the protein

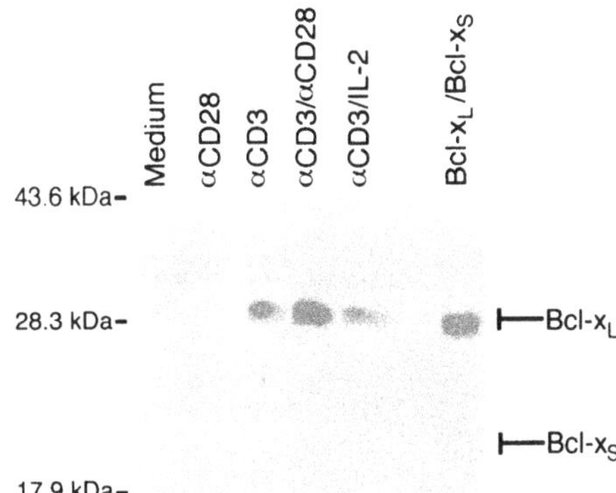

Figure 1. Expression of Bcl-x_L following T cell activation. Human T cells were activated for 24 hours under each of the indicated conditions. Cytoplasmic extracts were prepared from each sample and equal amounts of protein were subjected to SDS-PAGE and Western blot analysis with a polyclonal Bcl-x antisera[8].

was strongly induced, over those levels seen with anti-CD3 alone, when costimulation was provided in anti-CD3 activated cultures. Moreover, this induction is not further enhanced by the addition of exogenous IL-2. Furthermore, this induction is specific for CD28 since the addition of antibodies to CD2, CD5, CD11a, CD18 or MHC class I to anti-CD3 stimulated cultures did not induce Bcl-x_L above those levels seen with anti-CD3 alone. Additionally, in murine T cells blockade of CD28/B7 interactions reduces Bcl-x_L to levels comparable to those seen in anti-CD3 activated cultures alone[14]. These data indicate that specific CD28/B7 signaling events, independent of lymphokine induction, lead to the upregulation of Bcl-x_L which, in turn, is likely to play a role in preventing programmed cell death (PCD) in activated lymphocytes.

4. Bcl-x_L PROTECTS AGAINST Fas INDUCED PCD

The Fas antigen is another major regulator of cell survival. Crosslinking of Fas leads to the induction of cell death in both T cell lines and primary T cell cultures[15–17]. In cell lines, the induction of Fas mediated PCD is very rapid[16,17]. However primary T cells, in spite of an upregulation of Fas on the cell surface following activation, are not susceptible to cell death until 72 hours after initial activation[15]. This time course correlates with the expression of Bcl-x_L which peaks at approximately 24 hours after activation and is declining by 72 hours. In order to directly test whether Bcl-x_L is capable of protecting against Fas induced PCD, Jurkat T cells were transfected with Bcl-x_L or with a Neo control and their ability to resist PCD induced by treatment with anti-Fas or anti-CD3 was determined. Under both treatment conditions, Bcl-x_L transfectants displayed significantly higher survival than control transfectants demonstrating that Bcl-x_L was able to protect against Fas induced apoptosis.

5. IS CD28 REQUIRED FOR T CELL SURVIVAL DURING ACTIVATION?

The studies discussed above suggest that CD28 costimulation plays a role in the survival of activated T cells through the regulation of both intrinsic and extrinsic cell survival factors such as the survival gene *bcl-x_L* and lymphokines such as IL-2. We have further investigated the role of CD28 in T cell survival by comparing the survival characteristics of T cells from wildtype and CD28-deficient mice. Previous work characterizing the immune responses of CD28-deficient mice revealed a profound proliferative defect in the T cells of

Table 1. Viability of wildtype and CD28-deficient T cells stimulated with anti-CD3 in the presence and absence of CD28 costimulation

	Stimulation	0 hr	24 hr	48 hr	72 hr
CD28 +/+	α-CD3	92%	65%	46%	62%
	α-CD3+α-CD28		75%	73%	88%
	α-CD3+CTLA4Ig		64%	29%	20%
CD28 –/–	α-CD3	93%	66%	36%	18%
	α-CD3+α-CD28		63%	37%	20%
	α-CD3+CTLA4Ig		66%	33%	18%

Bulk lymph node cells were isolated from wildtype and CD28-deficient animals and cultured with anti-CD3, anti-CD3 + anti-CD28 or anti-CD3 + CTLA4Ig. Cells were taken at the indicated time points and stained for the T cell marker, Thy1.2. Viability of the Thy1.2+ population was assessed by FACS analysis of propidium iodide exclusion[14].

these animals as compared to wildtype controls[4]. Two alternative mechanisms may be invoked to explain the data. It is possible that T cells from CD28-deficient animals have a higher threshold for activation in the absence of costimulation ultimately resulting in fewer activated cells. Alternatively the same number of cells may become activated but CD28-deficient cells, lacking the induction of survival factors, may be unable to survive initial activation and go on to proliferate. In support of this hypothesis, similar numbers of wildtype and CD28-deficient cells were found to be CD69 positive 6 hours after activation in the presence and absence of CD28 costimulation, suggesting that the activation threshold is similar in both cell types and, at least at early time points, CD28-deficient cells are activated as well as wildtype cells[14].

As initial activation appeared similar in wildtype and CD28-deficient animals, we examined lymph node T cell survival following activation with anti-CD3, anti-CD3 + anti-CD28 and anti-CD3 + CLTA4Ig, a soluble inhibitor of CD28/B7 interactions. Wildtype T cells had a significantly higher percentage viability than CD28-deficient cells when activated with either anti-CD3 or anti-CD3 + anti-CD28 (Table I). However, the protective effect of activation was completely reversed by CD28/B7 blockade and, in fact, viability was identical to that observed in the CD28-deficient cultures under all conditions. Thus, it appears that a lack of CD28 costimulation is directly responsible for the increased cell death seen in CD28-deficient cultures and in wildtype cultures under conditions of CD28/B7 blockade.

In order to more fully understand the mechanism by which CD28 acts to enhance survival, we sought to differentiate between an increased capacity of individual cells to survive versus an increased rate of cell division resulting in an expansion of the number of viable cells. To address this issue, we quantitated the total number of Thy1.2+ cells in wildtype and CD28-deficient lymph node cultures under the conditions described above 48 hours following activation, before significant proliferation has occurred. Although the total number of Thy1.2+ cells was comparable between wildtype and CD28-deficient cultures at this time point, a significantly greater number of live Thy1.2+ cells was present in the wildtype cultures. In contrast, there was a reproducibly higher number of dead Thy1.2+ cells in CD28-deficient cultures as compared to wildtype cultures. Inhibition of CD28/B7 interactions, by the addition of CTLA4Ig, in wildtype cultures again resulted in decreased survival of Thy1.2+ cells to a level comparable to that observed in the CD28-deficient cultures. These data demonstrate that CD28/B7 signaling is directly influencing the ability of individual T cells to survive activation rather than merely driving proliferation at an increased rate resulting in an increase in the number of viable cells through clonal expansion.

6. A LACK OF CD28 COSTIMULATION RESULTS IN THE INDUCTION OF PCD PATHWAYS

Previous work has demonstrated that cells activated through the T cell receptor alone are more likely to undergo PCD in culture [9,10]. Genomic DNA isolated from cultures of lymph node cells from CD28-deficient animals or cultures of wildtype lymph node cells in which CD28/B7 interactions had been blocked displayed a high degree of nucleosomal fragmentation that is characteristic of apoptotic cell death. DNA isolated from wildtype lymph node cultures stimulated with anti-CD3 or anti-CD3 + anti-CD28 displayed almost no fragmentation, consistent with the observation that there are far fewer dead cells in these cultures. Interestingly, resting cells from both wildtype and CD28-deficient lymph node cultures also displayed a high degree of nucleosomal fragmentation indicating the induction of PCD pathways once cells are removed from their local physiologic environment. This is significantly different from the human system where resting, peripheral blood T cells maintain their viability out to at least 96 hours in culture[8]. It is possible that there are important differences in apoptotic control between circulating lymphocytes and cells localized in lymphatic tissue.

7. IS Bcl-x_L EXPRESSED IN CD28-DEFICIENT MICE?

Studies in the human system have shown that one mechanism by which CD28 acts to increase the survival of activated T cells is through the induction of Bcl-x_L [8]. We have also found that in the murine model Bcl-x_L is induced in lymphocytes stimulated with anti-CD3 and that there is a further upregulation of the protein upon direct ligation of CD28 by the addition of monoclonal antibody to the cultures. As mentioned previously, under conditions of CD28/B7 blockade there is a significant reduction in the level of Bcl-x_L induction. When we examined CD28-deficient cultures for the expression of Bcl-x_L, we found a level of induction similar to that seen in wildtype cultures stimulated with anti-CD3 in the absence of CD28 costimulation. Both wildtype and CD28-deficient cells expressed similar levels of Bcl-2 in resting cells and in cells that had been activated in the presence of anti-CD3. Thus, the low level of Bcl-x_L induced in the absence of CD28 costimulation is insufficient to enhance cell survival, and normal levels of Bcl-2 are unable to substitute for Bcl-x_L in the rescue of cells from initiation of programmed cell death.

8. WHAT IS THE MECHANISM OF CELL DEATH INDUCED IN THE ABSENCE OF CD28?

Fas is known to be a direct mediator of lymphocyte cell death[18]. However, as discussed previously, Bcl-x_L can confer resistance to Fas induced PCD. In order to determine whether Fas-Fas ligand interactions induce apoptosis in the absence of sufficient levels of Bcl-x_L in CD28-deficient mice we determined the percent viability of lymphocytes from *lpr* and wildtype mice. Although Fas is expressed on resting murine T cells, we found that resting cells from *lpr* mice were as susceptible to PCD in culture as cells from wildtype mice. Moreover, lymphocytes from both *lpr* and wildtype mice displayed similar percent viabilities under conditions of CD28/B7 blockade. Thus, Fas-Fas ligand interactions do not play a role in PCD of resting T cells in culture or during primary activation with anti-CD3 in the absence of CD28 costimulation.

Although the initiating event in the cell death pathway remains to be determined, it appears that the cell death we have studied takes place through the activation of the ICE-related family of proteases. The addition of the ICE-related protease inhibitor, zVAD-FMK, to wildtype and CD28-deficient cultures resulted in increased viability in resting cultures as well as cultures in which CD28/B7 interactions had been blocked confirming that the cell death we observed is through the induction of programmed cell death pathways.

9. CTLA4 IS A NEGATIVE REGULATOR OF T CELL ACTIVATION

CTLA4 is a member of the Ig supergene family of receptors[19]. It was originally isolated during a screen for genes induced during cytotoxic T cell activation[20], and shares significant structural homology with CD28. These two genes are closely linked in the genome and are likely to have arisen through a gene duplication event. Unlike CD28, CTLA4 is not expressed on resting T cells[21,22]. Its expression is induced transiently on activated T cells with peak expression occurring approximately 48 hours after activation[23]. Furthermore, cell surface expression of CTLA4 is lower than that of CD28, however, it binds B7–1 with a 20-fold greater avidity than CD28[24,25]. Initially, because of its structural similarity to CD28 and its ability to bind the same cell surface counter-receptors it was postulated that CTLA4 would also play a role in the positive regulation of T cell activation. However, recent studies indicate a negative regulatory role for CTLA4 in the control of T cell activation.

A study by Walunas *et al.* demonstrated that a monoclonal antibody against CTLA4 was capable of enhancing the response of T cells in an allogeneic mixed lymphocyte reaction[23]. An enhancement of proliferation was also observed when Fabs were added to the cultures. These data are consistent with the idea that CTLA4 is delivering a negative signal upon B7 engagement which is blocked by the addition of either an intact anti-CTLA4 monoclonal antibody or Fabs (Fragment antigen binding) derived from the antibody. An additional study by Gribben and coworkers[26] found that a panel of monoclonal antibodies against CTLA4 were able to induce cell death in previously activated T cell clones and CD4+ T cell blasts. Induction of PCD required prior activation by either anti-CD3 or antigen. In contrast, CTLA4 Fab fragments were not able to induce cell death suggesting that signaling via direct crosslinking of CTLA4 is required. However, signaling through CD28 or addition of IL-2 was able to protect against induction of apoptosis, suggesting that the death mechanisms induced by CTLA4 can be overidden by survival factors in a manner reminiscent of the ability of Bcl-x_L to override Fas induced cell death.

Both of the above studies suggest that CTLA4 plays a negative regulatory role in T cell activation, possibly through the induction of cell death pathways. The recent characterization of CTLA4-deficient animals provides the first definitive evidence that CTLA4 is critical for the maintenance of immune homeostasis and appears to exert its effects in a negative regulatory manner[27]. Mice that are deficient in CTLA4 die at 3–4 weeks of age due to a severe lymphoproliferative disorder. There is an extensive accumulation of activated lymphocytes in the lymphoid tissues as well as massive lymphocyte infiltration in many of the major organs. T cells from CTLA4-deficient mice displayed an enhanced proliferative response as compared to wildtype cells when stimulated through the T cell receptor with anti-CD3 or anti-TCRαβ, *in vitro*, indicating less stringent control of proliferation initiated by signaling through the T cell receptor. CTLA4-deficient cells remain susceptible to cell death induced by Fas or by gamma irradiation demonstrating that CTLA4 is not required for the initiation of cell death induced by these agents.

10. SUMMARY

In this article we have reviewed recent studies concerning the roles of CD28 and CTLA4 in controlling T cell activation. CD28 appears to have a dual role promoting both T cell proliferation and survival of activated T cells (Figure 2). Signaling through CD28 results in the upregulation of both intrinsic and extrinsic survival factors which are capable of acting independently to directly enhance cell survival. One way in which this enhancement may be accomplished is through the direct suppression of cell death pathways. This mechanism is supported by the data indicating that induction of the survival gene, *bcl-x_L*, can override Fas induced cell death and that the kinetics of Bcl-x_L expression correlate well with the susceptibility of normal T cells to apoptosis initiated by Fas-Fas ligand interactions[8]. The down regulation of an immune response is also a critical control point in the maintenance of immune homeostasis. Fas and now CTLA4 have both been identified as crucial negative regulators of lymphocyte activation as demonstrated by the lymphoproliferative defects present in both *lpr* and CTLA4-deficient mice[18,27]. Fas exerts its effects through a programmed cell death pathway mediated by the ICE-like family of proteases. The mechanism by which CTLA4 acts to down regulate proliferation is less clear (Figure 2). One possibility suggested by the data from Gribben and colleagues is that signaling through CTLA4 may induce apoptosis in previously activated cells[26]. Whatever mechanism(s) CTLA4 utilizes to modulate proliferation, they are likely to become active 48–72 hours after activation when cell surface expression of CTLA4 peaks, again correlating with the decreased expression of Bcl-x_L. It will be interesting to determine if CTLA4 signaling is specifically required for the down regulation of cell survival factors such as Bcl-x_L and IL-2 and whether these effects

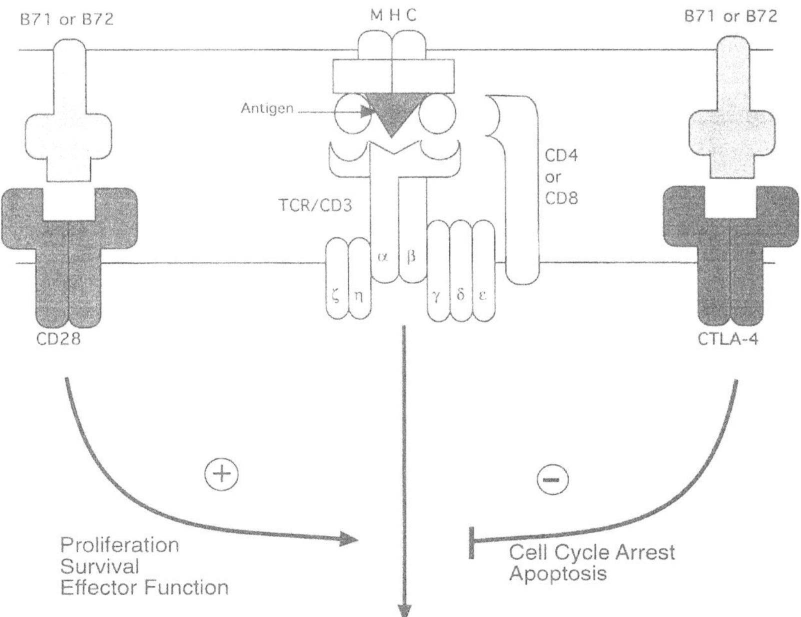

Figure 2. Regulation of T Cell Activation by CD28 and CTLA4. T cells are activated through the TCR/CD3 complex by antigen presented in the context of an MHC molecule on the surface of an APC. Costimulation via CD28/B7 interactions acts to enhance the production of lymphokines and Bcl-x_L resulting in cell survival and proliferation. Expression of CTLA4 and subsequent engagement of B71 and/or B72 then act to down regulate proliferation in the absence of continuing CD28 costimulation thus resulting in the deletion, in an antigen-specific manner, of inappropriately activated T cells.

are influenced directly through crosslinking of CTLA4 or indirectly as downstream effects of an inducible inhibitory pathway. Although the mechanistic details remain to be elucidated, CD28 and CTLA4 appear to play important and complex roles in the control of immune homeostasis.

ACKNOWLEDGMENTS

We would like to thank Carol Sampson for her assistance in preparing the manuscript. PJN is a fellow of the Cancer Research Institute. LHB is a fellow of the Leukemia Society of America. This work was supported in part from the National Institutes of Health grants PO1 AJ35294 and PO1 DK49799 and the National Cancer Institute grant R37 CA48023.

REFERENCES

1. R.H. Schwartz, A cell culture model for T Lymphocyte clonal anergy, *Science*. 248:1349 (1990).
2. C.D. Gimmi, G.J. Freeman, J.G. Gribben, K. Sugita, A.S. Freedman, C. Morimoto, and L.M. Nadler, B-cell surface antigen B7 provides a costimulatory signal that induces T cells to proliferate and secrete interleukin 2, *Proc. Natl. Acad. Sci. USA* 88:6575 (1991).
3. N.K. Damle, L.V. Doyle, L.S. Grosmaire, and J.A. Ledbetter, Differential regulatory signals delivered by antibody binding to the CD28 (Tp 44) molecule during the activation of human T lymphocytes, *J. Immunol.* 140:1753 (1988)
4. J.M. Green, P.J. Noel, A.I. Sperling, T.L. Walunas, G.S. Gray, J.A. Bluestone, and C.B. Thompson, Absence of B7-dependent responses in CD28- deficient mice, *Immunity*. 1:501 (1994).
5. Y. Shi, L.G. Radvanyi, A. Sharma, P. Shaw, D.R. Green, R.G. Miller, and G.B. Mills, CD28-mediated signaling *in vivo* prevents activation-induced apoptosis in the thymus and alters peripheral lymphocyte homeostasis, *The Journal of Immunology*. 155:1829 (1995).
6. T. Lindsten, C.H. June, J.A. Ledbetter, G. Stella, and C.B. Thompson, Regulation of lymphokine messenger RNA stability by a surface-mediated T cell activation pathway, *Science*. 244:339 (1989).
7. C.B. Thompson, T. Lindsten, J.A. Ledbetter, S.L. Kunkel, H.A. Young, S.G. Emerson, J.M. Leiden, and C.H. June, CD28 activation pathway regulates the production of multiple T-cell-derived lymphokines/cytokines, *Proc. Natl. Acad. Sci. USA*. 86:1333 (1989).
8. L.H. Boise, A.J. Minn, P.J. Noel, C.H. June, M. Accavitti, T. Lindsten, and C.B. Thompson, CD28 costimulation can promote T cell survival by enhancing the expression of Bcl-x_L, *Immunity*. 3:87 (1995).
9. D. Kabelitz and S. Wesselborg, Life and death of a superantigen-reactive human $CD4^+$ T cell clone: staphylococcal enterotoxins induce death by apoptosis but simultaneously trigger a proliferative response in the presence of $HLA\text{-}DR^+$ antigen-presenting cells, *International Immunology*. 4:1381 (1992).
10. H. Groux, D. Monte, B. Plouvier, A. Capron, and J.-C. Ameisen, CD3-mediated apoptosis of human medullary thymocytes and activated peripheral T cells: respective roles of interleukin-1, interleukin-2, interferon-γ and accessory cells, *Eur. J. Immunol.* 23:1623 (1993).
11. D.J. Veis, C.L. Sentman, E.A. Bach, and S.J. Korsmeyer, Expression of the Bcl- 2 protein in murine and human thymocytes and in peripheral T lymphocytes, *The Journal of Immunology*. 151:2546 (1993).
12. K-i. Nakayama, K. Nakayama, I. Negishi, K. Kuida, Y. Shinkai, M.C. Louie, L.E. Fields, P.J. Lucas, V. Stewart, F.W. Alt, and D.Y. Loh, Disappearance of the lymphoid system in Bcl-2 homozygous mutant chimeric mice, *Science*. 261:1584 (1993).
13. C.M. Chleq-Deschamps, D.P. LeBrun, P. Huie, D.P. Besnier, R.A. Warnke, R.K. Sibley, and M.L. Cleary, Topographical dissociation of BCL-2 messenger RNA and protein expression in human lymphoid tissues, *Blood*. 81:293 (1993).
14. P.J. Noel, L.H. Boise, J.M. Green, and C.B. Thompson, CD28 costimulation prevents cell death during primary T cell activation, *submitted*. (1995).
15. C. Klas, K.-M. Debatin, R.R. Jonker, and P.H. Krammer, Activation interferes with the APO-1 pathway in mature human T-cells, *International Immunology*. 5:625 (1993).
16. B.C. Trauth, C. Klas, A.M.J. Peters, S. Matzku, P. Möller, W. Falk, K.-M. Debatin, P.H. Krammer, Monoclonal antibody-mediated tumor regression by induction of apoptosis, *Science*. 245:301 (1989).

17. N. Itoh, S. Yonehara, A. Ishii, M. Yonehara, S.-I. Mizushima, M. Sameshima, A. Hase, Y. Seto, and S. Nagata, The polypeptide encoded by the cDNA for human cell surface antigen Fas can mediate apoptosis, *Cell*. 66:233 (1991).
18. S. Nagata and P. Golstein, The Fas death factor, *Science*. 267:1449 (1995).
19. K. Harper, C. Balzano, E. Rouvier, M.-G. Mattéi, M.-F. Luciani, and P. Golstein, CTLA-4 and CD28 activated lymphocyte molecules are closely related in both mouse and human as to sequence, message expression, gene structure, and chromosomal location, *The Journal of Immunology*. 147:1037 (1991).
20. J.-F. Brunet, F. Denizot, M.-F. Luciani, M. Roux-Dosseto, M. Suzan, M.-G. Mattei, and P. Golstein, A new member of the immunoglobulin superfamily---- CTLA-4, *Nature*. 328:267 (1987).
21. T. Lindsten, K.P. Lee, E.S. Harris, B. Petryniak, N. Craighead, P.J. Reynolds, D.B. Lombard, G.J. Freeman, L.M. Nadler, G.S. Gray, C.B. Thompson, and C.H. June, Characterization of CTLA-4 structure and expression on human T cells, *The Journal of Immunology*. 151:3489 (1993).
22. G.J. Freeman, D.B. Lombard, C.D. Gimmi, S.A. Brod, K. Lee, J.C. Laning, D.A. Hafler, M.E. Dorf, G.S. Gray, H. Reiser, C.H. June, C.B. Thompson, and L.M. Nadler, CTLA-4 and CD28 mRNA are coexpressed in most T cells after activation: expression of CTLA-4 and CD28 mRNA does not correlate with the pattern of lymphokine production, *The Journal of Immunology*. 149:3795 (1992).
23. T.L. Walunas, D.J. Lenschow, C.Y. Bakker, P.S. Linsley, G.J. Freeman, J.M. Green, C.B. Thompson, and J.A. Bluestone, CTLA-4 can function as a negative regulator of T cell activation, *Immunity*. 1:405 (1994).
24. P.S. Linsley, J.L. Greene, P. Tan, J. Bradshaw, J.A. Ledbetter, C. Anasetti, and N.K. Damle, Coexpression and functional cooperation of CTLA-4 and CD28 on activated T lymphocytes, *J. Exp. Med*. 176:1595 (1992).
25. P.S. Linsley, W. Brady, M. Urnes, L.S. Grosmaire, N.K. Damle, and J.A. Ledbetter, CTLA-4 is a second receptor for the B cell activation antigen B7, *J. Exp. Med*. 174:561 (1991).
26. J.G. Gribben, G.J. Freeman, V.A. Boussiotis, P. Rennert, C.L. Jellis, E. Greenfield, M. Barber, V.A. Restivo, Jr., X. Ke, G.S. Gray, and L.M. Nadler, CTLA4 mediates antigen-specific apoptosis of human T cells, *Proc. Natl. Acad. Sci. USA*. 92:811 (1995).
27. P. Waterhouse, J.M. Penninger, E. Timms, A. Wakeham, A. Shahinian, K.P. Lee, C.B. Thompson, H. Griesser, and T.W. Mak, Lymphoproliferative disorders with early lethality in mice deficient in *Ctla-4* , *Science*. 270:985 (1995).

23

GRANZYME B-INDUCED APOPTOSIS

Arnold H. Greenberg

Manitoba Institute of Cell Biology
Univeristy of Manitoba
Winnipeg, MB, Canada R3E OV9

1.0 INTRODUCTION

In the last several years the major mechanisms by which both cytotoxic T lymphocytes (CTL) and natural killer (NK) cells mediate cell death and apoptosis have been identified. Cytotoxic lymphocytes of both types induce apoptosis through granule- or Fas-dependent pathway[1–4] and perhaps TNF-α[5]. The granule exocytosis mechanism of CTL killing[1] appears to be primarily responsible for resistance to viruses and can mediate lysis of a wide variety of tumors and allogeneic cells[6,7]. The mechanism by which granulated cytotoxic cells lyse target cells requires their transient but intimate contact with the target[1,8]. Cytotoxic lymphocytes contain cytoplasmic granules that rapidly reorient to the area of target cell contact through a kinesin-based motor mechanism, then fuse with the plasma membrane and release their contents into the intercellular space formed between the lymphocyte and target cell[1,9,10]. The target cell then undergoes disintegration with the nucleus of the target cell rapidly breaking down in a manner similar to that seen in programmed cell death or apoptosis in which the nucleus collapses with severe chromatin condensation and oligonucleosomal-sized DNA fragments are released during DNA degradation[1]. Plasma membrane damage is also observed and is accompanied by the appearance of C9 complement-like pores and calcium permeability[1]. Morphologically the target cells exhibit the features of apoptosis with typical nuclear changes, as well as necrosis with evidence of plasma membrane damage[8,10,11].

2.0 CYTOTOXIC T CELL GRANULE-MEDIATED CELL DEATH

CTL and NK cell killing through granule exocytosis is the result of the action of two types of molecules, the pore forming protein perforin and the lymphocyte specific family of granule serine esterases known as granzymes which together can reproduce all of the features of CTL killing[8,10–12] and in whose absence CTL cytotoxicity does not proceed normally[6,7,13].

Mechanisms of Lymphocyte Activation and Immune Regulation VI, Edited by Gupta and Cohen
Plenum Press, New York, 1996

2.1 Perforin

Perforin is a 65–70 kDa glycoprotein that is synthesized only in CTL and NK cells and can be detected in lymphocytes at the site of viral infection and graft rejection[14]. Perforin is structurally related to members of the complement protein family and is thus thought to form the complement like pores seen in target cells after CTL attack[14]. Perforin unfolds a hydrophobic domain in the presence of calcium which then inserts into the plasma membrane of the target cell. The perforin monomers then form a tubular polymer in the membrane composed of 12–18 molecules that can act as an ion channel[14].

It has been clearly shown in several independently derived perforin-deficient mice that the lytic ability of CTL and NK cells is profoundly suppressed[3,4,6,7]. In targets cells not expressing Fas/Apo-1/CD95 (called Fas herein) no killing is observed, thus granule-based CTL killing appears to be completely dependent on perforin[3,4]. Purified perforin at high doses can induce lysis of virtually all cell types in the presence of calcium, however it cannot reproduce the apoptotic damage induced by CTL[15]. The granule serine proteases, known as granzymes, are required for apoptosis[13,15,16]. The exact role of perforin in apoptosis is not known and the commonly stated view that it acts as a pore through which granzymes pass into the target cell has not been established experimentally.

2.2 Granzymes

CTL/NK cell specific granule serine protease were identified and cloned long before their function was determined (see reviews[8,12,17]) . Although they are known by many alternate names (CCP, fragmentins, Hanukkah Factor, etc.) by common usage they have become known as granzymes. Ten different granzymes have been identified, four in humans, seven in mice, and six in rats. Because it has proved to be exceptionally difficult to express active granzymes in bacteria, most of our current knowledge is based on purified proteases from CTL and NK cells lines[16,18–20]. Four granzymes have been purified and the enzymatic activity characterized[8,12,16,18,19]. The proteases that have been purified to homogeneity in the human and mouse are granzyme A and B, granzyme 3/human tryptase-2/granzyme K and granzyme M (met-ase-1). Granzyme A and 3 are tryptases that cleave after arginine or lysine, granzyme B is an asp-ase that cleaves after aspartic acid or glutamine, and granzyme M is a met-ase and cleaves after methionine or leucine. The remaining six granzymes have not been purified and only a predicted proteolytic specificity based on deduced sequences is available[8]. Granzyme C is thought to cleave after asparagine/serine while D to G may cleave after phenylalanine or leucine. This rather large family of lymphocyte granule serine proteases thus carry proteolytic specificities that cover a potentially enormous range of substrates. The existence of such a large family of granzymes representing a diverse range of proteolytic specificities does not yet have a functional explanation and will be discussed later.

In humans granzyme A and 3 map to 5q11-q12, granzyme M (Met-ase-1) to 19p13.3 and granzyme B to 14q11-q12. The other granzymes that have been mapped are detected only in mice and rats on chromosome 14 near the murine granzyme B locus (see review[12]). Thus, their appears to be three different loci that cluster the granzymes, a tryptase locus, a met-ase locus and a locus for the remaining granzymes.

Except for granzyme A, all other granzymes show considerable homology in their protein structure. They initiate with a hydrophobic signal peptide followed by a short propeptide which is processed to release the mature enzyme. Granzyme A is a homodimer with two active catalytic sites, while all the other granzymes are monomeric. The family is most closely related to the rat mast cell protease II and cathepsin G. The primary amino acid sequence of this family are highly conserved, particularly in the first 16 amino acids of the mature enzyme although significant regions of variation have been detected, the most prominent of which is from amino acid 17 to 29.

All granzymes are expressed in CTL and granzymes A, B, H, 3 and M are detected in NK cells[12]. Granzymes A and B have also been found in $\gamma\delta^+$ T cells and thymocytes[12].

2.3 Granzyme Function in Apoptosis

The importance of granzymes in CTL/NK mediated killing has come from several types of experiments. Purified granzymes A, B and 3/K which by themselves are unable to kill, induce apoptosis in target cells in the presence of sublytic amounts of perforin[15]. Granzyme B is by far the most potent in specific activity and kinetics, while the tryptase granzymes, A and 3 are slow acting[16]. Of importance to in vivo experiments, the granzymes can synergistically interact suggesting they operate on complementary cell death pathways[15, 20]. Transfection of cDNAs encoding granzyme A or B alone or together, with perforin into mast cells will convert these cells to killers that are nearly as efficient as CTL[20]. Neither granzymes nor perforin when expressed alone in mast cells are effective in allowing these cells to induce apoptosis[21,22].

Another type of experiment to test the granzyme hypothesis is the creation of mice deficient in specific granzymes. Granzyme B homozygous null mutants exhibit a predicted phenotype of loss of apoptosis in CTL killing[13], and loss of apoptosis and plasma membrane damage in NK cell killing[23]. However, the phenotype was not one of total suppression of CTL function. Apoptosis was most affected in short term assays, while longer assays exhibited increasing effectiveness of the CTL[23]. The partial defect is not surprising in the face of alternative CTL killing mechanisms in short term in vitro assays, particularly Fas ligand/Fas interaction which would not be impaired in these mice. An alternate possibility is that other granzymes are responsible for the residual apoptosis. Candidates might be the tryptase granzymes A and 3 which are slow acting. Recently, a granzyme A deficient mouse has been reill express granzyme B, which is much more potent that granzyme A[16], as well as the alternate tryptase granzyme 3, this might explain why the loss of granzyme A would not have much impact. This does not mean that granzyme A is not a participant in CTL-induced apoptosis, as the 'knockout' strategy of evaluating gene function has the serious problem of how to interpret the phenotype produced by the loss of a single member of a redundant enzyme family such as the granzymes. Another test of the granzyme A hypothesis may be to examine double knockouts of granzyme B and A, although this still carries the risk that other granzymes may produce a compensatory effect to the loss of granzyme A (eg. granzyme 3). If this mouse exhibits a more profound defect in apoptosis than the granzyme B KO, then this would suggest that granzymes A interacts synergistically with granzyme B as observed in vitro[15] and in transfected mast cells[20].

3.0 APOPTOSIS PATHWAYS INDUCED BY GRANZYMES

In the last two years great advances have been made in identifying the molecules that have a regulatory role in apoptosis in mammalian cells. Much of this work has evolved by identifying homologs of cell death regulatory genes in the nematode C. elegans first discovered in the pioneering studies of Robert Horvitz and colleagues[25,26]. In this work, the ced-9 gene was found to be homologous to the bcl-2 family of genes[27], while the ced-3 gene led to the identification of the importance of the ICE family of proteases in cell death[28,29]. Another family of proteins that have been implicated in apoptosis are the cyclin dependent kinases(CDKs)[8,30,31]. These kinases control G1/S and G2/M transition. The G2/M kinase $p34^{cdc2}$ can induce cell damage similar to apoptosis when expressed in an unregulated manner in vertebrate cells[32,33]. Both the ICE and CDK families of molecules have been implicated in apoptosis induced by granzymes[34–36].

3.1 Cdc2 and Granzyme-Induced Apoptosis

Cell cycle progression is controlled by a group of highly conserved protein serine/threonine kinases referred to as cyclin dependent kinases. To ensure precise regulation for transition through G1 to S phase and through G2 to M phase the activity of the kinases is controlled through interaction with a cyclin regulatory protein and by phosphorylation and dephosphorylation (see reviews[30,31]). Cyclins exist in classes that interact with specific CDKs and are coordinately synthesized at different points in the cell cycle then degraded before the onset of the next cycle. Entry into mitosis is regulated by phosphorylation of thr-161 and dephosphorylation of tyr-15 and thr-14 of cdc2 kinase and binding to cyclin B or, to a lesser extent, cyclin A[8,31]. Overexpression of cdc2 with mutations at thr-14 and tyr-15[32], or cotransfection of cdc2 and cyclin A or B[33] induce premature mitosis and mitotic catastrophe, a phenomenon which at least morphologically resembles apoptosis.

Cdc2 was shown to be important in granzyme-induced apoptosis by the demonstration that granzyme Bor 3, in the presence of perforin, activates cdc2 prematurely and that reduced cdc2 expression in a temperature sensitive mutant prevents granzyme B or 3-induced apoptosis[34]. We have recently found that, in cycling cells, granzyme B is able to induce apoptosis during all phases of the cell cycle examining Jurkat cells separated by centrifugal elutriation (L. Shi et al, submitted for publication). However, cells entering G0 as a result of withdrawal of growth factors are less susceptible to killing by CTL and CTL granules[37,38] indicating the preference of granzymes for killing cycling target cells. The resistance of non-cycling cells may be a result of the lack of cdc2 expression in G0.

Another approach we have used to test this hypothesis is to transfect cells with the wee1 kinase[35]. Wee1 phosphorylates cdc2 on Tyr-15 and inactivates kinase activity[39,40]. Transient transfection into BHK cells of wee1 kinase with a c-myc epitope tag to identify successfully transfected and expressing cells, resulted in rescue from granzyme B induced apoptosis and rephosphorylation of cdc2[35]. The human p50 wee1 used for these experiments appears to be truncated wee1 based on recent experiments showing the full length protein, but the shortened form used in these experiments contains the catalytic region[40].

Cdc2 activation during apoptosis is not limited to granzyme B. Activation of cyclin A associated Cdc2 and Cdk2 was observed in HIV-1 Tat-induced apoptosis by Li *et al*[41]. They were able to inhibit the apoptotic effects of Tat on T cells with antisense cyclin oligonucleotides. Interestingly, the HIV-1 harbors a gene that suppresses cdc2 kinase activity and arrests cells at G2/M called Vpr[42,43]. These authors propose that by suppressing cdc2 Vpr may function to rescue the infected cell temporarily from either its own apoptotic effects induced by HIV-1 Tat and gp120, or that of CTL and thus allow viral replication before the cell dies[42].

Recent data indicates that the cyclin primarily associated with kinase activity in granzyme induced apoptosis is cyclin A, and that cdk2 kinase in addition to cdc2 activity is increased following granzyme B treatment (Shi, L. and Greenberg, A., submitted for publication). T cell receptor crosslinking with anti-CD3 antibodies of the A1.1 hybridoma also results in apoptosis and elevated Cdc2 activity but with a cyclin B rather than a cyclin A partner. In this model, apoptosis can be blocked with cyclin B antisense oligonucleotides[44]. Why there is preference for one rather another cyclin partner or CDK in different forms of apoptosis is not immediately obvious and will require a better understanding of the pathways leading to initiation of CDK/cyclin activation. One hypothesis is that inactivation of wee-1 is critical to initiation of apoptosis and is a prequisite for cdk disregulation. Wee-1 is required to maintain mitotic timing and can suppress both cdc2 and cdk2 activity[45]. Wee-1 activity might be disrupted if its phosphorylation is altered[46] or if it is hydrolysed by a protease such as granzyme B or a member of the ICE family.

Cdc2 or the closely related cdk2 are not the only members of the CDK family that are activated during apoptosis. Another CDK related kinase, p58 PITSLRE, was found to initiate apoptosis when overexpressed, and was activated during Fas-induced apoptosis[47]. Of further interest, the protein is activated by proteolytic cleavage and apoptosis and p58 PTSLRE cleavage can be blocked by serine protease inhibitors[47]. Thus it is possible that multiple CDKs are activated by apoptosis and these may be linked to induction pathways for granzyme and fas induced apoptosis, as well as apoptosis induced by HIV-1 and T cell receptor activation.

Having discussed the instances in which cdc2 and other CDKs are important for apoptosis, it must be noted that some models of drug and UV-induced apoptosis do not appear to be regulated by cdc2 kinase. Martin et al.[48] examined the FT210 cell line which carries a temperature sensitive cdc2, and found that when cdc2 was suppressed, the cells still underwent apoptosis to agents such as H_2O_2, actinomycin D, UV irradiation and ceramide. This experiment cannot exclude the participation of other CDKs such as cdk2[49], but one can say that cdc2 is not a universal regulator of apoptosis. This is not surprising considering that non-cycling cells that do not express cdc2 can still die a typical apoptotic death[50].

3.2 Granzymes and Interleulin-1 Beta Converting Enzyme (ICE) Family Proteases

During development, genetic analysis of programmed cell death in C. elegans identified three genes, ced-3, ced-4 and ced-9, that play important roles in this process[25]. Recessive loss-of-function mutations in ced-3 prevent the death of cells that normally undergo programmed cell death. The deduced product of ced-3 is homologous to mammalian ICE, a cysteine protease that cleaves pro-interleukin-1beta (IL-1β) to generate mature IL-1β. Expression of ICE induces programmed cell death in a variety of cell lines[51,52]. Expression of crmA, a specific inhibitor of ICE encoded by cowpox virus, inhibits the death of chicken dorsal root ganglion neurons induced by trophic factor deprivation[53]. Five additional members of the ICE family have been identified; Ich-1, CPP32/apopain, Tx/ ICErel-II, ICErel-III and Mch-2[52,54–59] with more to come (see Martin et al[54] for review) .

Granzyme B hydrolyses peptide substrates after aspartic acid and tripeptide aspartyl chloromethyl ketones irreversibly block granzyme B activity[16,18]. The activation of ICE family members requires cleavage after several aspartic acid residues which then releases subunits that form the active enzymes which like granzyme B are aspartyl proteases[60,61]. This suggests that granzyme B might act by processing ICE family members or by mimicking the effects of ICE in activating an apoptosis pathway. We have observed the processing of ICE to p20 and the conversion of pro-IL-1 beta to IL-1 beta after granzyme B treatment (L. Shi et al, submitted for publication), however, ICE is not directly activated by granzyme B[62] indicating that other substrates are important for its action. One ICE family member that has been recently found to be a substrate for granzyme B is CPP32/Apopain[63,64]. *In vitro* cleavage by granzyme B to the p17 subunit and activation of CPP32 through its ability to process poly ADP ribose polymerase (PARP) were also demonstrated[65]. Evidence for the direct *in vivo* activation of CPP32 by granzyme B is more difficult to establish. The processing of CPP32 *in vivo* after CTL induce apoptosis was shown[65] but it is not known if this is the result of the direct action of granzyme B on CPP32 in the cell cytoplasm or whether granzyme B activates another protease which initiates CPP32 processing. Furthermore, it is not known whether the CPP32 activation was necessary for death of the target cell. These issues will need to be addressed in the future to be sure that activation of CPP32 is on the effector pathway for granzyme B-induced apoptosis.

We have recently examined ICE protease deficient mice and find that some but not all cell types are resistant to granzyme B-induced apoptosis (L. Shi et al, submitted for publication). Interestingly, cells that are resistant to granzyme B are often not resistant to apoptosis by granzyme 3 or Fas suggesting that both ICE-dependent and independent apoptosis pathways exist in the same cell and that they may be activated by different initiating signals.

4.0 THE MULTIPLICITY OF GRANZYME AND ICE PROTEASES

The identification of the aspartyl protease granzyme B as a principle mediator of CTL induced apoptosis together with the discovery of the ICE proteases have given a major impetus to the hypothesis that proteases lie at the centre of the biochemical pathway that leads to apoptosis. Attempts to confirm the importance of the granzyme and ICE family members by creating deficient mouse mutants make it clear that these systems are far more complex than originally imagined. ICE deficiency produces no developmental defect[66,67] arguing that this ICE is not the ced-3 homologue of mammals. It has been suggested that this is the role of CPP32[63], however we must await the development of the CPP32 knockout for a direct test of this hypothesis. On the other hand, cells from ICE deficient mice exhibit resistance to either granzyme B, granzyme 3 (L. Shi et al, submitted for publication) or Fas killing[67] suggesting that induced apoptosis in adult cells may be regulated on different pathways than those that are important for development. Since both the granzyme and ICE proteases are highly redundant, other family members may be able to substitute for a deficiency of any one protease. Redundant ICE family proteases may substitute functionally for one another making the interpretation of knockout experiments difficult unless one identifies a clear phenotype. Double and triple knockouts in the ICE family may be necessary to observe widspread resistance to CTL or the anticipated developmental defects equivalent to ced-3 deficiency. Similar concerns about the impact of redundancy in the interpretation of experiments apply to the granzyme family. For example, CTL from granzyme B deficient mice are suppressed in their ability to induce rapid but not long term apoptosis[13], thus indicating the existence of other apoptosis inducers in the CTL such as other granzymes. Granzyme A deficiency has no observable effect on CTL cytotoxicity[24] which again may be due to the presence of the powerful granzyme B in these CTL.

Why do multiple granzymes and ICE proteases exist? It has been suggested that the immune system in vertebrates evolved in response to the appearance of viruses, and recent evidence indicates that some viruses produce proteins that are inhibitory to granzyme and ICE. The cowpox protease inhibitor crmA[63,64,68] and baculovirus p35[69] inhibit several different ICE family members. We have observed that crmA which is an ICE inhibitor[63,64,68] can block the asp-ase granzyme B but not the tryptase granzyme 3 (L. Shi et al, unpublished data). The CTL also contains granzymes of other proteolytic specificities including a met-ase and chymase which presumably will not be inhibitable by crmA[12]. Thus CTL may have evolved granzymes that can potentially avoid the inhibitory effect of crmA-like protease inhibitors. However this viral protease inhibitor can also suppress the cellular death pathway through its effects on ICE proteases thus potentially blocking the alternate granzymes if they act through the same ICE pathway. Thus, different ICE molecules may have evolved in parallel with the granzymes creating independent cell death pathways. These alternate cell death pathways through which the other granzymes can act would thus prevent the devastating effects of uncontrolled viral replication by allowing CTL to rapidly kill the infected cell expressing the protease inhibitor (Figure 1).

Although an important place for the proteases in world of cell medited cytotoxicity and apoptosis seems assured, the major task facing the granzyme-induced apoptosis field in

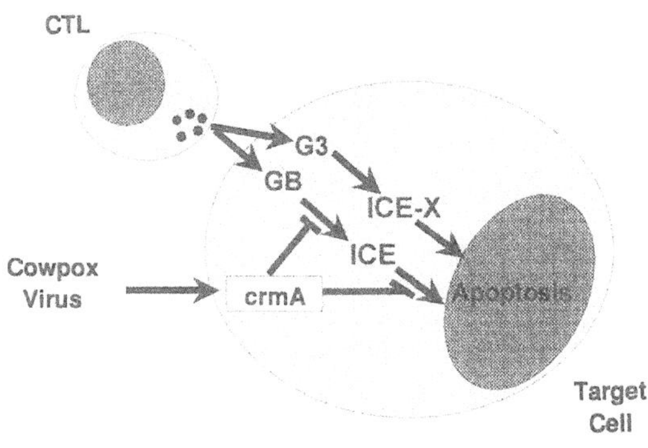

Figure 1. Model for the existence of multiple granzymes and ICE proteases in CTL induced cytotoxicity of virally infected cells. The cowpox virus crmA protease inhibitor is capable of blocking the activity of both granzyme B and ICE protease (see text). To bypass this viral block to apoptosis which would be lethal for the host, the CTL simultaneously release a second granzyme located in their granules with a tryptase specificity, shown here as granzyme 3. This granzyme is not blocked by the crmA gene and it induces the activation of another ICE family protease which is also not crmA inhibitable resulting in the initiation of an apoptotic pathway and cell death thus bypassing the viral strategy to block cell death and allow viral replication. It has been suggested that HIV-1 uses the Vpr gene that blocks cdc2 activation as a means of preventing apoptosis and thus allow viral replication before the cell dies or is killed by CTL[42]. It is not known whether or not the cdc2 and ICE regulated apoptosis pathways are identical.

the future is to determine the function and relationship of members of the large granzyme and ICE families. We will need to answer such questions as: What initiates the processing of different ICE proteases by granzymes? How do granzymes cause cell death? Do granzymes of different proteolytic specificity act independently and on different ICE pathways? How does the processing of the ICE proteases result in the activation of the CDKs? And many other intriguing questions.

5.0 REFERENCES

1. P.A. Henkart, Mechanism of lymphocyte-mediated cytotoxicity. *Annu. Rev. Immunol.* 3:31 (1985).
2. E. Rouvier, M.F. Luciani, and P. Golstein, Fas involvement in Ca(2+)-independent T cell-mediated cytotoxicity. *J. Exp. Med.* 177:195 (1993).
3. D. Kaegi, F. Vignaux, B. Ledermann, K. Buerki, V. Depraetere, S. Nagata, H. Hengartner, and P. Golstein, Fas and perforin pathways as major mechanisms of T cell-mediated cytotoxicity. *Science* 265:528 (1994).
4. B. Lowin, M. Hahne, C. Mattmann, and J. Tschopp, Cytolytic T-cell cytotoxicity is mediated through perforin and Fas lytic pathways. *Nature* 370:650 (1994).
5. L.X. Zheng, G. Fisher, R. E. Miller, J. Peschon, D. H. Lynch, and M. J. Lenardo, Induction of apoptosis in mature T cells by tumour necrosis factor. *Nature* 377:348 (1995).
6. D. Kaegi, B. Ledermann, K. Buerki, P. Seiler, B. Odermatt, K. J. Olsen, E. R. Podack, R. M. Zinkernagel, and H. Hengartner, Cytotoxicity mediated by T cells and natural killer cells is greatly impaired in perforin-deficient mice. *Nature* 369:31 (1994).
7. C.M. Walsh, M. Matloubian, Liu, C.C. Ueda, C.G. Kurahara, J.L. Christensen, M.T.F. Huang, J.D.E. Young, R. Ahmed, and W.R. Clark, Immune function in mice lacking the perforin gene. *Proceedings. of. the. National. Academy. of. Sciences. of. the. United. States. of. America.* 91:10854 (1994).
8. A.H. Greenberg, and D.W. Litchfield. Granzymes and apoptosis: Targeting the cell cycle. In Pathways for cytolysis. G.M. Griffiths and J. Tschopp, editors. *Current Topics in Microbiology and Immunology.* 198:95 (1995).
9. G.M. Griffiths, and Y. Argon, Structure and biogenesis of lytic granules. In Pathways for cytolysis. G.M. Griffiths and J. Tschopp, editors. *Current Topics in Microbiology and Immunology* 198:39 (1995).
10. P.A. Henkart, M.S. Williams, and H. Nakajima, Degranulating cytotoxic lymphocytes inflict multiple damage pathways on target cells. In Pathways for cytolysis. G.M. Griffiths and J. Tschopp, editors. *Current Topics in Microbiology and Immunology* 198:75 (1995).
11. B. Lowin, M.C. Peitsch, and J. Tschopp, Perforin and Granzymes: Crucial effector moleucles in cytolytic T lymphocyte and ntural killer cell-mediated cytotoxicity. In Pathways for cytolysis. G.M. Griffiths and J. Tschopp, editors. *Current Topics in Microbiology and Immunology* 198:1 (1995).

12. M.J. Smyth, and J. A. Trapani, Granzymes: Exogenous proteinases that induce target cell apoptosis. *Immunology Today* 16:202 (1995).
13. J.W. Heusel, R.L. Wesselschmidt, S. Shresta, J.H. Russell, and T. J. Ley, Cytotoxic lymphocytes require granzyme B for the rapid induction of DNA fragmentation and apoptosis in allogeneic target cells. *Cell* 76:977 (1994).
14. C.C Liu, C.M. Walsh, and J.D.E. Young , Perforin: Structure and function. *Immunology Today* 16:194 (1995).
15. L. Shi, R.P.Kraut, R. Aebersold, and A.H. Greenberg, A natural killer cell granule protein that induces DNA fragmentation and apoptosis. *J. Exp. Med.* 175:553 (1992).
16. L. Shi, C.M. Kam, J.C. Powers, R. Aebersold, and A.H. Greenberg, Purification of three cytotoxic lymphocyte granule serine proteases that induce apoptosis through distinct substrate and target cell interactions. *J. Exp. Med.* 176:1521 (1992).
17. D. Masson, and J. Tschopp, A family of serine esterases in lytic granules of cytolytic T lymphoyctes. *Cell* 49:679 (1987).
18. M. Poe, J.T. Blake, D.A. Boulton, M. Gammon, N.H. Sigal, J.K. Wu, and H.J. Zweerink, Human cytotoxic lymphocyte granzyme B. Its purification from granules and the characterization of substrate and inhibitor specificity. *J. Biol. Chem.* 266:98 (1991).
19. S. Odake, C.M. Kam, L. Narasimhan, M. Poe, J.T. Blake, O. Krahenbuhl, J. Tschopp, and J.C. Powers, Human and murine cytotoxic T lymphocyte serine proteases: subsite mapping with peptide thioester substrates and inhibition of enzyme activity and cytolysis by isocoumarins. *Biochemistry* 30:2217 (1991).
20. H. Nakajima, H.L. Park, and P.A. Henkart, Synergistic roles of granzymes A and B in mediating target cell death by rat basophilic leukemia mast cell tumors also expressing cytolysin/perforin. *J. Exp. Med.* 181:1037 (1995).
21. J.W. Shiver, and P. A. Henkart, A noncytotoxic mast cell tumor line exhibits potent IgE-dependent cytotoxicity after transfection with the cytolysin/perforin gene. *Cell* 64:1175 (1991).
22. J.W. Shiver, L. Su, and P.A. Henkart, Cytotoxicity with target DNA breakdown by rat basophilic leukemia cells expressing both cytolysin and granzyme A. *Cell* 71:315 (1992).
23. S. Shresta, D.M. MacIvor, J.W. Heusel, J.H. Russell, and T.J. Ley, Natural killer and lymphokine-activated killer cells require granzyme B for the rapid induction of apoptosis in susceptible target cells. *Proceedings. of. the. National. Academy. of. Sciences. of. the. United. States. of. America.* 92:5679 (1995).
24. K. Ebnet, M. Hausmann, F. Lehmann-Grube, A. Muellbacher, M. Kopf, M. Lamers, and M.M. Simon, Granzyme A-deficient mice retain potent cell-mediated cytotoxicity. *EMBO Journal.* 14:4230 (1995).
25. R.E. Ellis, J.Y. Yuan, and H.R. Horvitz, Mechanisms and functions of cell death. *Annu. Rev. Cell Biol.* 7:663 (1991).
26. M.O. Hengartner, and H.R. Horvitz, Programmed cell death in Caenorhabditis elegans. *Current. Opinion. in Genetics and. Development* 4:581 (1994).
27. M.O. Hengartner, and H.R. Horvitz, C. elegans cell survival gene ced-9 encodes a functional homolog of the mammalian proto-oncogene bcl-2. *Cell* 76:665 (1994).
28. J.Y. Yuan, and H.R. Horvitz, The Caenorhabditis elegans genes ced-3 and ced-4 act cell autonomously to cause programmed cell death. *Dev. Biol.* 138:33 (1990).
29. J. Yuan, S. Shaham, S. Ledoux, H.M. Ellis, and H.R. Horvitz, The C. elegans cell death gene ced-3 encodes a protein similar to mammalian interleukin-1Beta-converting enzyme. *Cell* 75:641 (1993).
30. C. Norbury, and P. Nurse, Animal cell cycles and their control. *Annu. Rev. Biochem.* 61:441 (1992).
31. D.O. Morgan, Principles of CDK regulation. *Nature* 374:131 (1995).
32. W. Krek, and E.A. Nigg, Mutations of p34cdc2 phosphorylation sites induce premature mitotic events in HeLa cells: evidence for a double block to p34cdc2 kinase activation in vertebrates. *EMBO J.* 10:3331 (1991).
33. R. Heald, M. McLoughlin, and F. McKeon, Human wee1 maintains mitotic timing by protecting the nucleus from cytoplasmically activated Cdc2 kinase. *Cell* 74:463 (1993).
34. L. Shi, W.K. Nishioka, J. Th'ng, E.M. Bradbury, D.W. Litchfield, and A.H. Greenberg, Premature p34cdc2 activation required for apoptosis. *Science* 263:1143 (1994).
35. G. Chen, L. Shi, D.W. Litchfield, and A.H. Greenberg, Rescue from granzyme B-induced apoptosis by Wee1 kinase. *Journal. of. Experimental. Medicine* 181:2295 (1995).
36. A.J. Darmon, D.W. Nicholson, and R.C. Bleackley, Activation of the apoptotic protease CPP32 by cytotoxic T-cell-derived granzyme B. *Nature* 377:446 (1995).
37. N. Khalil, S.C. Battistuzzi, R.P. Kraut, L.C. Schwarz, and A.H. Greenberg, Growth factor-initiated proliferation of mouse embryonic fibroblasts induces cytotoxicity by natural killer cells and by a non-cytolysin cytotoxin in natural killer granules. *J. Immunol.* 145:1286 (1990).

38. W.K. Nishioka, and R.M. Welsh, Susceptibility to cytotoxic T lymphocyte-induced apoptosis is a function of the proliferative status of the target. *J. Exp. Med.* 179:769 (1994).
39. L.L. Parker, and H. Piwnica Worms, Inactivation of the p34cdc2-cyclin B complex by the human WEE1 tyrosine kinase. *Science* 257:1955 (1992).
40. C.H. McGowan, and P. Russell, Cell cycle regulation of human WEE1. *EMBO Journal.* 14:2166 (1995).
41. C.J. Li, D.J. Friedman, C. Wang, V. Metelev, and A.B. Pardee, Induction of apoptosis in uninfected lymphocytes by HIV-1 Tat protein. *Science* 268:429 (1995).
42. J.L. He, S. Choe, R. Walker, P. Di Marzio, D.O. Morgan, and N.R. Landau, Human immunodeficiency virus type 1 viral protein R (Vpr) arrests cells in the G2 phase of the cell cycle by inhibiting p34cdc2 activity. *Journal. of. Virology* 69:6705 (1995).
43. F. Re, D. Braaten, E.K. Franke, and J. Luban, Human immunodeficiency virus type 1 Vpr arrests the cell cycle in G2 by inhibiting the activation of p34cdc2-cyclin B. *Journal. of. Virology* 69:6859 (1995).
44. R. Fotedar, J. Flatt, S. Gupta, R.L. Margolis, P. Fitzgerald, H. Messier, and A. Fotedar, Activation-induced T-cell death is cell cycle dependent and regulated by cyclin B. *Molecular. and. Cellular. Biology.* 15:932 (1995).
45. N. Watanabe, M. Broome, and T. Hunter, Regulation of the human WEE1 Hu CDK tyrosine 15-kinase during the cell cycle. *EMBO Journal.* 14:1878 (1995).
46. L. Wu, and P. Russell, Nim1 kinase promotes mitosis by inactivating Wee1 tyrosine kinase. *Nature* 363:738 (1993).
47. J.M. Lahti, J. Xiang, L.S. Heath, D. Campana, and V.J. Kidd, PITSLRE protein kinase activity is associated with apoptosis. *Molecular. and. Cellular. Biology.* 15:1 (1995).
48. S.J. Martin, A.J. McGahon, W.K. Nishioka, D. La Face, X. Guo, J. Th'ng, E.M. Bradbury, and D.R. Green, p34cdc2 and apoptosis. *Science* 269:106 (1995).
49. A.H. Greenberg, and D.W. Litchfield, p34cdc2 and apoptosis. Response. *Science* 269:107 (1995).
50. R.S. Freeman, S. Estus, and E.M. Johnson,Jr, Analysis of cell cycle-related gene expression in postmitotic neurons: Selective induction of cyclin D1 during programmed cell death. *Neuron* 12:343 (1994).
51. M. Miura, H. Zhu, R. Rotello, E.A. Hartwieg, and J. Yuan, Induction of apoptosis in fibroblasts by IL-1Beta-converting enzyme, a mammalian homolog of the C. elegans cell death gene ced-. *Cell* 75:653 (1993).
52. L. Wang, M. Miura, L. Bergeron, H. Zhu, and J. Yuan, Ich-1, an Ice/ced-3-related gene, encodes both positive and negative regulators of programmed cell death. *Cell* 78:739 (1994).
53. V. Gagliardini, P.A. Fernandez, R.K. Lee, H.C. Drexler, R.J. Rotello, M.C. Fishman, and J. Yuan, Prevention of vertebrate neuronal death by the crmA gene [see comments]. *Science* 263:826 (1994).
54. S.J. Martin, and D.R. Green, Protease activation during apoptosis: Death by a thousand cuts?. *Cell* 82:349 (1995).
55. S. Kumar, M. Kinoshita, M. Noda, N.G. Copeland, and N.A. Jenkins, Induction of apoptosis by the mouse Nedd2 gene, which encodes a protein similar to the product of the Caenorhabditis elegans cell death gene ced-3 and the mammalian IL-1Beta-converting enzyme. *Genes and. Development* 8:1613 (1994).
56. T. Fernandes-Alnemri, G. Litwack, and E.S. Alnemri, CPP32, a novel human apoptotic protein with homology to Caenorhabditis elegans cell death protein Ced-3 and mammalian interleukin-1Beta-convert ng enzyme. *Journal. of. Biological. Chemistry.* 269:30761 (1994).
57. C. Faucheu, A. Diu, A.W.E. Chan, A.M. Blanchet, C. Miossec, F. Herve, V. Collard-Dutilleul, Y. Gu, R.A. Aldape, J.A. Lippke, C. Rocher, M.S.S. Su, D.J. Livingston, T. Hercend, and J.L. Lalanne, A novel human protease similar to the interleukin-1Beta converting enzyme induces apoptosis in transfected cells. *EMBO Journal.* 14:1914 (1995).
58. N.A. Munday, J.P. Vaillancourt, A. Ali, F.J. Casano, D.K. Miller, S.M. Molineaux, T.T. Yamin, V.L. Yu, and D.W. Nicholson, Molecular cloning and pro-apoptotic activity of ICErelII and ICErelIII, members of the ICE/CED-3 family of cysteine proteases. *J. Biol. Chem.* 270:15870 (1995).
59. T. Fernandes Alnemri, G. Litwack, and E.S. Alnemri, Mch2, a new member of the apoptotic Ced-3/Ice cysteine protease gene family. *Cancer Res.* 55:2737 (1995).
60. N.A. Thornberry, H.G. Bull, J.R. Calaycay, K.T. Chapman, A.D. Howard, M.J. Kostura, D.K. Miller, S.M. Molineaux, J.R. Weidner, J. Aunins, and et al, A novel heterodimeric cysteine protease is required for interleukin-1 beta processing in monocytes. *Nature* 356:768 (1992).
61. D.K. Miller, J.M. Ayala, L.A. Egger, S.M. Raju, T.T. Yamin, G.J. Ding, E.P. Gaffney, A.D. Howard, O.C. Palyha, A.M. Rolando, and et al, Purification and characterization of active human interleukin-1 beta-converting enzyme from THP.1 monocytic cells. *J. Biol. Chem.* 268:18062 (1993).
62. A.J. Darmon, N. Ehrman, A. Caputo, J. Fujinaga, and R.C. Bleackley, The cytotoxic T cell proteinase granzyme B does not activate interleukin-1Beta-converting enzyme. *Journal. of. Biological. Chemistry.* 269:32043 (1994).

63. D.W. Nicholson, A. Ali, N.A. Thornberry, J.P. Vaillancourt, C.K. Ding, M. Gallant, Y. Gareau, P.R. Griffin, M. Labelle, Y.A. Lazebnik, N.A. Munday, S.M. Raju, M.E. Smulson, T.T. Yamin, V. L. Yu, and D.K. Miller, Identification and inhibition of the ICE/CED-3 protease necessary for mammalian apoptosis. *Nature* 376:37 (1995).
64. M. Tewari, L.T. Quan, K. O'Rourke, S. Desnoyers, Z. Zeng, D.R. Beidler, G.G. Poirier, G.S. Salvesen, and V.M. Dixit, Yama/CPP32Beta, a mammalian homolog of CED-3, is a CrmA-inhibitable protease that cleaves the death substrate poly(ADP-ribose) polymerase. *Cell* 63. 81:801 (1995).
65. A.J. Darmon, D.W. Nicholson, and R.C. Bleackley, Activation of the apoptotic protease CPP32 by cytotoxic T-cell-derived granzyme B. *Nature* 377:446 (1995).
66. P. Li, H. Allen, S. Banerjee, S. Franklin, L. Herzog, C. Johnston, J. McDowell, M. Paskind, L. Rodman, J. Salfeld, E. Towne, D. Tracey, S. Wardwell, F. Wei, W. Wong, R. Kamen, and T. Seshadri, Mice deficient in IL-1β-converting enzyme are defective in production of mature IL-1β and resistant to endotoxic shock. *Cell* 80:401 (1995).
67. K. Kuida, J.A. Lippke, G. Ku, M.W. Harding, D.J. Livingston, M.S.S. Su, and R.A. Flavell, Altered cytokine export and apoptosis in mice deficient in interleukin-1Beta converting enzyme. *Science* 267:2000 (1995).
68. C.A. Ray, R.A. Black, S.R. Kronheim, T.A. Greenstreet, P.R. Sleath, G.S. Salvesen, and D.J. Pickup, Viral inhibition of inflammation: cowpox virus encodes an inhibitor of the interleukin-1 beta converting enzyme. *Cell* 69:597 (1992).
69. N.J. Bump, M. Hackett, M. Hugunin, S. Seshagiri, K. Brady, P. Chen, C. Ferenz, S. Franklin, T. Ghayur, P. Li, P. Licari, J. Mankovich, L.F. Shi, A.H. Greenberg, L.K. Miller, and W.W. Wong, Inhibition of ICE family proteases by baculovirus antiapoptotic protein p35. *Science* 269:1885 (1995).

MATURE T LYMPHOCYTE APOPTOSIS IN THE HEALTHY AND DISEASED IMMUNE SYSTEM

Lixin Zheng, Galen Fisher, Behazine Combadiere, Felicita Hornung, David Martin, Clara Pelfrey, Jin Wang, and Michael Lenardo

Laboratory of Immunology
National Institute of Allergy and Infectious Diseases
National Institutes of Health
Bethesda, Maryland 20892

1. INTRODUCTION

We have been studying a feedback regulatory mechanism termed propriocidal regulation that induces mature peripheral T cells to undergo apoptosis during an antigen response. We proposed this mechanism in order to account for two observations. First, we observed that mature T lymphocytes can be triggered to undergo apoptosis following T cell receptor (TCR) ligation [1],[2], for a reviews see [3], [4]. This finding challenged the prevailing idea that only immature T cells in the thymus could exhibit an apoptotic response to TCR stimulation. Second, we made the surprising observation that interleukin-2 (IL-2) was the principal influence that caused peripheral T lymphocytes to become susceptible to TCR-induced apoptosis[1]. This led to the notion that antigen-induced death was the end-result of a feedback loop. We conceived that strong, repeated antigen stimulation would first initiate a robust proliferative reaction and then induce the death of T cells following their effector response. This would decrease the number of specifically-responding T cells and the magnitude of the response to subsequent antigen challenge. The consequence of this feedback loop is that an unregulated geometric expansion of activated T lymphocytes and their attendant toxicities would be prevented. Thus, propriocidal regulation entails immune regulation by the programmed cell death of mature T cells. This provides a means for antigen-specific, extrathymic immunological tolerance.

We report progress in understanding the molecular basis of the propriocidal mechanism and define the molecular basis for a human disease syndrome comprising dysregulated lymphoproliferation and autoimmunity.

Mechanisms of Lymphocyte Activation and Immune Regulation VI, Edited by Gupta and Cohen
Plenum Press, New York, 1996

2. RESULTS

2.1 A Key Role for IL-2 in Mature T Cell Apoptosis

We have shown in a number of settings using different populations of mature T cells that respond to different TCR ligands, that apoptosis is directly related to the level of IL-2-induced proliferation and to the concentration of TCR ligand used for re-stimulation[4]. To determine the importance of IL-2 during T cell deletion in vivo, we administered the bacterial superantigen, *Staphylococcus Aureus B* (SEB), together with an IL-2-blocking antibody [1](Figure 1). This experiment showed that repeated infections of SEB which is a powerful mitogenic TCR ligand, will cause deletion of 60–70% of the specifically responding T cells. This experiment also revealed that deletion of the SEB-reactive Vß8 T cells was substantially abrogated when IL-2 was blocked. Recently, this result was affirmed by the genetic experiment of Willerford et al.[5]. These investigators prepared a homozygous gene deficiency of the α chain of the high affinity IL-2 receptor. They found that B and T lymphocyte development was apparently normal but as mice grew older, they exhibited progressive polyclonal lymphocyte expansion as autoimmune hemolytic anemia and inflammatory bowel disease. These mice showed defective T lymphocyte deletion in response to SEB administration. Similar results were found in mice that carried homozygous deficiencies of the IL-2 gene itself [6]. Therefore, loss of the IL-2 component of the propriocidal pathway leads to dysregulation of peripheral T cell homeostasis and autoimmune disease.

2.2 Contribution of the TCR ζ Chain to Mature T Cell Apoptosis

We also investigated the participation of various TCR chains in the induction of apoptosis and found a particularly important role for the TCR ζ chain. We found that

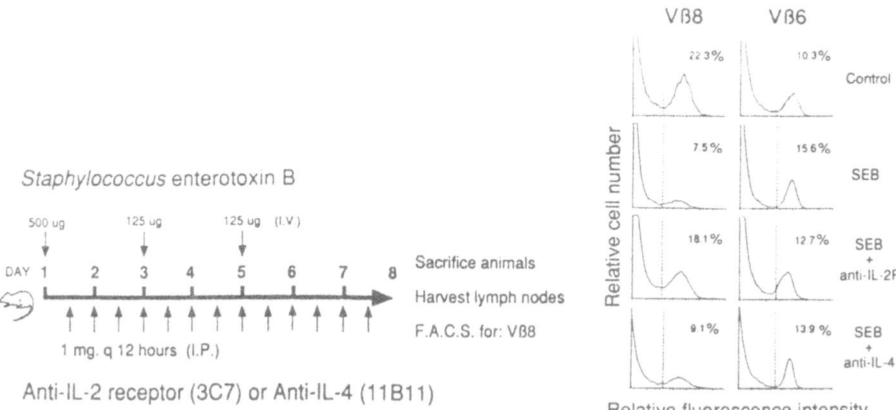

Figure 1. Antibody blockade of IL-2 abrogates deletion of Vß8 T cells by repetitive immunization with SEB. (Left panel) Illustrated Is the time course of injections used to stimulate deletion using the bacterial superantigen, *Staphylococcus Aureus* enterotoxin B (SEB). Amounts and timing of SEB injections that were given intravenously are shown above the timeline. The amounts and timing of monoclonal antibody (mAb) injections are shown below the timeline. One group of mice received SEB injections together with the 3C7 mAb that binds to the α chain of the IL-2 receptor and blocks IL-2 signals and another group of mice received SEB injections together with the 11B11 mAb that binds to and neutralizes the IL-4 molecule. (Right panel) Flow cytometric analysis of the lymph node T cells to quantitate the fraction of SEB-responsive cells (Vß8) or a population of control cells that do not respond to SEB (Vß6). A one-week course of injections of the 3C7 or 11B11 mAbs without SEB had no effect. Further experimental details can be found in Lenardo (1). Previously published data used with permission of MacMillan Press Ltd.

cross-linking a single chain chimera containing the cytoplasmic domain of the ζ chain that fused to the extracellular portion of the Tac (CD25) molecule caused T cells to undergo apoptosis (B. Combadiere and M. Lenardo, data not shown). Furthermore, we found that T cell hyridomas in which the TCR contained a truncated ζ chain lacking the cytoplasmic domain exhibited reduced apoptosis in response to antibody or peptide TCR stimulation. Also lymph node cells from a genetically-altered mouse with a similar truncation of the ζ cytoplasmic domain showed a defect in apoptosis. By contrast, these alterations in the TCR ζ chain had little effect on IL-2 production after TCR cross-linking. In further experiments, we carried out site-specific mutagenesis in order to address the contributions of the immunoreceptor tyrosine-based activation motifs (ITAMs) which are believed to be the principal signalling elements in the cytoplasmic domains of the TCR/CD3 chains. Mutations in the most membrane proximal ITAM of the three found in the TCR ζ chain causes the largest decrease in apoptosis. We conclude from these studies that the TCR ζ chain makes both qualitative and quantitative contributions to the efficiency of TCR-induced apoptosis.

2.3 FasL/Fas and TNFR/TNF Interaction Are Both Involved in TCR-Induced Mature T Cell Apoptosis

Another important insight into the molecular mechanism of propriocidal regulation comes from the study of mice that are homozygous for the *lpr* (lymphoproliferation, a Fas mutation) and *gld* (generalized lymphoproliferative disease) alleles [7,8]. These strains of mice exhibit severe lymphoproliferation coupled with autoimmune manifestations including vasculitis, hemolytic anemia, and glomerulonephritis[7,8]. Molecular genetic studies revealed that *lpr* was due to a mutation in the Fas/Apo-1/CD95 molecule that is a surface receptor which bears a close resemblance to the apoptosis-inducing p55 TNF receptor. Subsequent work showed that *gld* was a mutant allele of the gene encoding Fas ligand (FasL). In the case of both *lpr* and *gld*, the mutations abrogated the ability of the Fas:FasL interaction to induce cell death. In order to explore the autoimmune phenotype of *lpr* and *gld* mice, we examined the sensitivity of mature peripheral T cells from these mutant mice as well as TNF receptor mutant mice to apoptosis induced by TCR/CD3 crosslinking.

We first investigated whether FasL/Fas and TNF/TNFR interactions were essential for TCR-induced mature T cell apoptosis using T cells from wild-type mice [9]. T cells were freshly isolated from lymph nodes and stimulated with the lectin concanavalin A (Con A). Con A upregulates the expression of high affinity IL-2 receptor which is not present on resting T cells. The activated T cells were then treated with 50 IU/ml of IL-2 for 2 to 5 days. These IL-2 treated LN cells are >98% TCR+, CD4+ or CD8+ single-positive T cells, and generally in log-phase of proliferation as demonstrated by the 3H-thymidine incorporation assay. When we rechallenged these proliferating lymph node T cells (LNTC) in vitro by TCR/CD3 crosslinking, we observed a dramatic loss of viable cells accompanied by an increase in apoptotic cells (Fig. 2A, B). Previously, it was postulated -induced apoptosis [10, 11, 12]. If this assumption was correct, one would expect that TCR-triggered apoptosis could not take place in T cells derived from either *lpr* or *gld* mice. On the contrary, we found that proliferating *lpr* T cells, although less sensitive than wild-type controls, manifested substantial apoptosis in response to TCR stimulation (Fig. 2 A, B). The death of *lpr* T cells did not seem to be due to 'leakiness' of the Fas mutation as *lpr* cells had almost no surface expression of Fas compared to wild-type T cells (data not shown). The loss of cells was not due to cell cycle blockade because of the prominent increase in cells with a clearly apoptotic morphology. Similar data was obtained with LNTCs from *gld* mice (data not shown). From these experiments, we conclude that T cell apoptosis is still partly inducible in Fas and FasL mutant systems, indicating that there are both Fas-dependent as well as Fas-independent mechanisms.

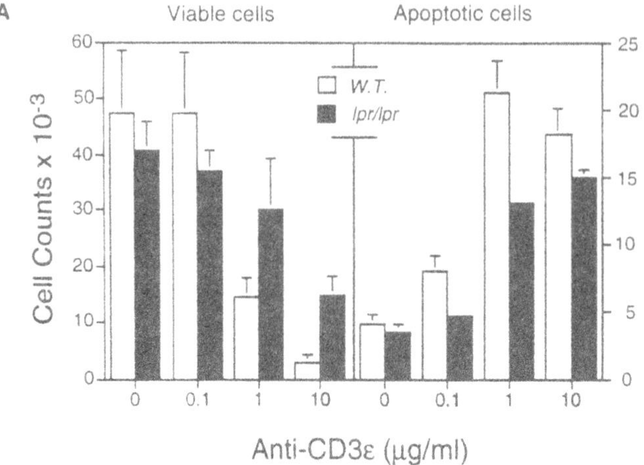

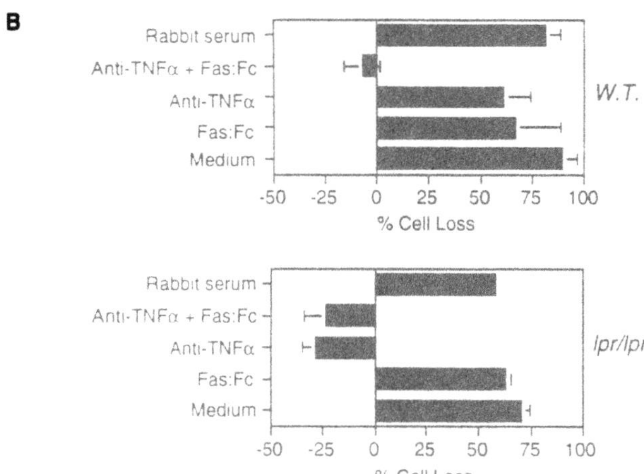

Figure 2. Both Fas and Tumor Necrosis Factor (TNF) cause mature T cell apoptosis. (A) IL-2 treated lymph node T cells from either wild-type or *lpr* mice show the indicated loss of viable T cells and increase in apoptotic cells after rechallenge with various concentrations of platebound anti-CD3ε. (B) The degree of wild-type or *lpr* T cell loss after TCR rechallenge either alone or in the presence of Fas:Fc (FasL blocking reagent), anti-TNFα or both. Further experimental details can be found in Zheng et al.[9]. Previously published data used with permission of MacMillan Press Ltd.

Tumor necrosis factor (TNF) and lymphotoxin (LT) are apoptosis-inducing cytokines that share amino acid homology with FasL [13], [14]. Two distinct TNF receptors have been characterized which can bind either TNF or LT: a 55-kDa TNFR1 and a 75-kDa TNFR2 [13]. TNFR1 but not TNFR2 contains a "death domain" in its intracytoplasmic portion that is required for apoptosis signalling and highly homologous with a similar region of Fas [13]. The TNF/TNFR1 interaction is able to induce apoptosis in various types of cells [14], [15]. The role of TNFR2 in apoptosis has been controversial, although certain evidence supports such a role [15], [16], [17]. We and other groups have noticed that T cells can secrete large amounts of TNF following TCR stimulation. Therefore, we examined the role of TNF in the induction of apoptosis of T cells derived from *lpr* and *gld* mice. Under conditions in which TCR stimulation causes the death of IL-2 treated T cells, TNF mRNA and protein expression were greatly increased in wild-type as well as *lpr* and *gld* T cells (data not shown)[9] . However, direct evidence confirming the contribution of both FasL/Fas and TNFR/TNF to TCR-induced apoptosis was obtained from blocking experiments (Fig.2B). After 48 hours of TCR stimulation, Fas-Fc (a Fas extracellular domain and human IgG Fc fragment fusion protein) slightly reduced apoptosis in wild-type LNTC but not in *lpr* LNTC, whereas anti-TNF blocking antibody partly rescued wild-type cells but completely prevented *lpr* T cell death

(Fig. 3B). The combination of Fas-Fc and anti-TNF together blocked all cell loss in both wild-type and *lpr* T cells. Similar effects were observed using T cells derived from *gld* mice. Control experiments showed that apoptosis was not prevented by control serumnteractions alone could not abrogate TCR-induced wild-type T cell apoptosis, suggesting that neither of the pathways is sufficient for all T cell death. However, most, if not all, T cell apoptosis was abrogated by blocking TNF in *lpr* or *gld* T cells indicating that the Fas and TNF pathways together played the most prominent role in propriocidal apoptosis of mature T cells.

Kinetic analyses showed that Fas-Fc, but not TNFR-Fc, could protect the TCR-induced apoptosis at 24 hours [9]. Nevertheless, either Fas-Fc or TNFR-Fc could substantially decrease, while both blocking agents together could almost completely abrogate, the TCR-induced death at 48 or 72 hours. The overall apoptosis was greater at these later stages. Interestingly, only partial protection could be obtained with either Fas-Fc or TNFR-Fc alone, which means that certain T cell populations may be sensitive only to FasL or TNF but not both.

2.4 The p75 TNFR2 Can Mediate TCR-Induced Apoptosis Signal Transduction

TNFs exert their function by binding either p55 (TNFR1) which has an intracytoplasmic signaling "death domain" highly homologous with that of Fas, or p75 (TNFR2), which signals through a distinct intracytoplasmic domain [18]. Stimulation of TNFR1 by either TNF or anti-p55 crosslinking could transduce death signals in various kinds of cells, while p75 TNFR2 occupancy by TNF could induce NF-kB and other cytokine activities which usually accompanied with cellular activation processes [19], [20]. Evidence has been put forward both for and against a role for p75 in apoptosis (15–20). It has been postulated that TNFR2 does not signal itself but recruits TNF for p55 and enhances p55-mediated signaling [21]. However, several reports have described mAbs against p75 could initiate signaling directly and there is evidence that p75 can directly initiate apoptosis [22], [23], [24]. To definitively assess the contributions of the p55 and p75 TNFRs to T cell death we constructed mice bearing homozygous deficiencies in each TNFR and bred them to obtain mice with deficiencies in both TNFRs (Fig. 3). In each of the TNFR-deficient strains, TCR-induced T cell death at 24 hrs could be blocked by Fas:Fc, but barely by TNFR:Fc, confirming that Fas causes rapid apoptosis. However, after 48 hrs, the fractions of apoptotic cells from $(B6x129)F_1$ or $p55^{-/-}$ mice were 51.8% and 41.6%, and were decreased to 37.9% and 25.5% by adding TNFR:Fc alone; to 19.4 and 12.5% with Fas:Fc alone; and to 14.4% and 9.0% with both TNFR:Fc and Fas:Fc. Thus, both TNF and FasL cause death in T cells from wild-type or $p55^{-/-}$ mice. In T cells from $p75^{-/-}$ or double deficient $(p55^{-/-}/p75^{-/-})$ mice, apoptosis was decreased by Fas:Fc but not TNFR:Fc , suggesting that a deficiency of the p75 TNFR essentially abolished TNF-mediated T cell apoptosis.

We also assessed surface expression of TNFRs on T cells. Flow cytometry revealed binding of biotinylated TNF to both wild-type and $p55^{-/-}$ T cells, but almost no binding to $p75^{-/-}$ or double deficient T cells (data not shown). By using specific anti-TNFR1 and anti-TNFR2 monoclonal antibodies, we found that p75 can be readily detected on activated wild-type T cells whereas there is no detectable p55 TNFR despite the fact that control L929 fibroblast cells can be readily stained with the same mAb against the p55 TNFR (Fig. 3B). Despite poor surface expression, p55 mRNA was detected in activated T cells (data not shown). Thus, although we cannot exclude a role for p55 in the induction of apoptosis in wild-type T cells, we find that it is not necessary and may not be sufficient. In summary, we conclude: 1) TNF interactions with TNFRs participate in TCR-induced apoptosis, 2) the p75 receptor, which has no "death domain" or intracytoplasmic homology to Fas, is sufficient to

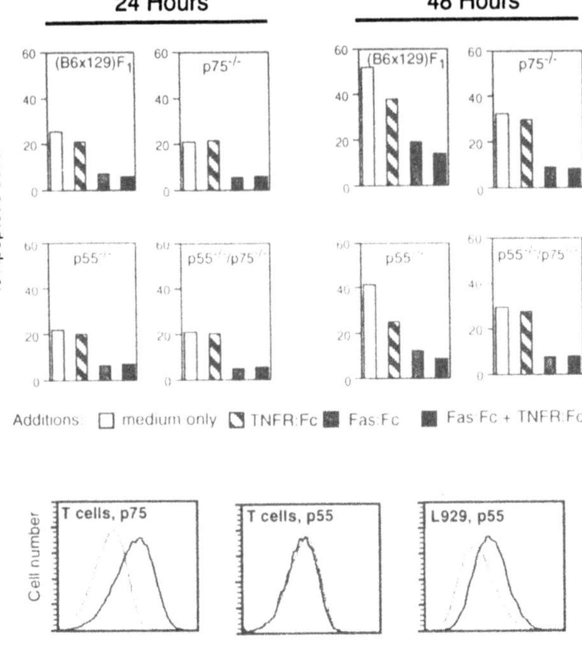

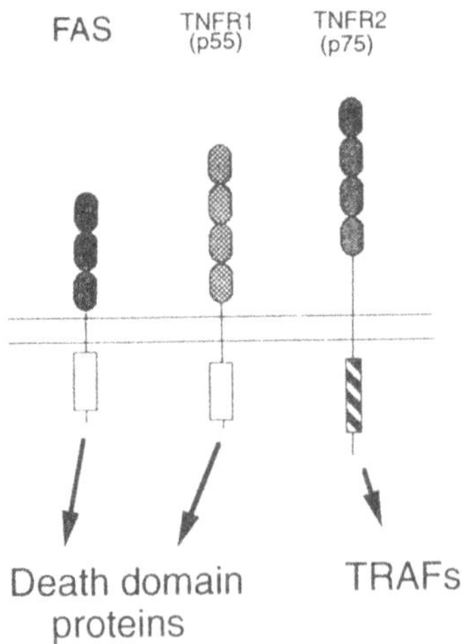

Figure 3. Mature T cell apoptosis induced b y TNF is mediated by the Type 2 TNFR. (Top) Percent apoptotic cells after TCR cross-linking in lymph node T cells derived from mouse strains that are either wild-type or homozygously deficient in the p55 or p75 TNFR with either Fas:Fc or TNFR:Fc blocking agents as indicated. At the bottom is staining T cells or L929 cells with mAbs for the p55 and p75 TNFRs. (Bottom) Schematic of Fas and the TNFRs. Shaded ovals shown the cysteine-rich domains that are involved in ligand binding and define membership in the TNFR superfamily. Open rectangle is the intracytoplasmic death domain and hatched rectangle is the TRAF-binding domain. Further experimental details can be found in Zheng et al.[9]. Previously published data used with permission of MacMillan Press Ltd.

mediate T cell death, and 3) Fas has a quantitatively greater effect on T cell apoptosis at earlier times than TNF. It will be interesting to understand the molecular pathway that TNFR2 employs to initiate apoptosis of T cells. Recently, Rothe et al. reported two proteins, cIAP1 and cIAP2 that could bind TRAF1/TRAF2 heterodimeric complex that associates with p75 and that have homology with genes that control apoptosis in viruses, Drosophila, and mammalian neurons [25]. The involvement of cIAP proteins in signal transduction for either T cell activation or apoptosis remains to be clarified.

2.5 Susceptibility of TCR-Induced Apoptosis in T Cell Subsets

As mentioned above, results from blocking experiment showed that by using either Fas-Fc or TNFR-Fc alone, substantial number of T cells could be protected from TCR-induced apoptosis, suggesting that certain T cell subpopulations may be only sensitive to FasL or TNF but not both (Fig. 2). To explore this phenomenon further, we analyzed the susceptibility of CD4+ and CD8+ T cell subsets to apoptosis induced by TCR crosslinking [9]. In wild-type controls, both CD4+ and CD8+ subsets underwent dramatic apoptosis although CD8+ cells were a little less sensitive than CD4+ cells. Interestingly, we found the *gld* mutation of FasL almost completely blocked the sorted CD4+ T cells from TCR-induced apoptosis but was incapable of preventing apoptosis for approximately 70% of CD8+ T cells (Fig.4A). Similar results were obtained in *lpr* system. By contrast, anti-TNF had little effect on CD4+ cells, but substantially protected CD8+ cells (Fig. 4B). These results suggest that T cell subpopulations may employ mechanisms that are kinetically and molecularly distinct to sense and respond to the feedback control signals by undergoing apoptosis. The biological significance of this observation is not completely clear at the present time. However, we would like to propose a model in which CD4+ helper T cells, may be activated and carry out their physiological functions first, and need to be controlled earlier and faster than the forthcoming effector cells. The rapid FasL/Fas-mediated apoptosis pathway could provide optimal regulation in this case. On the other hand, by using TNF-mediated apoptosis pathway which seems to be kinetically delayed as compared to that of FasL, a time lag is available for the CD8+ effector T cells, to ensure their proliferation and exertion of cytolytic functions before self-regulatory apoptosis is ignited.

2.6 Genetic Abnormalities in Fas Cause an Autoimmune/ Lymphopro-Liferative Syndrome

The association of Fas/FasL defects with an autoimmunity and lymphoid hyperplasia in *lpr* and *gld* mice prompted consideration of the role of these molecules in human diseases. Based on studies carried out by Sneller, Strober, and Straus, a group of 5 unrelated patients was identified that exhibited massive enlargement of the secondary lymphoid tissues and autoimmune symptoms such as vasculitis, autoantibodies, and immune-mediated blood cell destruction [26] This findings were present in children between the ages of 2 months and 5 years of age. Each of the patients exhibited large increases of circulating $\alpha\beta$T lymphocytes that lacked expression of either the CD4 or the CD8 co-receptor. When the T cells derived from such patients were treated with IL-2 and then rechallenged with TCR ligation, it was found that the patients' cells exhibited substantially less apoptosis compared to controls (Fig. 5) [27]. To determine the cause of this apoptotic defect, genetic analyses were carried out in collaboration with the laboratory of Dr. Jennifer Puck of the National Center for Human Genome Research. In her laboratory, it was found that each of the abnormal children had various mutations of the Fas gene that would be predicted to impair the function of this receptor [27].

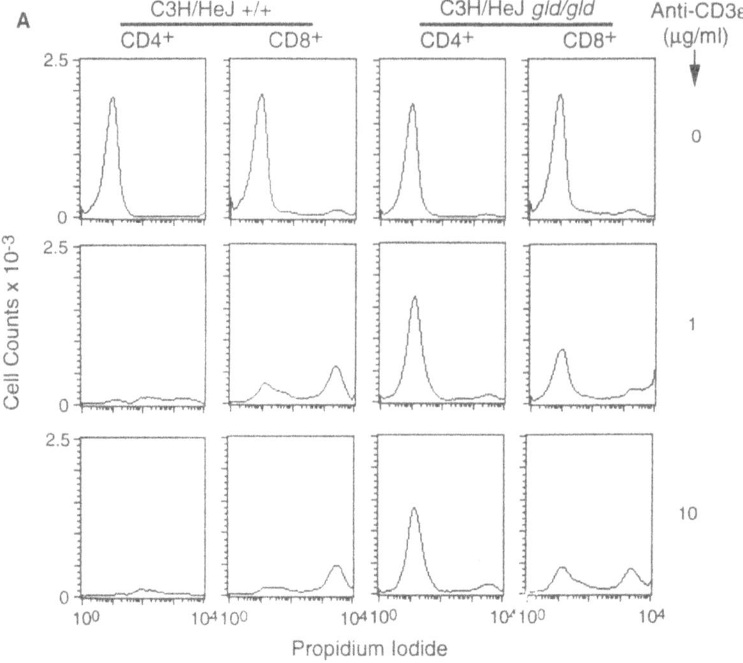

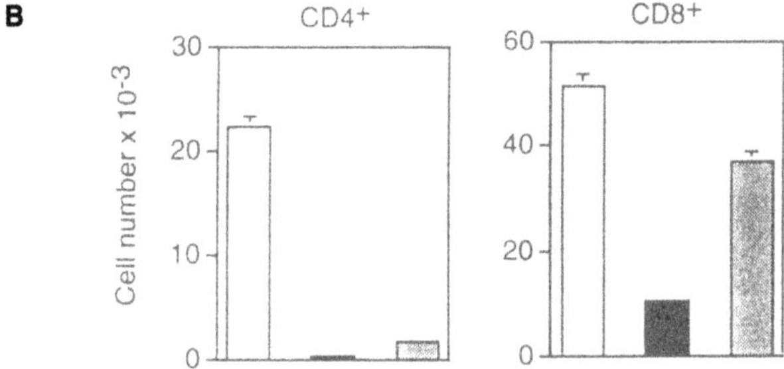

Figure 4. Fas mediates mostly CD4$^+$ T cell death whereas TNF mediates mostly CD8$^+$ T cell death. (A) Viable cell recovery of CD4$^+$ and CD8$^+$ T cells from wild-type or *gld* mice after TCR cross-linking. (B) Viable cell recovery of CD4$^+$ and CD8$^+$ T cells from wild-type mice after no treatment (open bars), TCR re-stimulation (black bars), or TCR re-stimulation together with anti-TNF (grey bars). Further experimental details can be found in Zheng et al.[9]. Previously published data used with permission of MacMillan Press Ltd.

We next investigated the function of the abnormal Fas alleles by transfecting expression constructs containing cDNA versions of the mutants into a mouse cell line and determining the level of apoptosis induced by anti-Fas antibody. These experiments showed that each of the identified mutations impaired the death-inducing function of the Fas molecules (Fig.6). Further genetic testing by Dr. Puck's laboratory unveiled the surprising finding that in all five patients, the mutant alleles were heterozygous with a normal allele. This lead to the hypothesis that the Fas mutations might operate in a dominant negative fashion in four of the patients. In the fifth patient, the mutation created an early translational stop in the Fas coding sequence and therefore no dominant-interfering Fas protein could be

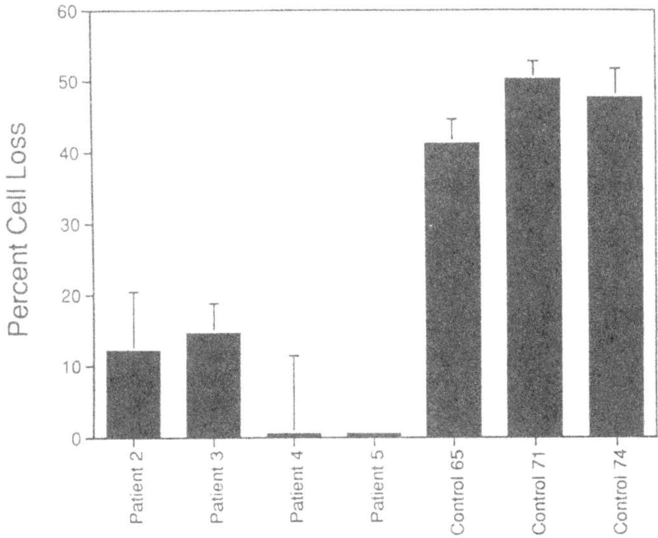

Figure 5. Peripheral blood T cells from ALPS patients show reduced apoptosis after TCR cross-linking. IL-2 treated peripheral blood T cells were re-stimulated with anti-CD3 for 18–24 hours and the percent cell loss was determined flow cytometrically. Further experimental details can be found in Fisher et al. [9]. Previously published data used with permission of Cell Press.

made. We confirmed dominant negative effects by co-transfecting expression plasmids for mutant Fas alleles and wild-type Fas alleles into mouse cells. After cross-linking with anti-Fas antibody, we observed that the mutant Fas interfered with death signalling through normal Fas in a dose-response manner. Given the trimeric structure of Fas, only 1/8 of the Fas complexes would contain only wild-type Fas subunit, if it can be assumed that both the wild-type and mutant Fas alleles are expressed equally. This could account for the large

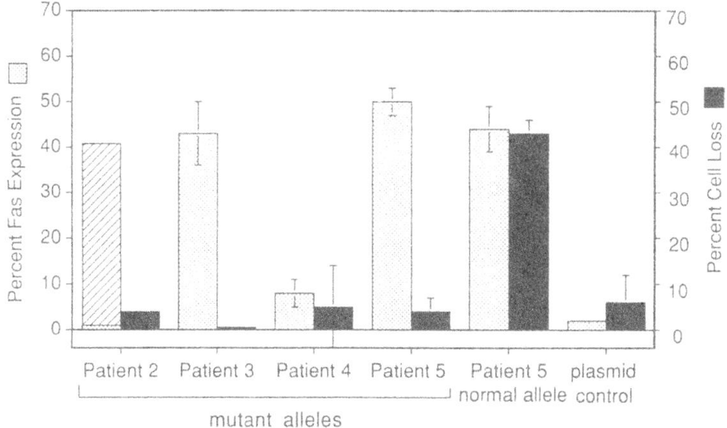

Figure 6. Transfection of constructs expressing the mutant Fas alleles reveal that the Fas mutations are defective for apontal details can be found in Fisher et al. [9]. Previously published data used with permission of Cell Press.

reduction of TCR-mediated apoptosis in the patients T cells. This constellation of clinical, pathological, and molecular findings have been named the autoimmune/lymphoproliferative syndrome (ALPS)[27] and may account for earlier case reports of pediatric diseases comprising autoimmunity and lymphoproliferation [28].

3. CONCLUSIONS

The involvement of the same antigen receptor complex in T cell proliferation and death is dictated by the requirement for clonal specificity of these processes. But the dual nature of this receptor's response raises a paradox. Under what conditions should a strong antigenic stimulus, for example that due to an infectious agent, induce the programmed death of the T cells that are specifically designed to respond to that infectious agent? The involvement of the powerful T cell proliferative cytokine, IL-2, as an essential component of the death pathway sheds light on this question. It suggests that the death pathway is designed to limit T cell proliferation to avoid the toxicities and potential autoimmune complications associated with a large expansion of activated T cells. Recent genetic experiments in mice together with the molecular analyses of ALPS patients have shown that the propriocidal mechanism play a critical role in mature T lymphocyte homeostasis as well as peripheral tolerance. Further elucidation of the molecular pathway from the TCR to apoptosis should shed light on how this receptor complex can control the diametrically opposed outcomes of proliferation and death. Moreover, molecular genetic analyses of more autoimmune/lympho-proliferative illnesses in man should reveal critical molecular controls of the lymphocyte homeostasis. Finally, for nearly 2 centuries, vaccination has been used to elicit specific protective immune responses. We can now foresee employing our understanding of antigen-specific T cell apoptosis in an attempt to generate antigen-specific immunological tolerance to alleviate autoimmune, allergic, and allograft rejection diseases.

ACKNOWLEDGMENTS

We would like to thank our close collaborators and colleagues who generously contributed to the studies described in this summary: David Lynch, Jacques Peschon, Robert Miller for the work on TNF and Stephen Straus, Warren Strober, Jennifer Puck, Michael Sneller, Fred Rosenberg, Lindsay Middleton, and Janet Dale on the ALPS patients. Galen Fisher and David Martin were supported by the Howard Hughes Medical Institutes-NIH Research Scholars Program.

REFERENCES

1. Lenardo MJ. Interleukin-2 programs mouse alpha beta T lymphocytes for apoptosis. Nature 1991;353(6347):858–61.
2. Critchfield JM, Racke MK, Zuniga-Pflucker JC, et al. T cell deletion in high antigen dose therapy of autoimmune encephalomyelitis. Science 1994;263(5150):1139–43.
3. Critchfield JM, Boehme SA, Lenardo MJ. The regulation of antigen-induced apoptosis in mature T lymphocytes. Apoptosis and the Immune Response 1995;Wiley-Liss, Inc.:55–114.
4. Lenardo MJ, Boehme SA, Chen L, et al. Autocrine feedback death and the regulation of mature T lymphocyte antigen responses. Intern. Rev. Immunol. 1995;13:115–134.
5. Willerford DM, Chen J, Ferry JA, Davidson L, Ma A, Alt FW. Interleukin-2 receptor alpha chain regulates the size and content of the peripheral lymphoid compartment. Immunity 1995;3(4):521–30.

6. Kneitz B, Herrmann T, Yonehara S, Schimpl A. Normal clonal expansion but impaired Fas-mediated cell death and anergy induction in interleukin-2-deficient mice. Eur J Immunol 1995;25(9):2572–7.
7. Adachi M, Watanabe-Fukunaga R, Nagata S. Aberrant transcription caused by the insertion of an early transposable element in an intron of the Fas antigen gene of lpr mice. Proc Natl Acad Sci U S A 1993;90(5):1756–60.
8. Takahashi T, Tanaka M, Brannan CI, et al. Generalized lymphoproliferative disease in mice, caused by a point mutation in the Fas ligand. Cell 1994;76(6):969–76.
9. Zheng L, Fisher G, Miller RE, Peschon J, Lynch DH, Lenardo MJ. Induction of apoptosis in mature T cells by tumour necrosis factor. Nature 1995;377(6547):348–51.
10. Dhein J, Walczak H, Baumler C, Debatin KM, Krammer PH. Autocrine T-cell suicide mediated by APO-1/(Fas/CD95) [see comments]. Nature 1995;373(6513):438–41.
11. Brunner T, Mogil RJ, LaFace D, et al. Cell-autonomous Fas (CD95)/Fas-ligand interaction mediates activation-induced apoptosis in T-cell hybridomas [see comments]. Nature 1995;373(6513):441–4.
12. Ju ST, Panka DJ, Cui H, et al. Fas(CD95)/FasL interactions required for programmed cell death after T-cell activation [see comments]. Nature 1995;373(6513):444–8.
13. Smith CA, Farrah T, Goodwin RG. The TNF receptor superfamily of cellular and viral proteins: activation, costimulation, and death. Cell 1994;76(6):959–62.
14. Nagata S, Golstein P. The Fas death factor. Science 1995;267(5203):1449–56.
15. Beyaert R, Fiers W. Molecular mechanisms of tumor necrosis factor-induced cytotoxicity. What we do understand and what we do not. Febs Lett 1994;340(1–2):9–16.
16. Reid T, Louie P, Heller RA. Mechanisms of tumor necrosis factor cytotoxicity and the cytotoxic signals transduced by the p75-tumor necrosis factor receptor. Circ Shock 1994;44(2):84–90.
17. Grell M, Zimmermann G, Hulser D, Pfizenmaier K, Scheurich P. TNF receptors TR60 and TR80 can mediate apoptosis via induction of distinct signal pathways. J Immunol 1994;153(5):1963–72.
18. Rothe M, Wong SC, Henzel WJ, Goeddel DV. A novel family of putative signal transducers associated with the cytoplasmic domain of the 75 kDa tumor necrosis factor receptor. Cell 1994;78(4):681–92.
19. Tartaglia LA, Weber RF, Figari IS, Reynolds C, Palladino M, Jr., Goeddel DV. The two different receptors for tumor necrosis factor mediate distinct cellular responses. Proc Natl Acad Sci U S A 1991;88(20):9292–6.
20. Rothe M, Sarma V, Dixit VM, Goeddel DV. TRAF2-mediated activation of NF-kappa B by TNF receptor 2 and CD40. Science 1995;269(5229):1424–7.
21. Tartaglia LA, Pennica D, Goeddel DV. Ligand passing: the 75-kDa tumor necrosis factor (TNF) receptor recruits TNF for signaling by the 55-kDa TNF receptor. J Biol Chem 1993;268(25):18542–8.
22. Vandenabeele P, Declercq W, Vanhaesebroeck B, Grooten J, Fiers W. Both TNF receptors are required for TNF-mediated induction of apoptosis in PC60 cells. J Immunol 1995;154(6):2904–13.
23. Grell M, Scheurich P, Meager A, Pfizenmaier K. TR60 and TR80 tumor necrosis factor (TNF)-receptors can independently mediate cytolysis. Lymphokine Cytokine Res 1993;12(3):143–8.
24. Bigda J, Beletsky I, Brakebusch C, et al. Dual role of the p75 tumor necrosis factor (TNF) receptor in TNF cytotoxicity. J Exp Med 1994;180(2):445–60.
25. Rothe M, Pan M-G, Henzel WJ, Ayres TM, Goeddel DV. The TNFR2-TRAF signaling complex contains two novel proteins related to baculoviral inhibitor of apoptosis proteins. Cell 1995;83(7):1243–1252.
26. Sneller MC, Straus SE, Jaffe ES, et al. A novel lymphoproliferative/autoimmune syndrome resembling murine lpr/gld disease. The Journal of Clinical Investigation 1992;90:334–341.
27. Fisher GH, Rosenberg FJ, Straus SE, et al. Dominant interfering Fas gene mutations impair apoptosis in a human autoimmune lymphoproliferative syndrome. Cell 1995;81(6):935–46.
28. Weisdorf SA, Krivit W. Paucity of splenic germinal centers: A new and unique splenomegaly syndrome including dysfunctional immune system. Clincal Immunology and Immunopathology 1982;23:492–500.

25

AUTOIMMUNITY DUE TO DEFECTIVE NUR77, FAS, AND TNF-RI APOPTOSIS

John D. Mountz,[1,2] Carl K. Edwards III,[3] Jianhua Cheng,[1,2] Pingar Yang,[1] Zheng Wang,[1] Changdan Liu,[1] Xiao Su,[1] Horst Bluethmann,[4] and Tong Zhou[1]

[1] University of Alabama at Birmingham, Department of Medicine Division of Clinical Immunology and Rheumatology, Birmingham, Alabama 35294
[2] The Birmingham Veterans Administration Medical Center Birmingham, Alabama 35233
[3] Amgen Boulder, Inc., Department of Inflammation Boulder, CO 80301
[4] F. Hoffmann-LaRoche, Ltd., Pharmaceutical Research Gene Technologies Basel, Switzerland

1. PATHWAYS FOR NEGATIVE SELECTION OF THYMOCYTES BY APOPTOSIS*

1.1. Fas and p53 Mutant Mice

Several molecules and pathways known to be of importance in apoptosis have been described in the thymus; however, their contribution to clonal deletion and tolerance induction remains controversial[1–4]. Although knockout of p53 leads to decreased sensitivity of murine thymocytes to radiation-induced apoptosis, negative selection remains intact[5–7]. Fas is a cell surface receptor that mediates apoptosis by interaction with a specific ligand and is expressed on most murine thymocytes[8–11]. Although mutant Fas antigen and Fas ligand cause autoimmune disease in *lpr/lpr* and *gld/gld* mice, respectively[10–12], no major negative selection defects have been found in *lpr/lpr* mice[13–17]. Therefore, it is unlikely that Fas

*Abbreviations used in this paper: Tg71: TCR transgenic mice expressing the D^b/HY TCR; M33: D^b/HY TCR anticlonotypic Mab; BrdU: bromodeoxyuridine; TUNEL: Terminal digoxigenin nucleotide end labeling; NGFI-B: nerve growth factor I-B; NBRE: NGFI-B response element; ΔNur77: Dominant negative mutation of Nur77; TINUR: Transcriptionally inducible nuclear receptor; AICD: activation induced cell death; 7-AAD: 7-amino actinomycin D; FasFP: Fas fusion protein; TNF-RI: p55 TNF receptor; $Tnfrl^{0/0}$: homozygous p55 TNF receptor mutant mice; $Tnfrl^{0/0}$-*lpr/lpr*: homozygous p55 TNF receptor mutant *lpr/lpr* mice.

Mechanisms of Lymphocyte Activation and Immune Regulation VI, Edited by Gupta and Cohen
Plenum Press, New York, 1996

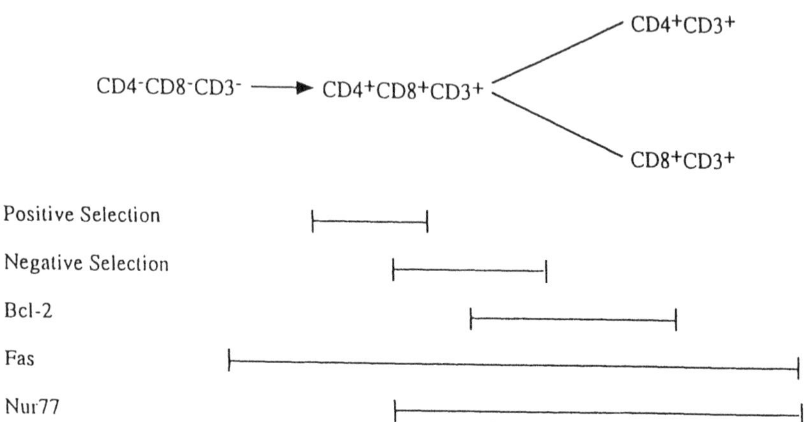

Figure 1. Expression of Nur77, Fas and *bcl-2* relative to positive and negative selection. Positive selection begins after CD3 expression on $CD4^+CD8^+$ thymocytes. Negative selection is thought to occur after thymocytes express high levels of CD3. Fas is expressed at low levels on $CD4^-CD8^-CD3^-$ thymocytes, but is upregulated on the earliest $CD4^+CD8^+CD3^+$ thymocytes. This is associated with high sensitivity of thymocytes to Fas apoptosis. Nur77 is expressed on $CD4^+CD8^+CD3^+$ thymocytes and is further increased after engagement of the TCR/CD3 molecule by MCH + Ag. *Bcl-2* is expressed on later stage $CD4^+CD8^+CD3^{bright}$ thymocytes and is correlated with resistance of thymocytes to apoptosis.

antigen is directly involved in negative selection in the thymus, but may be involved in apoptosis during early T cell development in the thymus. We have previously proposed that Fas expression duirng early thymocyte development plays a role in positive selection or pre-positive selection of thymocytes (Figure 1).

1.2. *Bcl-2* Mutant Mice

Bcl-2 prevents thymocyte apoptosis that is induced by radiation, steroids, and other chemicals[18–20]. Expression of *bcl-2* has been reported to be decreased in $CD4^+8^+$ thymocytes, but not in mature thymocytes, and has been proposed to play a role in inhibition of negative selection in the thymus[21]. However, *bcl-2* knockout mice do not exhibit excessive clonal deletion in the thymus[22] and, conversely, *bcl-2* transgenic mice do not exhibit a major defect in negative selection or T cell tolerance[23–25]. Double transgenic *bcl-2*+D^b/HY TCR mice show that constitutive expression of *bcl-2* increases the survival of thymocytes in the absence of positive selection[26–28]. The *bcl-2* transgene also reduces the efficiency of negative selection, but the mature peripheral T cells which appear in increased numbers were not autoreactive. Thus, although *bcl-2* can play a role in both positive and negative selection, tolerance is maintained by a mechanism that can bypass *bcl-2*.

1.3. Nur77 and Thymocyte Apoptosis

NGFI-B/Nur77 is a growth factor-inducible member of the steroid/thyroid hormone receptor superfamily originally identified in nerve growth factor (NGF)-treated P12 pheochromocytoma cells[29] and in serum-stimulated fibroblasts[30]. NGFI-B/Nur77 is transcriptionally regulated as an immediate-early gene and is rapidly activated by phosphorylation after stimulation with serum or nerve growth factor[31, 32]. NGFI-B/Nur77 has a centrally located, highly conserved DNA binding domain containing two zinc-fingers and a transcriptional trans-activating domain[33–37]. NGFI-B/Nur77 gene is expressed in thymic medulla and is

Table I. Classification of NGFI-B/Nur77 Family of Genes[42]

Type	Human	Mouse	Rat
α	NAK-1 (TR3)	Nur77 (N10)	NGFI-B (TIS1)
β	TINUR	Nurr1	RNF-1

rapidly upregulated in T cell hybridomas and thymocytes after treatment with anti-CD3 or anti-TCR and this expression has been correlated with induction of apoptosis[38, 39]. Blocking of NGFI-B with either a dominate negative truncated[38] or antisense[39] NGFI-B/Nur77 gene prevented TCR/CD3 signaling-mediated apoptosis in T cell hybridomas. There are at least two gene families with an identical NGFI-B response element (NBRE) AAAGGTCA[40]. The first member of this family, referred to as Nur77 (mouse), NGFI-B (rat), and NAK-1 (human) peaks 1 hr after stimulation of the PEER T cell line. The second member, referred to as Nurr1 (mouse), RNR-1 (rat), and TINUR (human) peaks 24 hrs after stimulation and correlates with apoptosis (Table I).

1.4. Steroids and Thymocyte Apoptosis

Glucocorticoids are a potent means of inducing thymocyte apoptosis. The thymocytes are most sensitive to the lethal effects of glucocorticoids are the same that undergo selection: $CD4^+CD8^+$ immature cells located in the cortex; medullary $CD4^+CD8^-$ and $CD4^-CD8^+$ cells, like peripheral T cells, are relatively resistant. Both TCR/CD3 and glucocorticoid signaling are lethal by themselves, but glucocorticoids and TCR-mediated signals prevent each other's induction of T cell hybridoma and thymocyte programmed cell death *in vitro* (mutual antagonism). Cultured thymic non-T cells have been shown to produce soluble pregnenolone and deoxycorticosterone, and immunohistochemistry demonstrated steroidogenic enzymes in radioresistant thymic epithelial cells but not in thymocytes[41, 42] Inhibition of thymic corticosterone production or blockade of the glucocorticoid receptor with RU-486 resulted in enhanced TCR-mediated, antigen-specific deletion of immature thymocytes. These data indicate that locally produced glucocorticoids, because of their antagonism of TCR-mediated signaling for death, may be one of the key elements of antigen-specific thymocyte selection.

1.5. Nur77 Mutant Mice

Although Nur77 has been shown to be important in T cell signaling and apoptosis[38, 39], other signaling proteins can also contribute to T cell maturation and apoptosis. This was recently demonstrated in a Nur77 mutant mouse, in which T cells did not exhibit defective apoptosis after anti-CD3 crosslinking and exhibit normal T cell development and apoptosis in D^b/HY TCR and in class II MHC-restricted pigeon cytochrome TCR transgenic mice[43]. The results indicate that the truncated form of ΔNur77 gene binds to the NBRE and inhibits TCR/CD3 signaling-mediated apoptosis and also interferes with selection and clonal deletion in the thymus of ΔNur77-D^b/HY TCR double transgenic mice.

2. TRANSGENIC MICE EXPRESSING A DOMINANT NEGATIVE MUTATION OF NUR77 (ΔNUR77)

2.1. Production of ΔNur77 Transgenic Mice

Six lines of mice carrying ΔNur77 transgenic DNA were produced. These mice express a truncated ΔNur77 RNA transcript under the regulation of a TCR Cβ1 minigene[44,

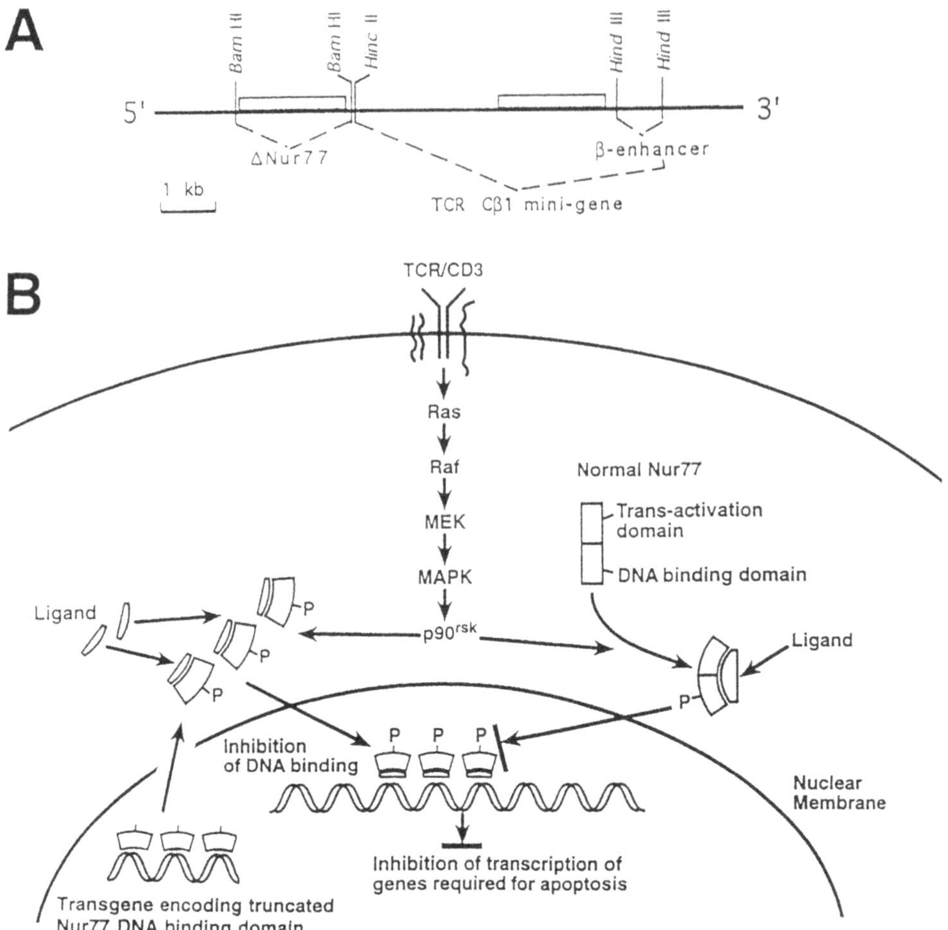

Figure 2. Production of transgenic mice expressing the DNA binding domain of Nur77. **A)** A 1.3 kb DNA binding portion of the Nur77 gene was cloned upstream of the TCR CTβ minigene, including the CTβ promoter and enhancer. This resulted in high levels of expression of the ΔNur77 RNA in TCR-α/β^{+} T cells. **B)** Inhibition of binding of normal Nur77 by the truncated DNA binding domain of Nur77. The transgene encoding the Nur77 DNA binding domain produces high levels of the truncated Nur77 DNA binding domain which is activated after phosphorylation and also binds an unidentified steroid ligand. After translocation to the nucleus, the DNA binding domain binds to the NGFI-B response element (NBRE), AAAGGTCA. This prevents binding of the functional Nur77 activated peptide and prevents transactivation and gene expression by Nur77.

[45] (Figure 2A). The translation product in the ΔNur77 DNA-binding protein is specifically expressed in the T cells of the transgenic mice, and that the truncated ΔNur77 protein produced in ΔNur77 transgenic mice competitively inhibits the binding of endogenous Nur77 to the NBRE (Figure 2B).

2.2. Inhibition of Anti-TCR/CD3 Mediated Thymocyte Apoptosis in ΔNur77 Transgenic Mice

There was no significant abnormality in the development of thymocytes in ΔNur77 transgenic mice (Figure 3). There were normal numbers of cells in each thymocyte popula-

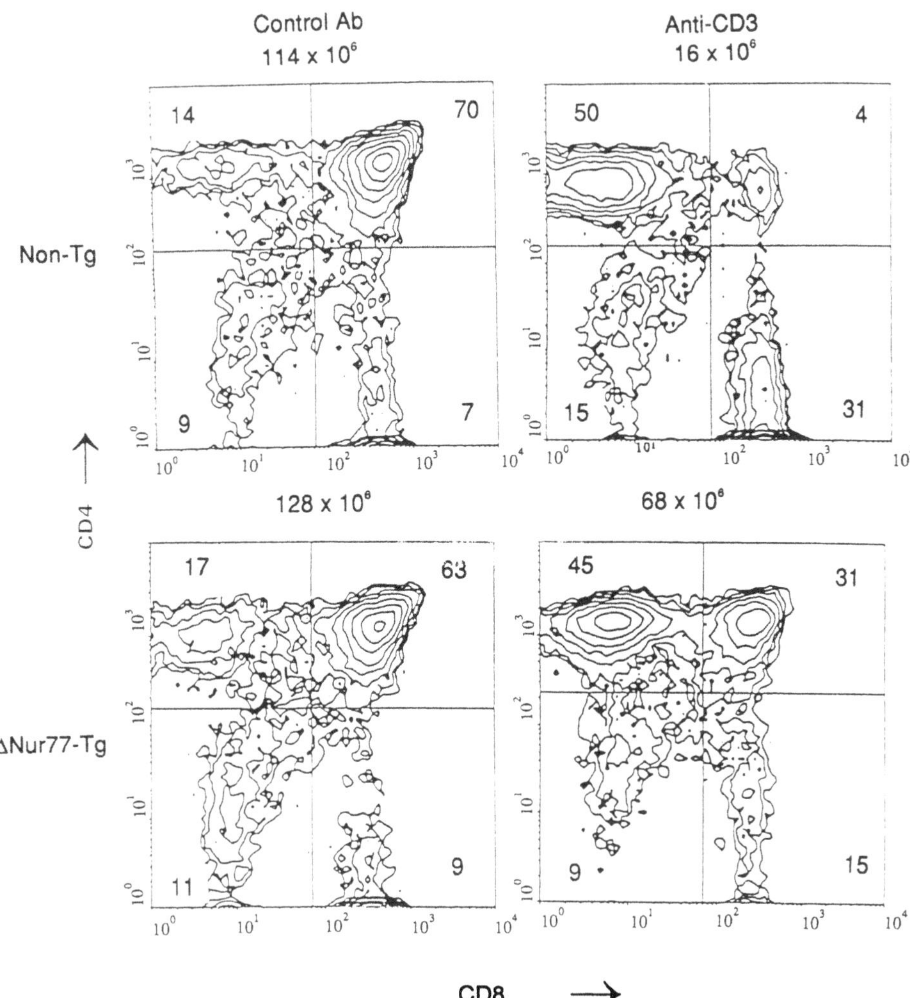

Figure 3. Inhibition of anti-CD3 apoptosis in ΔNur77 transgenic mice. Non-transgenic or ΔNur77 transgenic mice were treated with either control antibody or anti-CD3 *in vivo*. Thymocyte single-cell suspensions were analyzed by flow cytometry analysis for expression of CD4 and CD8. In non-transgenic mice there was a decrease in total number of thymocytes as indicated above the flow cytometry contour figure. There was a depletion of the $CD4^+CD8^+$ thymocytes. ΔNur77 transgenic mice exhibited a lessor decrease in total number of thymocytes. The $CD4^+CD8^+$ thymocytes were relatively resistant to depletion after anti-CD3 treatment.

tions in ΔNur77-transgenic mice after treatment with control antibody. Anti-CD3 antibody treatment led to depletion of 86% of the total and 99% of the double positive (DP) thymocytes in non-transgenic mice and reduced the percentage of DP thymocytes to 4%. In contrast, anti-CD3-induced deletion of thymocytes was less efficient in ΔNur77 transgenic mice, with depletion of 50% of total and 63% of DP thymocytes and reduced the percentage of DP thymocytes to 31%. Depletion of $CD4^+$ and $CD8^+$ thymocytes was also inhibited after CD3 antibody treatment of ΔNur77 transgenic mice, whereas deletion of double negative (DN) thymocytes was equivalent in both non-transgenic and ΔNur77-transgenic mice. The preferential inhibition of apoptosis of $CD4^+8^+$ thymocyte by blocking the NBRE, combined with previous results that Nur77 knockout does not inhibit apoptosis of $CD4^+8^+$ thymocytes support the conclusion that other factors that interact with the NBRE such as Nurr1/TINUR plays a role in apoptosis of thymocytes.

2.3. Decreased Negative Selection of Thymocytes in ΔNur77-D^b/HY TCR Double Transgenic Male Mice

The D^b/HY TCR transgenic male mouse has been extensively analyzed as a model for analysis of positive and negative selection[46–50]. The D^b/HY reactive thymocytes exhibit extensive positive selection in D^b/HY TCR transgenic female mice and are extensively deleted at the DP stage of thymocyte development in D^b/HY TCR transgenic male mice. To determine whether interruption of Nur77/Nurr1 function in ΔNur77 transgenic mice leads to defective negative selection of thymocytes, H-2^b ΔNur77 transgenic mice were back-crossed to D^b/HY TCR transgenic C57BL/6 mice.

D^b/HY TCR transgenic male mice exhibited a 10-fold decrease in total thymocyte number and nearly complete deletion of DP and $CD8^{bright}$ thymocytes due to negative selection of thymocytes bearing the transgenic TCR. Several changes were observed in the thymus of the C57BL/6 ΔNur77-D^b/HY TCR double transgenic male mice compared to that of D^b/HY TCR transgenic male mice. First, there was a 4-fold increase in total thymocytes that comprised a 20-fold increase in $CD4^{low}CD8^{low}$ DP thymocytes and a 5-fold increase in CD8 thymocytes (Figure 4). Second, increased numbers of DP and CD8 thymocytes in ΔNur77-D^b/HY TCR transgenic male mice expressed the transgenic TCR. 70% of $CD4^+CD8^+$ thymocytes were also M33 positive in the ΔNur77-Tg71 double transgenic male compared to 17% in D^b/HY TCR transgenic male mice. Third, in ΔNur77-D^b/HY TCR

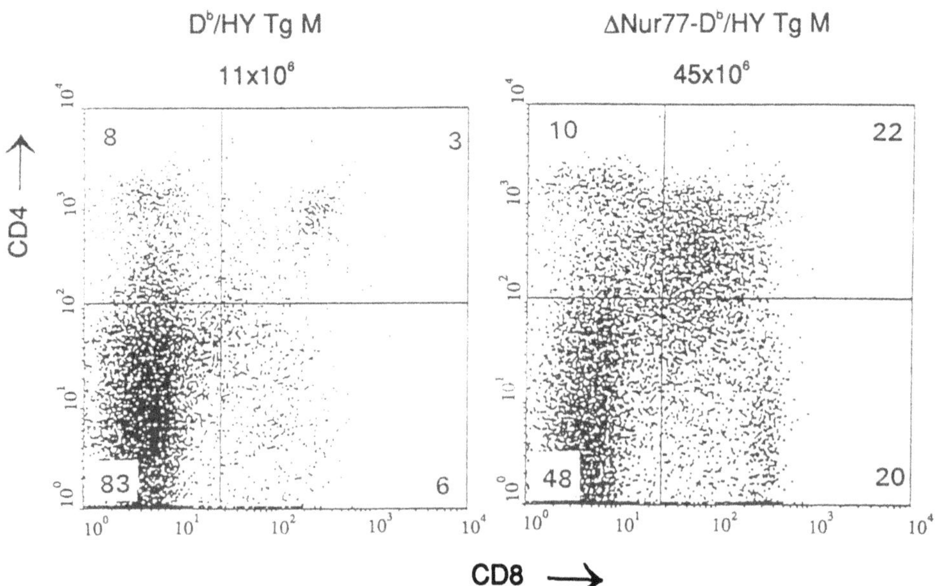

Figure 4. Defective negative selection in ΔNur77-D^b/HY transgenic male mice. D^b/HY T cell receptor transgenic mice were crossed with the ΔNur77 transgenic mice. Thymocytes were analyzed for expression of CD4 and CD8. In D^b/HY T cell receptor transgenic male mice, most thymocytes were $CD4^-CD8^-$ and very few thymocytes progressed to the $CD4^+CD8^+$ stage of development. In the ΔNur77-D^b/HY T cell receptor transgenic male mice, there was a significant increase in $CD4^{dull}CD8^{dull}$ thymocytes. In addition, there was a significant increase in the percentage of $CD4^+$ and $CD8^+$ mature thymocytes.

transgenic male mice, there was the appearance of a substantial number of CD8 thymocytes that exhibited high levels of expression of CD8; 70–80% of these thymocytes expressed the transgenic TCR. This population was absent in the D^b/HY male mice. Taken together, these results indicate that negative selection of the autospecific D^b/HY reactive thymocytes was inhibited in ΔNur77-D^b/HY TCR transgenic male mice compared to D^b/HY TCR transgenic male mice and that the family of DNA binding proteins including Nurr1 and Nur77 are directly involved in signaling of negative selection during T cell development in the thymus. It is highly likely that the NBRE might be one factor leading to regulation of apoptosis-related genes resulting in induction of genes such as *Bcl-Xs* and Bax that down-regulate apoptosis and decreased transcription of other genes such as *bcl-2* that inhibit apoptosis[51–54].

Another important pathway of apoptosis for thymocytes is dexamethasone-induced apoptosis[41, 42]. There is no defect of dexamethasone-induced apoptosis in the ΔNur77 transgenic mice which indicates that there is no competitive inhibition between the Nur77/Nurr1 orphan steroid receptors either at the level of cytoplasmic steroid binding or at the receptor-DNA binding site with glucocorticoids. This lack of inhibition between dexamethasone apoptosis and Nur77/Nurr1 apoptosis indicates that apoptosis is mediated by separate independent pathways. Taken together, there are multiple independent pathways of thymocyte apoptosis related to TCR/CD3 signaling that may be affected by inhibition of Nurr1/Nur77 interaction with NBRE.

2.4. Model for Inhibiton of Negative Selection of $CD8^+M33^+$ T Cells in the Thymus of ΔNur77 D^b/HY Transgenic Mice.

In C57BL/6-+/+ D^b/HY TCR transgenic male mice, thymocyte selection occurs under the influence of H2-D^b + the H-Y male antigen (Figure 5). In male mice, negative selection results in the failure to develop $CD4^+CD8^+$ cells and CD8 single positive cells in the thymus. In female mice, there is efficient selection of $CD8^+$ thymocytes expressing the

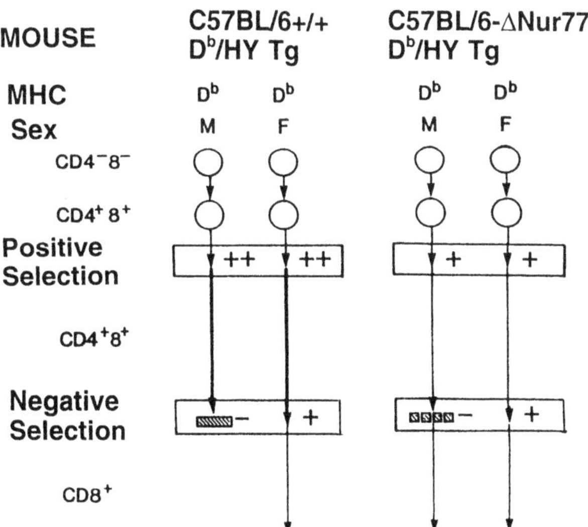

Figure 5. Model for positive and negative selection of $CD8^+$ T cells in the thymus of ΔNur77-D^b/HY transgenic mice. In C57BL/6-+/+ D^b/HY TCR transgenic male mice, thymocyte selection occurs under the influence of H2-D^b + the H-Y male antigen. In female mice, there is efficient selection of $CD8^+$ thymocytes expressing the D^b/HY T cell receptor. There is a significant defect in negative selection in C57BL/6-ΔNur77-D^b/HY transgenic male mice resulting in the development of $CD4^+CD8^+$ thymocytes and also mature $CD4^+M33^+$ and $CD8^+M33^+$ autoreactive thymocytes that escape to the periphery.

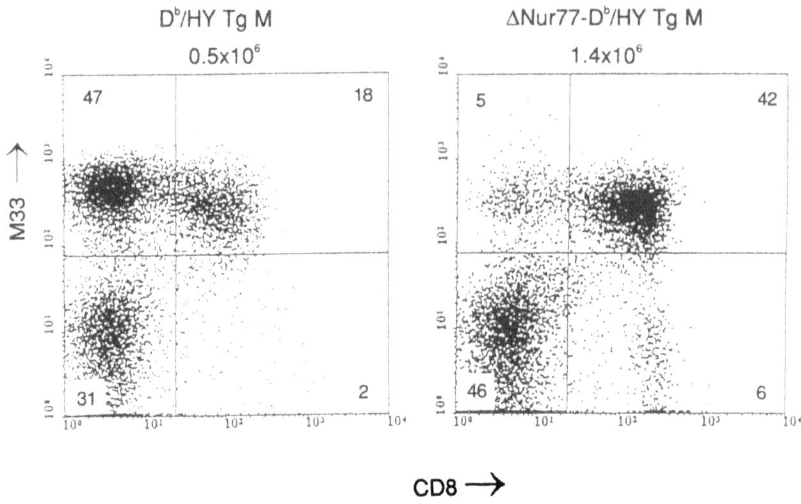

Figure 6. Increased autoreactive CD8$^+$M33$^+$ T Cells in LN of ΔNur77-D^b/HY transgenic male mice. Most of the T cells in D^b/HY transgenic male mice express the autospecific TCR recognized by the anti-clonotypic antibody M33. Approximately one-third of these T cells express low levels of CD8 and two-thirds downmodulate CD8 as a tolerance mechanism. In ΔNur77-D^b/HY transgenic male mice, almost all of the M33$^+$ T cells also express high levels of CD8.

D^b/HY T cell receptor. In C57BL/6-ΔNur77-D^b/HY transgenic mice, there is only a minimal decrease in positive selection despite inhibition of functional Nur77-DNA binding due to inhibition by the ΔNur77 transgene product. This result indicates that Nur77-DNA binding is not critical for positive selection. There is a significant defect in negative selection in C57BL/6-ΔNur77-D^b/HY transgenic male mice resulting in the development of CD4$^+$CD8$^+$ thymocytes and also mature CD4$^+$M33$^+$ and CD8$^+$M33$^+$ autoreactive thymocytes that escape to the periphery.

2.5. Increased M33$^+$CD8$^+$ Lymph Node T Cells in the ΔNur77-D^b/HY TCR Double Transgenic Male Mice

To determine whether defective thymic clonal deletion leads to the escape of autospecific T cells to the periphery, the phenotype of LN T cells was analyzed. In the D^b/HY male mice, 47% of lymph node cells were CD4$^-$CD8$^-$ but expressed equivalent levels of the M33 TCR transgene as observed in female mice. In these mice, peripheral lymph node cells expressed down modulated levels of CD8 and also expressed the D^b/HY TCR transgene (Figure 6). In the ΔNur77 D^b/HY TCR double transgenic male mice, there was a significant increase in the total number of cells, and M33$^+$CD8$^+$ lymph node T cells, but a decreased number of M33$^+$CD8$^-$ T cells. Compared to D^b/HY transgenic male mice, most M33$^+$ T cells expressed intermediate to high levels of CD8.

2.6. Tolerance of Lymph Node T Cells in ΔNur77-D^b/HY TCR Transgenic Male Mice

In D^b/HY T cell receptor transgenic male mice, less than 5% of thymocytes are M33$^+$CD8$^+$. These cells undergo clonal deletion and are highly tolerant in the thymus and undergo weak or no stimulation in the LN (Figure 7). Weak stimulation in the LN results in

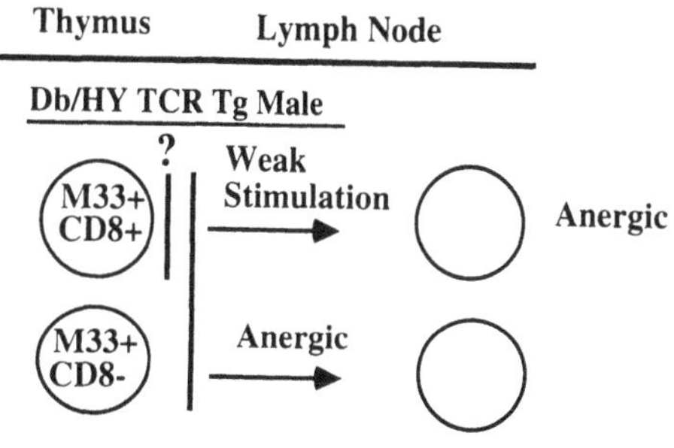

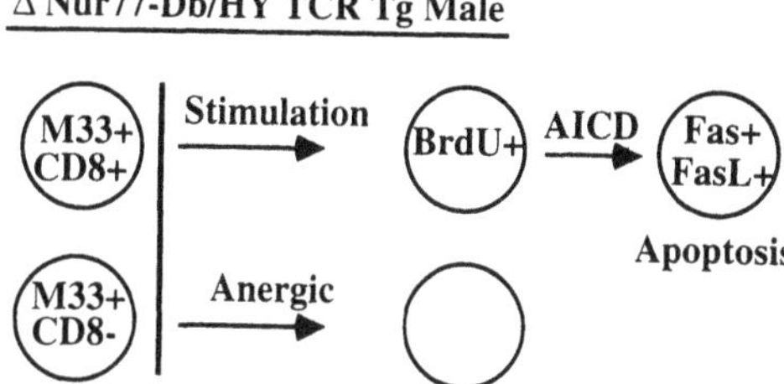

Figure 7. Model for T cell tolerance in ΔNur77-D^b/HY T cell receptor transgenic male mice. Proposed model which depicts T cells that escape CD3/Nur77-mediated apoptosis in the thymus become tolerant by the Fas and Fas ligand-mediated activation induced cell death in the lymph node.

an 18% population of $M33^+CD8^+$ cells with low BrdU uptake. Fas and Fas ligand is not expressed by these cells indicating tolerance is maintained in the lymph node. There is a larger population of $M33^+CD8^-$ thymocytes that are anergic in the thymus and in the lymph node. These cells do not uptake BrdU in the lymph node and do not express Fas and Fas ligand. In the ΔNur77-D^b/HY TCR transgenic male there is increased production of $M33^+CD8^+$ cells in the thymus which escape to the lymph node. These cells are not anergic and undergo stimulation with incorporation of BrdU and an increase to 45% of lymph node cells. These cells subsequently express high levels of Fas and Fas ligand and undergo activation-induced cell death as a tolerance mechanism. In contrast, the $M33^+CD8^-$ thymocytes, which comprise the majority of thymocytes are functionally anergic and are not stimulated or undergo apoptosis upon arrival to the lymph node. Fas and Fas ligand interaction has been shown to be an important mechanism for activation-induced apoptosis[55–60]. Several lines of evidence indicate that this AICD in ΔNur77-D^b/HY TCR transgenic male mice was due to Fas/Fas ligand interaction. First, apoptosis *in vitro* could be blocked by the FasFP. Second, Fas expression was increased in the $M33^+$, $CD8^+$ T cells, but not in the $M33^+$, $CD8^-$ T cells, of Nur77-D^b/HY TCR transgenic male mice, consistent with ongoing AICD using a Fas/Fas ligand pathway. Finally, there was increased Fas ligand production by LN T cells from Nur77-D^b/HY TCR transgenic male but not female mice. Taken together, these results indicate that T cells that escape CD3/Nur77-mediated apoptosis in the thymus become tolerant by the Fas and Fas ligand-mediated activation induced cell death in the lymph node.

3. INCREASED TNF AND SEPTIC SHOCK IN *LPR/LPR* MICE

3.1. Increased TNF-Mediated Septic Shock in *lpr/lpr* Mice

The expression of TNF-α by macrophages is increased in *lpr/lpr* compared to +/+ mice, which suggests that in the absence of Fas, the production TNF-α is upregulated[61, 62]. Furthermore, *lpr* mice produce high levels of TNF and are highly susceptible to septic shock

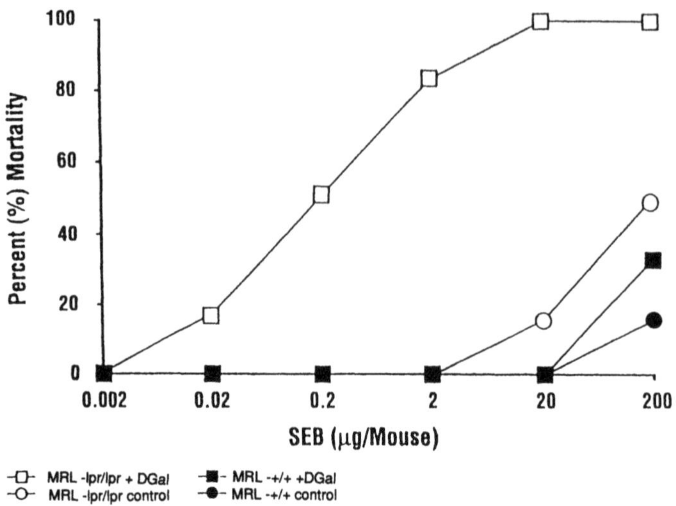

Figure 8. SEB-induced mortality in non-transgenic MRL-+/+ and MRL-*lpr/lpr* mice. Non-transgenic MRL-+/+ or MRL-*lpr/lpr* mice (8 mice per group) were either untreated or treated with D-GalNH2 (25 mg/mouse) in varying concentrations of SEB. Mortality was determined at 18 h.

indicating that in the absence of the Fas-mediated apoptosis system, the TNFR system does not function as an efficient down-modulator of certain pro-inflammatory cytokines including IL-1 and TNF-α that have been implicated in septic shock[62–64].

3.2. Increased Susceptibility to SEB-Induced Lethal Shock in *lpr/lpr* Mice

MRL-+/+ and MRL-*lpr/lpr* mice were treated with D-galNH$_2$ (25 mg i.p.) and then challenged with increasing doses of SEB. MRL-+/+ mice that were not treated with D-galNH$_2$ exhibited an LD$_{50}$ after treatment with SEB of greater than 100 μg per mouse (Figure 8). D-galNH$_2$ sensitization resulted in an approximately 10-fold increase in the sensitivity of MRL-+/+ mice to SEB. MRL-*lpr/lpr* mice that were not sensitized with D-galNH$_2$ exhibited an LD$_{50}$ of 100 μg of SEB. Sensitization with D-galNH$_2$ resulted in an approximately 1,000-fold increase in the sensitivity of MRL-*lpr/lpr* mice to SEB, reducing the LD$_{50}$ to 0.2 μg per mouse.

Similar results were observed in V$_\beta$8 TCR transgenic MRL-+/+ and MRL-*lpr/lpr* mice. The presence of the V$_\beta$8 transgene results in expression of the V$_\beta$8 TCR on nearly 100% of the T cells. This increased T cell receptor expression from approximately 20–30% to 100% of the T cells greatly increases the mortality after treatment with SEB. In V$_\beta$8 TCR transgenic MRL-+/+ mice that were not sensitized with D-galNH$_2$ the LD$_{50}$ is greater than 100 μg (Figure 9). Sensitization of these mice with D-galNH$_2$ reduces the LD$_{50}$ by approximately 1,000-fold to 0.1 μg of SEB. The effect of the transgene is even greater in the MRL-*lpr/lpr* mice. V$_\beta$8 TCR transgenic MRL-*lpr/lpr* mice that have not been sensitized by D-galNH$_2$ exhibited an LD$_{50}$ of 1 μg per mouse of SEB and sensitized mice exhibited a 1,000-fold decrease in LD$_{50}$ of approximately 2 ng per mouse.

3.3. Protection Against SEB-Induced Lethal Shock by Anti-TNF and Anti-IL-2 mAbs

In other strains of mice, the main cytokines responsible for SEB-induced lethal shock have been found to be TNF-α, and to a lessor extent IL-1, IL-2 and IL-6 (13–16). V$_\beta$8 TCR

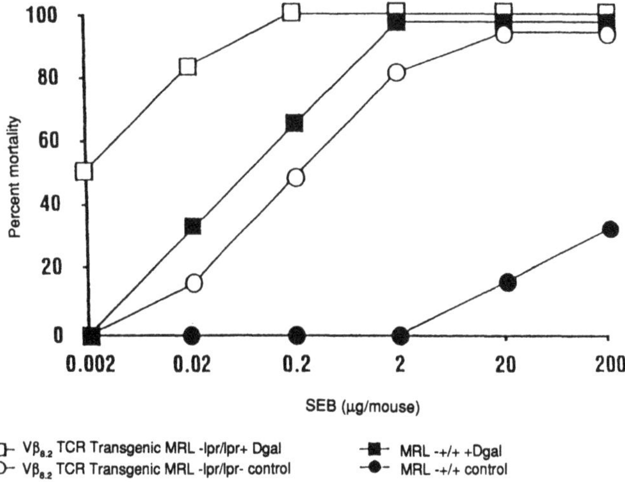

Figure 9. SEB-induced mortality in Vβ8 TCR transgenic mice. Vβ8 TCR Transgenic MRL-+/+ or Vβ8 TCR Transgenic MRL-*lpr/lpr* Mice were monitored for mortality in response to SEB treatment. Results shown are the mortality determined at 18 h for groups of 8 mice.

transgenic MRL-*lpr/lpr* mice exhibit a dramatically heightened sensitivity to SEB-induced lethal shock. The cytokine responsible for this lethal effect has not been determined, and it is possible that the defect in Fas expression and result of autoimmune features are associated with a novel cytokine profile in induction of SEB-induced lethal shock. This was investigated using anti-cytokine antibodies *in vivo* in $V_{\beta}8$ TCR transgenic MRL-*lpr/lpr* mice that had been pre-sensitized with D-$galNH_2$ and then treated with SEB (0.25 μg/mouse) (Table II). Treatment with a monoclonal anti-mouse TNF-α antibody resulted in protection in almost all mice against the SEB-induced lethal shock (Table II). Treatment of mice with anti-IL-2 protected approximately 50% of the mice from SEB-induced lethal shock. In contrast, anti-mouse IL-1β, anti-mouse IL-6, or anti-mouse IFN-γ did not significantly prevent SEB-induced lethal shock (Table II). These results indicate that SEB-induced lethal shock

Table II. Protection against SEB-induced mortality in $V_{\beta}8$ TCR transgenic MRL-*lpr/lpr* female mice by pretreatment with anti-cytokine antibodies[1]

		Expt. 1[3]		Expt. 2[3]	
Superantigen	Antibody[2]	Deaths/total	% Protection	Deaths/total	% Protection
SEB	rat IgG1	8/8	0.0	6/6	0.0
SEB	anti-IL-1β	8/8	0.0	6/6	0.0
SEB	anti-TNF	1/8	87.5	0/6	100.0
SEB	anti-IL-2	4/8	50.0	2/6	66.7
SEB	anti-IFN-γ	7/8	12.5	6/6	0.0
SEB	anti-IL-6	8/8	0.0	6/6	0.0

[1]Two separate groups (Expt. 1 = 8 mice/group; Expt. 2 = 6 mice/group) of $V_{\beta}8$ TCR transgenic MRL-*lpr/lpr* female mice (10 wk old) were first treated with D-$galNH_2$ (25mg) and then with SEB (250 ng i.p.) and assayed for lethality after 18 hr.

[2]Mice were either untreated or pretreated 1 hr before SEB with the indicated anti-mouse cytokine or a control Ab (100 μg/mouse i.p.).

[3]Results are presented as number of dead mice/total mice per group (deaths/total) and surviving mice/total mice per group (% protection).

in the $V_{\beta}8$ TCR transgenic MRL-*lpr/lpr* mice is similar to that observed in other strains of mice in that it is mediated primarily by TNF-α and, to a lessor extent, by IL-2. However, the time course kinetic expression and relative levels of these cytokines upon SEB challenge has not been compared with other strains of inbred mice such as C57Bl/6 mice backcrossed with the *lpr* phenotype. We are actively pursuing these questions in our laboratories.

4. COMPENSATORY ROLE OF FAS AND TNF FOR APOPTOSIS OF MACROPHAGES

4.1. Activation-Induced Cell Death (AICD) of Macrophages

In T cells, AICD is highly dependent on the interaction of Fas with its ligand[55–60]. AICD is defective in T cells obtained from Fas-deficient *lpr/lpr* mice and Fas ligand-deficient *gld* mice[8–12]. TNF-α has also been shown to mediate the AICD of T cells from *lpr/lpr* mice[65]. Fas and TNF-RI share sequence homology at the "death domain" signaling region[66–72]. However, TNF-RI and Fas deliver different signals with different physiologic consequences[73,74]. Ligation of Fas activates production of DNase and proteases in a process known as apoptosis or programmed cell death, whereas ligation of TNF-RI activates lipases and proteases, interferes with mitochondrial metabolism, and results in cell necrosis. The expression of Fas ligand and the role of the interaction of Fas and its ligand in the regulation of macrophage AICD has not been analyzed.

4.2. 7-AAD Analysis of Activation-Induced Cell Death of C57BL/6-+/+ Macrophages

Cell death was analyzed by flow cytometry after stimulation of C57BL/6-+/+ peritoneal macrophages with LPS in RPMI-1640–10% FCS for 24 h (Figure 10). Unstimulated macrophages cultured in RPMI-1640–10% FCS under similar conditions were used as controls. After 24 h of culture with LPS, there was significantly increased cell death as indicated by the increase in the population of 7-AAD^{+} small macrophages. DNA fragmentation analysis indicated that at least some of the macrophage cell death after LPS stimulation occurred through an apoptotic mechanism. Most of the unstimulated macrophages exhibited a stellate appearance on phase contrast microscopy, but after culture with LPS for 24 h, approximately 40% of the macrophages exhibited a condensed nuclear appearance and loss of adherence consistent with the flow cytometry findings. These data indicate that macrophages obtained from C57BL/6-+/+ mice undergo AICD after stimulation with LPS and that at least some of the cell death is mediated by an apoptotic mechanism.

4.3. Increased Expression of Fas, Fas Ligand, TNF-RI, and TNF-α in Activated Macrophages

Both Fas and TNF receptor have been shown to be involved in AICD of T cells. To determine whether Fas and TNF receptor are also involved in AICD of macrophages, we examined the expression of Fas and TNF-RI in resting and activated macrophage from C57BL/6-+/+ and *lpr/lpr* mice. Low levels of Fas were expressed by the majority of resting C57BL/6-+/+ macrophages, whereas 58% of these macrophages expressed increased level of Fas expression after 12 h of LPS stimulation (Figure 11). As expected, neither resting nor activated macrophage from C57BL/6-*lpr/lpr* mice expressed Fas. Although most resting macrophages from C57BL/6-+/+ mice did not express TNF-RI, LPS stimulation led to

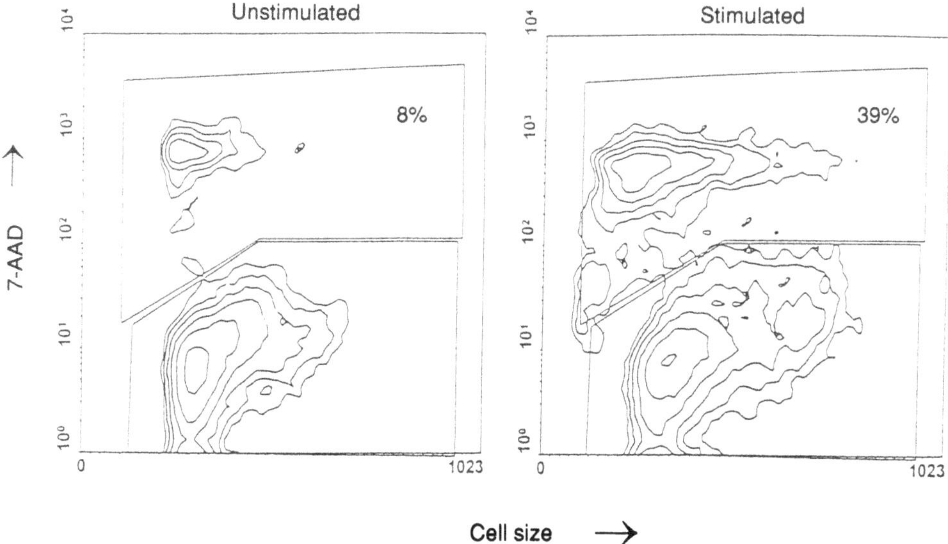

Figure 10. Activation-induced cell death determined by 7-amino actinomycin D (7-AAD) staining. Resident macrophages obtained from the peritoneal cavity of C57BL/6-+/+ mice were either unstimulated or stimulated *in vitro* with LPS for 18 h. Unstimulated peritoneal macrophages exhibited little increase in cell size and only 8% of macrophages were 7-AAD$^+$. Stimulated macrophages exhibited an increase in cell size. There was an increase in activation-induced cell death indicated by 39% of macrophages as 7-AAD$^+$. These cells also exhibited a small cell size consistent with cell shrinkage and apoptosis. Apoptosis of these cells was verified by TUNEL staining and intranucleosomal chromosomal cleavage resulting in DNA fragmentation into the DNA ladder of multiples of 180 bp.

increased expression of TNF-RI by 42% of these macrophages. In contrast, a small percentage of resting macrophages from C57BL/6-*lpr*/*lpr* mice expressed TNF-RI and, after activation with LPS, 66% of macrophages from these mice expressed high level of TNF-RI.

Fas ligand production was determined by the ability of macrophages to lyse Fas ligand sensitive TNFR$^-$ A20 cells at different times after LPS stimulation. In macrophages obtained from C57BL/6-+/+ mice, Fas ligand activity was upregulated after LPS stimulation and the expression remained high for 72 h after stimulation. In macrophages obtained from C57BL/6-*lpr*/*lpr* mice, Fas ligand activity was rapidly up-regulated after LPS stimulation and remained elevated for 72 h. This specific lysis of A20 cells was mediated by Fas ligand as the addition of FasFP led to a greater than 70% reduction of cytotoxicity. There was no significant production of Fas ligand by macrophages incubated in media alone (Figure 11). These results indicate that activation of macrophages obtained from C57BL/6-+/+ mice induces increased expression of Fas, Fas ligand and TNF-RI. Activation of macrophages obtained from C57BL/6-*lpr*/*lpr* mice does not induce expression of Fas, but does induce an increased expression of Fas-ligand activity and TNF-RI.

4.4. Both Fas and TNF-RI Mediate Cell Death in Activated C57BL/6-+/+ Macrophages, Whereas TNF-RI Mediates Cell Death in Activated *lpr/lpr* Macrophages

To determine the roles of Fas and TNF-RI in AICD of macrophages, peritoneal macrophages from C57BL/6-+/+ mice were stimulated with LPS in the presence or absence of anti-TNF antibody or Fas-Fc fusion protein (FasFP). Unstimulated C57BL/6-+/+ macro-

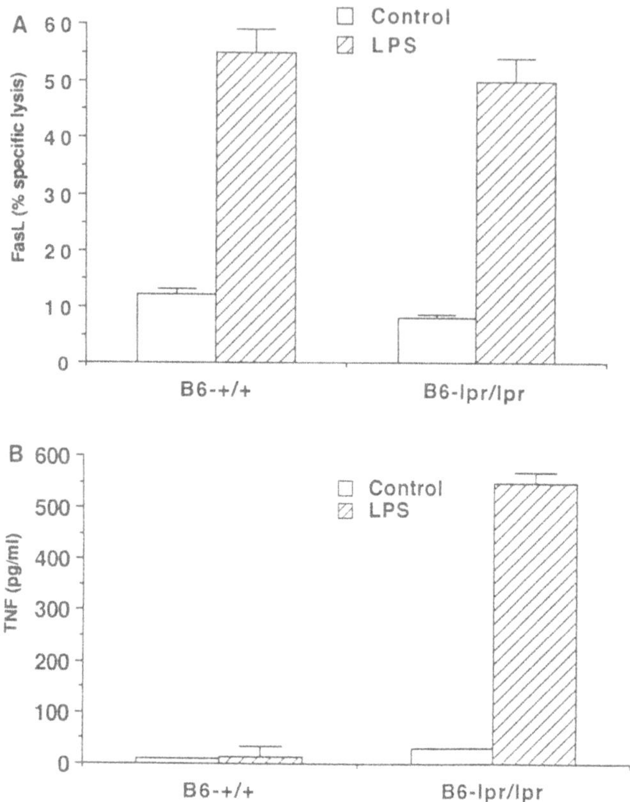

Figure 11. TNF and Fas ligand production after macrophage activation with LPS. Macrophages were activated with LPS and the supernate was assayed for TNF by ELISA assay or for Fas ligand by percent specific lysis of a Fas-sensitive A20 target cell. TNF production was increased in C57BL/6-*lpr/lpr* mice compared to C57BL/6-+/+ mice. Fas ligand production was nearly equivalent by LPS stimulated macrophages from C57BL/6-*lpr/lpr* mice compared to C57BL/6-+/+ mice. A. FasL after macrophage activation; B. TNF after macrophage activation.

phages showed only low levels of cell death during 72 h culture (Figure 12). LPS stimulation of macrophages induced a time-dependent increase in cell death over the three days of culture. Both FasFP and anti-TNF antibody reduced the AICD to approximately 50% of that observed in the LPS-stimulated control. The time course indicates that both TNF-α-mediated cell death, which occurred in the presence of Fas fusion protein, and Fas-induced cell death, which occurred in the presence of anti-TNF-α, continued to increase in an additive fashion for the first 72 h of culture. These results indicate that activation-induced cell death in macrophages is mediated by Fas as well as TNF-α.

To further determine the role of Fas and TNF-α in AICD of macrophages, peritoneal macrophages of C57BL/6-*lpr/lpr* mice were stimulated under identical conditions. A greater number of the macrophages of C57BL/6-*lpr/lpr* mice underwent AICD after LPS stimulation compared to those obtained from C57BL/6-+/+ mice. This increased AICD of the C57BL/6-*lpr/lpr* macrophages was not affected by addition of Fas fusion protein, but was markedly inhibited by addition of anti-TNF-α antibody. These results indicate that, in spite of defective Fas, the AICD of C57BL/6-*lpr/lpr* macrophages is increased compared to that of C57BL/6-+/+ macrophages and is apparently mediated by TNF-α.

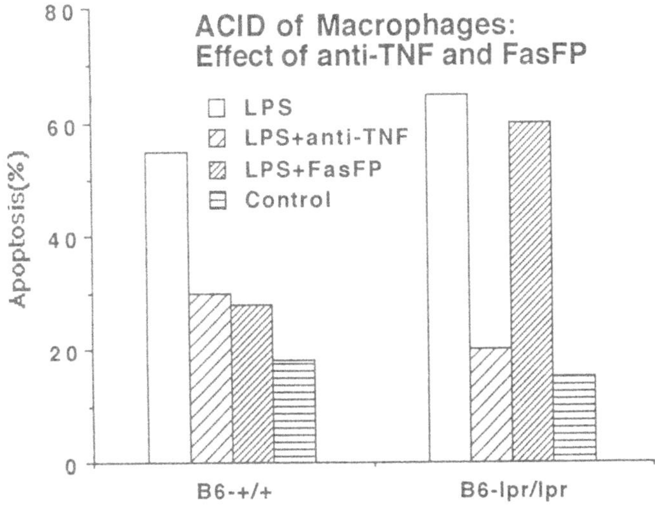

Figure 12. Activation-induced cell death (AICD) of macrophages after stimulation with LPS. Freshly isolated peritoneal macrophages were stimulated with LPS *in vitro* and apoptosis was determined by 7-AAD staining. Macrophages were stimulated by LPS alone or by LPS in the presence of anti-TNF or a Fas fusion protein capable of neutralizing Fas ligand. In C57BL/6-+/+ mice there was high AICD of macrophages after LPS stimulation which was reduced by approximately 50%, either by blocking of TNF with anti-TNF or by blocking of Fas ligand with the Fas fusion protein. LPS stimulated macrophages from C57BL/6-*lpr/lpr* mice underwent increased AICD compared to C57BL/6-+/+ mice, and AICD could be totally blocked by addition of anti-TNF, whereas neutralization of Fas ligand by Fas fusion protein have no effect.

4.5. FasL and TNF Regulate the Inflammatory Response by Induction of Apoptosis

Macrophages are known to express Fas and TNF-RI and produce TNF-α[73, 74]. The present results show that, in addition, macrophages are capable of expressing a functional Fas ligand. This was demonstrated by the ability of activated macrophages to induce apoptosis of a Fas^+ cell line and by the detection of Fas/Fas ligand mediated AICD in a purified macrophage population. Macrophage production of Fas ligand as an AICD mechanism may be due to an autocrine or paracrine signaling mechanism that serves to kill macrophages with high expression of TNF-α. The present data suggest that there are two independent pathways for macrophage AICD, one mediated by Fas/Fas ligand interaction and one mediated by TNF-RI/TNF-α interaction. The macrophage population that expresses Fas and produces TNF-α is highly susceptible to Fas-mediated apoptosis.

The Fas and TNF-RI signaling pathways are distinct. Trimerization of Fas leads to binding of a different set of cytoplasmic proteins than does trimerization of TNF-RI[66–72]. The downstream signaling pathways and secondary mediators induced after Fas and TNF-RI signaling are also different[73, 74]. This was demonstrated using transfection of the TNF sensitive L929 cell with the Apo-1/Fas gene in a panel of cell lines that can be induced to undergo death by either anti-Fas-induced or anti-TNF-RI-induced signaling pathways[73]. Most of the cell lines studied expressed both TNF-RI and Fas, but only a minority of cell lines were susceptible to induction of cell death in a synergistic fashion by both pathways. TNF-α mediated cell death was more susceptible to modulation as labile proteins are involved in cell death mediated by TNF-RI and inducible proteins can protect against TNF-RI-mediated cell death, but not against Fas-mediated cytotoxicity[74].

AICD of macrophages is the result of TNF-RI/TNF-α signaling and Fas/Fas ligand signaling. In C57BL/6-+/+ normal mice, there is an approximately equal contribution of both pathways since activation-induced apoptosis is reduced by approximately 50% by blocking either TNF-α with anti-TNF-α or Fas ligand with Fas fusion protein. The present results indicate that the two pathways can augment each other in the AICD of macrophages, or can replace each other if one is defective. Activation-induced cell death in macrophages is higher in Fas-deficient C57BL/6-*lpr/lpr* mice compared to normal mice. Thus, AICD in macrophages can be completely attributed to TNF-α/TNFRI signaling, Fas/Fas ligand signaling, or combined signaling. The data suggest that there is an orderly sequence of events leading to activation-induced cell death of macrophages. First, there is upregulation of Fas, TNF-RI, TNF-α, and possibly Fas ligand by a majority of macrophages. Second, macrophages expressing high levels of Fas begin to undergo cell death by either an autocrine or paracrine interaction with Fas ligand. We favor a paracrine mechanism of regulation leading to apoptosis of macrophages producing high levels of TNF-α by adjacent macrophages that produce high levels of Fas ligand. Defective Fas-mediated apoptosis led to persistence of macrophages with high levels of cytoplasmic TNF-α and continued production of TNF-α into the supernate. These results indicate that AICD mediated by Fas/Fas ligand interactions helps to eliminate macrophages producing high levels of TNF-α, thereby eliminating the immune response.

5. BI-FUNCTIONAL CONTROL OF INFLAMMATION AND AUTOIMMUNITY BY TNF

5.1. Production of *Tnfr1$^{0/0}$-lpr/lpr* Mice

TNF is involved in activation-induced apoptosis in T cells[65]. This observation is consistent with a cytotoxic role for TNF-RI and TNF-RII[75–78]. The role of TNF-RI or TNF-RII in AICD as an inhibitor of an inflammatory response or autoimmune disease has not been established. Neither TNF-RI nor TNF-RII knockout mice exhibit an apoptosis defects and these mice are not predisposed to develop autoimmune disease[79–81]. Moreover, TNF is a proinflammatory cytokine and has been demonstrated to play a major role in inflammation and tissue damage in autoimmune disease[82–90]. The administration of soluble TNF-R or anti-TNF antibody reduces inflammation in patients with rheumatoid arthritis and other autoimmune diseases[90–92]. To investigate the role of TNF-RI-mediated apoptosis in autoimmune disease and T cells apoptosis, we generated Fas deficient *lpr/lpr* mice lacking the TNF-RI by backcrossing *Tnfr1$^{0/0}$* mice with C57BL/6-*lpr/lpr* mice.

5.2. Accelerated Lymphadenopathy in *Tnfr1$^{0/0}$ lpr/lpr* Mice

Tnfr1$^{0/0}$-lpr/lpr mice developed visibly enlarged LN as early as 6 weeks of age whereas C57BL/6-*lpr/lpr* mice exhibited equivalently enlarged lymph nodes at 12 weeks of age (Figure 13). There was a 5–10 fold increase in LN weight and total cell number of *Tnfr1$^{0/0}$-lpr/lpr* mice compared to C57BL/6-*lpr/lpr* mice at all ages studied. The accumulation of Thy1.2$^+$ B220$^+$ T cells rapidly increased and reached 80% in 12-week-old *Tnfr1$^{0/0}$-lpr/lpr* mice. There was no significant difference in lymphoproliferative disease in C57BL/6-*Tnfr1$^{0/0}$* mice compared to control C57BL/6-+/+ mice. These results indicate that defective TNF-RI signaling can greatly accelerate lymphoproliferative disease when combined with the Fas defect in *lpr/lpr* mice, but does not cause lymphadenopathy in the absence of defective Fas apoptosis.

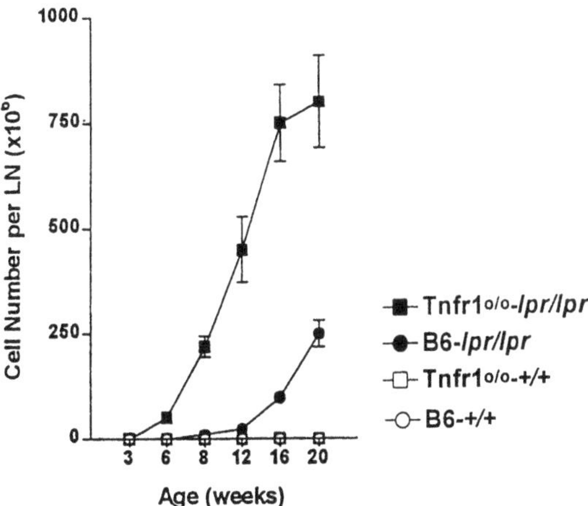

Figure 13. Accelerated lymphoproliferative disease in Tnfr$^{0/0}$-*lpr/lpr* double mutant mice. The mean number of cells per lymph node was determined at different ages. There was no lymphoproliferative disease in C57BL/6-+/+ mice or in C57BL/6-Tnfr1$^{0/0}$ knockout mice. Lymphoproliferative disease was evident between 12 and 16 wks of age in C57BL/6-*lpr/lpr* mice. There is greatly accelerated lymphoproliferative disease in C57BL/6-Tnfr1$^{0/0}$-*lpr/lpr* mice.

5.3. Increased Production of IgG Immunoglobulins and Autoantibodies in *Tnfr1$^{0/0}$-lpr/lpr* Mice

Total IgG1 and IgG2a was significantly increased by 6 weeks of age in *Tnfr1$^{0/0}$-lpr/lpr* compared to 6-week-old C57BL/6-*lpr/lpr* mice (Figure 14). Similarly, IgG1 and IgG2a anti-DNA antibodies were also increased in *Tnfr1$^{0/0}$-lpr/lpr* mice compared to C57BL/6-*lpr/lpr* mice. In contrast, the levels of IgM and IgM RF were less affected by the TNF-RI deficiency. There was no significant increase in Ig or anti-DNA production in C57BL/6-*Tnfr1$^{0/0}$* mice compared to control C57BL/6-+/+ mice. Increased Ig and autoantibody production could be due to an apoptosis defect of B cells or due to increased T cell-derived help for Ig production by B cells. IgG1 and IgG2a autoantibodies, but not IgM autoantibodies were elevated in the C57BL/6-Tnfr$^{0/0}$-*lpr/lpr* mice suggesting that TNF-RI-mediated apoptosis might function predominantly in the autoreactive T cells but not B cells in *lpr/lpr* mice. Massive lymphocyte infiltration in the diseased organs further support the concept that there is a profound T cell apoptosis defect in C57BL/6-*Tnfr$^{0/0}$-lpr/lpr* mice. Upregulation of cytokine production due to defective AICD of T cells would lead to increased expression of IFN-γ, IL-2, and IL-6. The expression of all of these cytokines is known to be increased in *lpr* mice[61–64]. IFN-γ and IL-4 would lead to increased production of IgG1 and IgG2a as observed in the *Tnfr1$^{0/0}$-lpr/lpr* mice. Thus, the additive defects in the *Tnfr1$^{0/0}$*-

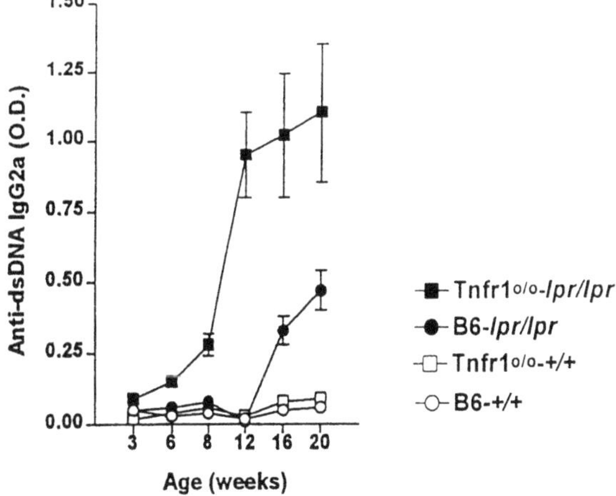

Figure 14. Increased anti-DNA production in Tnfr$^{0/0}$-*lpr/lpr* double mutant mice. Anti-double stranded DNA of the IgG2a isotype was determined at different ages in the different strains of mice. There was no significant production of anti-double stranded DNA in C57BL/6-+/+ mice or in C57BL/6-Tnfr1$^{0/0}$ mice. IgG2a anti-double stranded DNA was detectable in C57BL/6-+/+-*lpr/lpr* mice after 12 wks of age. Increased production of anti-double stranded DNA was observed as early as 3–6 wks of age in the C57BL/6-Tnfr1$^{0/0}$-*lpr/lpr* mice.

lpr/lpr mice contribute to increased numbers of activated T cells, B cells and macrophages with resultant increased lymphadenopathy and tissue infiltration by mononuclear cells.

5.4. Dual and Functionally Opposite Role of TNF-RI in Apoptosis and Inflammation

The predominate histopathologic lesion observed in the *Tnfr1*$^{o/o}$-*lpr/lpr* mice was mononuclear cell infiltration into tissues and immune complex deposition. The arthritis of *Tnfr1*$^{o/o}$-*lpr/lpr* mice was characterized by synovial cell hyperplasia and mononuclear cell infiltration but these changes did not progress to cartilage erosion and joint damage by 5 months of age suggesting that cartilage erosion is not accelerated in the absence of TNF-RI. There was increased accumulation of mononuclear cells, including CD3$^+$ T cells, in the liver, kidney, lung, and joint. Immune complex deposition in the renal glomerulus was the predominant feature of the glomerulonephritis in the *Tnfr1*$^{o/o}$-*lpr/lpr* mice. Taken together, these data suggest that TNF-RI exhibits a bi-directional function in the process of autoimmune disease. During an early phase, TNF-RI may mediate AICD in the mononuclear cells and prevent tissue infiltration. At a later phase of the disease, TNF-RI has been reported to promote production of collagenase and stromyelysin that results in tissue damage. These results are consistent with previous observations that TNF can both inhibit and accelerate autoimmune disease in autoimmune mice.

In *lpr/lpr* mice, low expression of Fas results in increased expression of Fas ligand and TNF. Decreased expression of Fas results in inefficient apoptosis of inflammatory cells

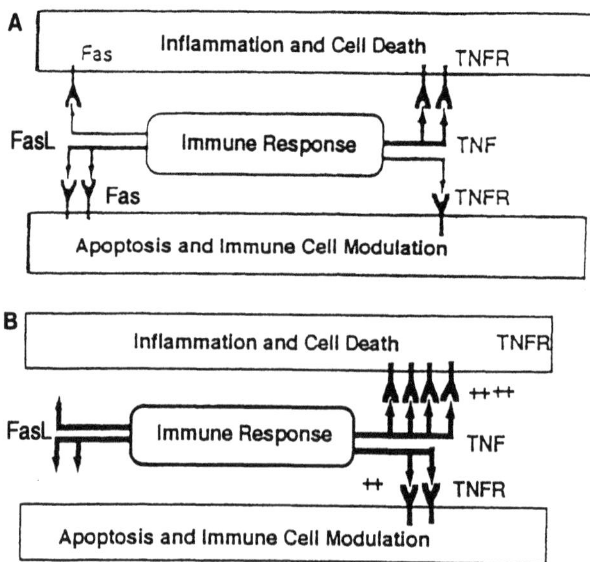

Figure 15. Regulation of the immune response by TNF and Fas ligand in normal and *lpr/lpr* Mice. Normal balance regulation of the immune response in apoptosis requires interaction of Fas with Fas ligand and is facilitated by TNF/TNF receptor apoptosis. TNF also has a proinflammatory cytokine effect leading to inflammation and tissue destruction when it interacts with its receptor. In some cases, interaction of Fas ligand with Fas can lead to inflammation and cell death. In Fas deficient *lpr* mice, there is inefficient down regulation of the immune response by Fas ligand production. This is partially compensated for by overproduction of TNF which interacts with the TNF receptor leading to apoptosis. This apoptosis effect acts to limit the development of lymphoproliferative disease and autoantibody production. However, increased TNF also results in increased inflammation, cell death and septic shock in *lpr/lpr* mice. A. Balanced regulation of immune response and apoptosis by TNF and FasL; B. overproduction of TNF regulates T cell apoptosis but causes septic shock in *lpr* mice.

of the immune system including T cells, B cells and macrophages. Low expression of Fas also results in high expression of TNF and the predisposition toward development of septic shock and autoimmune disease. A model for imbalanced production of TNF and Fas in *lpr/lpr* mice is shown in Figure 15. These results help explain previous findings that TNF can act as both a pro and anti-inflammatory cytokine[93].

ACKNOWLEDGMENTS

This work is supported in part by VA Career Development and Merit Review Award, National Institutes of Health grants P60 AR20614, P50 AI23694, P01 AR03555, R01 AI30744 and a grant from Sankyo Co., Ltd. Tong Zhou is the recipient of an Arthritis Foundation Investigator Award.

REFERENCES

1. G.J.V. Nossal, Negative selection of lymphocytes. *Cell* 76:229 (1994).
2. P. Golstein, D.M. Ojcius, and J.D. Young, Cell death mechanisms and the immune system. *Immunol. Rev.* 121:29 (1991).
3. B. Lucas, F. Vasseur, and C. Penit, Production, selection, and maturation of thymocytes with high surface density of TCR. *J. Immunol.* 153:53 (1994).
4. J.A. Punt, B.A. Osborne, Y. Takahama, S.O. Sharrow, and A. Singer, Negative selection of $CD4^+CD8^+$ thymocytes by T cell receptor-induced apoptosis requires a costimulatory signal that can be provided by CD28. *J. Exp. Med.* 179:709 (1994).
5. A.R. Clarke, C.A. Purdie, D.J. Harrison, R.G. Morris, C.C. Bird, M.L. Hooper, and A.H. Wyllie, Thymocyte apoptosis induced by *p*53-dependent and independent pathways. *Nature*. 362:849 (1993).
6. S.W. Lowe, E.M. Schmitt, S.W. Smith, B.A. Osborne, and T. Jacks, *p*53 is required for radiation-induced apoptosis in mouse thymocytes. *Nature* 362:847 (1993).
7. J.M. Lee and A. Bernstein, *p*53 mutations increase resistance to ionizing radiation. *Proc. Natl. Acad. Sci. USA*. 90:5742 (1993).
8. T. Suda, T. Takahashi, P. Golstein, and S. Nagata, Molecular cloning and expression of the Fas ligand, a novel member of the tumor necrosis factor family. *Cell*. 75:1169 (1993).
9. S. Nagata and P. Golstein. The Fas death factor. *Science*. 267:1449 (1995).
10. R. Watanabe-Fukunaga, C.I. Brannan, N.G. Copeland, N.A. Jenkins, and S. Nagata, Lymphoproliferation disorder in mice explained by defects in Fas antigen that mediates apoptosis. *Nature*. 356:314 (1992).
11. D. Lynch, M. Watson, M.R. Alderson, P.R. Baum, R.E. Miller, T. Tough, M. Gibson, T. Davis-Smith, C.A. Smith, K. Hunter, D. Bhat, W. Din, R.G. Goodwin, and M.F Seldin, The mouse Fas-ligand gene is mutated in gld mice and is part of a TNF family gene cluster. *Immunity*. 1:131 (1994).
12. T. Takahashi, M. Tanaka, C.I. Brannan, N.A. Jenkins, N.G. Copeland, T. Suda, and S. Nagata, Generalized lymphoproliferative disease in mice, caused by a point mutation in the Fas ligand. *Cell*. 76:969 (1994).
13. B.L. Kotzin, S.K. Babcock, and L.R. Herron, Deletion of potentially self-reactive T cell receptor specificities in $L3T4^-$, $Lyt2^-$ T cells of *lpr* mice. *J. Exp. Med.* 168:2221 (1988).
14. P.A. Singer, R.S. Balderas, R.J. McEvilly, M. Bobardt, and A.N. Theofilopoulos, Tolerance-related Vβ clonal deletions in normal $CD4^-CD8^-$, TCR-α/β^+ and abnormal *lpr* and *gld* cell populations. *J. Exp. Med.* 170:1869 (1989).
15. J.D. Mountz, T.M. Smith, and K.S. Toth, Altered expression of self-reactive T cell receptor Vβ regions in autoimmune mice. *J. Immunol.* 144:2159 (1990).
16. T. Zhou, H. Bluethmann, J. Eldridge, K. Berry, and J.D. Mountz, Abnormal thymocyte development and production of autoreactive T cells in TCR transgenic autoimmune mice. *J. Immunol.* 147:466 (1991).
17. T. Zhou, J.D. Mountz, C.K. Edwards, III, K. Berry, and H. Bluethmann, Defective maintenance of T cell tolerance to a superantigen in MRL-*lpr*/*lpr*. *J. Exp. Med.* 176:1063 (1992).
18. R.P. Bissonnette, F. Echeverri, A. Mahboubi, and D.R Green, Apoptotic cell death induced by *c-myc* is inhibited by *bcl-2*. *Nature*. 359:552 (1992).
19. A. Fanidi, E.A. Harrington, and G.I. Evan, Cooperative interaction between c-*myc* and *bcl-2* proto-oncogenes. *Nature*. 359:554 (1992).

20. A. Strasser, A.W. Harris, and S. Cory, *Bcl-2* transgene inhibits T cell death and perturbs thymic self-censorship. *Cell.* 67:889 (1991).
21. N.C. Moore, G. Anderson, G.T. Williams, and J.J Owen, Developmental regulation of *bcl-2* expression in the thymus. *Immunology.* 81:115 (1994).
22. D. Veis, C.M. Sorenson, J.R. Shutter, and S.J. Korsmeyer, *Bcl-2*-deficient mice demonstrate fulminant lymphoid apoptosis, polycystic kidneys, and hypopigmented hair. *Cell.* 75:229 (1993).
23. M. Katsumata, R.M. Siegel, D.C. Louie, T. Miyashita, Y. Tsujimoto, P.C. Nowell, M.I. Greene, and J.C. Reed, Differential effects of *bcl-2* on T and B cells in transgenic mice. *Proc. Natl. Acad. Sci. USA.* 89:11376 (1992).
24. R.M. Siegel, M. Katsumata, T. Miyashita, D.C. Louie, M.I. Greene, and J.C. Reed, Inhibition of thymocyte apoptosis and negative antigenic selection in bcl-2 transgenic mice. *Proc. Natl. Acad. Sci., USA.* 89:7003 (1992).
25. C.L. Sentman, J.R. Shutter, D. Hockenbery, O. Kanagawa, and S. Korsmeyer, *Bcl-2* inhibits multiple forms of apoptosis but not negative selection in thymocytes. *Cell.* 67:879 (1991).
26. A. Strasser, A.W. Harris, H. von Boehmer, and S. Cory, Positive and negative selection of T cells in T-cell receptor transgenic mice expressing a *Bcl-2* transgene. *Proc. Natl. Acad. Sci. USA.* 91:1376 (1994).
27. K. Lundberg and K. Shortman, Small cortical thymocytes are subject to positive selection. *J. Exp. Med.* 179:1475 (1994).
28. W. Tao, S.J. Teh, I. Melhado, F. Jirik, S.J. Korsmeyer, and H.S. Teh, The T cell receptor repertoire of $CD4^-8^+$ thymocytes is altered by overexpression of the *bcl-2* protooncogene in the thymus. *J. Exp. Med.* 179:145 (1994).
29. J. Milbrandt, Nerve growth factor induces a gene homologous to the glucocorcoid receptor gene. *Neuron.* 1:183 (1988).
30. T.G. Hazel, D. Nathans, and L.F. Lau, A gene inducible by serum growth factors encodes a member of the steroid and thyroid hormone receptor superfamily. *Proc. Natl. Acad. Sci. USA.* 85:8444 (1988).
31. G.T. Williams and L.F. Lau, Activation of the inducible orphan receptor gene Nur77 by serum growth factors: dissociation of immediate-early and delayed-early responses. *Mol. Cell Biol.* 13:6124 (1993).
32. S.R. Abu-Shakra, A.J. Cole, and D.B. Drachman, Nerve stimulation and denervation induce differential patterns of immediate early gene mRNA expression in skeletal muscle. *Brain Res. Mol. Brain Res.* 18:216 (1993).
33. S.W. Law, O.M. Conneely, F.J. DeMayo, and B.W. O'Malley, Identification of a new brain-specific transcription factor, NurR1. *Mol. Endocrinol.* 6:2129 (1992).
34. I.J. Davis, T.G. Hazel, R.H. Chen, J. Blenis, and L.F. Lau, Functional domains and phosphorylation of the orphan receptor Nur77. *Mol. Endocrinol.* 7:953 (1993).
35. J.K. Yoon and L.F. Lau, Transcriptional activation of the inducible nuclear receptor gene Nur77 by nerve growth factor and membrane depolarization in PC12 cells. *J. Biol. Chem.* 268:9148 (1993).
36. T.E. Wilson, T.J. Fahrner, M. Johnston, and J. Milbrandt, Identification of the DNA binding site for NGFI-B by genetic selection in yeast. *Science.* 252:1296 (1991).
37. T.E. Wilson, R.E. Paulsen, K.A. Padgett, and J. Milbrandt, Participation of non-zinc finger residues in DNA binding by two nuclear orphan receptors. *Science.* 256:107 (1992).
38. Z-G. Liu, S.W. Smith, K.A. McLaughlin, L.M. Schwartz, and B.A. Osborne, Apoptotic signals delivered through the T-cell receptor of a T-cell hybrid require the immediate-early gene Nur77. *Nature.* 367:281 (1994).
39. J.D. Woronica, B. Cainan, V. Ngo, and A. Winoto, Requirement for the orphan steroid receptor Nur77 in apoptosis of T-cell hybridomas. *Nature.* 367:277 (1994).
40. T. Okabe, R. Takayanagi, K. Imasaki, H. Masafumi, H. Nawata, and T. Watanabe, cDNA cloning of a NGF1-B/Nur77-related transcription factor from an apoptotic human T cell line. *J. Immunol.* 154:3871 (1995).
41. Y. Yang, M. Mercep, C.F. Ware, and J.D. Ashwell, Fas and activation-induced Fas ligand mediate apoptosis of T cell hybridomas: inhibition of Fas ligand expression by retinoic acid and glucocorticoids. *J. Exp. Med.* 181:1673 (1995).
42. M.S. Vacchio, V. Papadopoulos, J.D. Ashwell, Steroid production in the thymus: implications for thymocyte selection. *J. Exp. Med.* 179:1835 (1994).
43. S.L. Lee, R.L. Wesselschmidt, G.P. Linette, O. Kanagawa, J.H. Russell, and J. Milbrandt, Unimpared thymic and peripheral T cell death in mice lacking the nuclear receptor NGFI-B Nur77. *Science.* 269:532 (1995).
44. J. Kaye, M.-L. Hsu, M.-E. Sauron, S.C. Jameson, N.R.J. Gascoigne, and S.M. Hedrick, Selective development of $CD4^+$ T cells in transgenic mice expressing a class II MHC-restricted antigen receptor. *Nature.* 341:746 (1989).

45. J. Kaye and S.M. Hedrick, Analysis of specificity for antigen, Mls, and allogeneic MHC by transfer of T-cell receptor α- and β-chain genes. *Nature*. 336:580 (1988).
46. P. Kisielow, H. Bluethmann, U.D. Staerz,, M. Steinmetz, and H. von Boehmer, Tolerance in T-cell receptor transgenic mice involves deletion of immature $CD4^{+}8^{+}$ thymocytes. *Nature*. 333:742 (1988).
47. P. Kisielow, H.S. Teh, H. Bluethmann, and H. von Boehmer, Positive selection of antigen-specific T cells in thymus by restricting MHC molecules. *Nature*. 335:730 (1988).
48. H. von Boehmer, Developmental biology of T cells in T cell receptor transgenic mice. *Annu. Rev. Immunol.* 8:531 (1990).
49. H-S. Teh, H. Kishi, B. Scott, and H. von Boehmer, Deletion of autospecific T cells in T cell receptor (TCR) transgenic mice spares cells with normal TCR levels and low levels of CD8 molecules. *J. Exp. Med.* 169:795 (1989).
50. T. Zhou, H. Bluethmann, J. Eldridge, K. Berry, and J.D. Mountz, Origin of $CD4^{-}CD8^{-}B220^{+}$ T cells in MRL-*lpr/lpr* mice. Clues from a T cell receptor β transgenic mouse. *J. Immunol.* 150:3651 (1993).
51. M.C. Kiefer, M.J. Brauer, V.C. Powers, J.J. Wu, S.R. Umansky, L.D. Tomei, and P.J. Barr, Modulation of apoptosis by the widely distributed *bcl-2* homologue Bak. *Nature*. 374:736 (1995).
52. E. Yang, J. Zha, J. Jockel, L.H. Boise, C.B Thompson, and S.J. Korsmeyer, Bad, a heterodimeric partner for Bcl-XL and *Bcl-2*, displaces Bax and promotes cell death. *Cell*. 80:285 (1995).
53. X.M. Yin, Z.N. Oltval, and S.J. Korsmeyer, BH1 and BH2 domains of *Bcl-2* are required for inhibition of apoptosis and heterodimerization with Bax. *Nature*. 369:321 (1994).
54. Z.N. Oltvai, C.L. Milliman, and S.J. Korsmeyer, *Bcl-2* heterodimerizes *in vivo* with a conserved homolog, Bax, that accelerates programmed cell death. *Cell*. 74:609 (1993).
55. S.T. Ju, D.J. Panka, H. Cui, R. Ettinger, M. el-Khatib, D.H. Sherr, B.Z. Stanger, and A. Marshak-Rothstein, Fas (CD95)/FasL interactions required for programmed cell death after T-cell activation. *Nature*. 373:444 (1995).
56. T. Brunner, R.J. Mogil, D. LaFace, N.J. Yoo, A. Mahboubi, F. Echeverri, S.J. Martin, W.R. Force, D.H. Lynch, and C.F. Ware, Cell-autonomous Fas (CD95)/Fas-ligand interaction mediates activation-induced apoptosis in T-cell hybridomas. *Nature*. 373:441 (1995).
57. J. Dhein, H. Walczak, C. Baumler, K.M. Debatin, and P.H. Krammer, Autocrine T-cell suicide mediated by APO-1/(Fas/CD95). *Nature*. 373:438 (1995).
58. M.R. Alderson, T.W. Tough, T. Davis-Smith, S. Braddy, B. Falk, K.A. Schooley, R.G. Goodwin, C.A. Smith, F. Ramsdell, and D.H. Lynch, Fas ligand mediates activation-induced cell death in human T lymphocytes. *J. Exp. Med.* 181:71 (1995).
59. D. Kabelitz, T. Pohl, and K. Pechhold, Activation-induced cell death (apoptosis) of mature peripheral T lymphocytes. *Immunol. Today*. 14:339 (1993).
60. D.R. Green and D.W. Scott, Activation-induced apoptosis in lymphocytes. *Curr. Opin. Immunol.* 6:476 (1994).
61. W.F. Davidson, C. Calkins, A. Hugins, T. Giese, and K.L. Holmes, Cytokine secretion by C3H-*lpr* and -*gld* T cells. Hypersecretion of IFN-γ and tumor necrosis factor-α by stimulated $CD4^{+}$ T cells. *J. Immunol.* 146:4138 (1991).
62. J.D. Mountz, T.J. Baker, D.R. Borcherding, H. Bluethmann, T. Zhou, and C.K. Edwards, III, Increased susceptability of fas mutant MRL-*lpr/lpr* mice to staphylococcal enterotoxin B-induced septic shock. *J. Immunol.* 155:4829 (1995).
63. T. Zhou, J.D. Mountz, C.K. Edwards, III, K. Berry, and H. Bluethmann, Defective maintenance of T cell tolerance to a superantigen in MRL-*lpr/lpr*. *J. Exp. Med.* 176:1063 (1992).
64. J.D. Mountz, T. Zhou, R.E. Long, H. Bluethmann, W.J. Koopman, and C.K. Edwards, III, T cell influence on superantigen-induced arthritis in MRL-*lpr/lpr* mice. *Arthritis Rheum.* 37:113 (1994).
65. L. Zheng, G. Fisher, R.E. Miller, J. Peschon, D. Lynch, and M.J. Lenardo, Induction of apoptosis in mature T cells by tumour necrosis factor. *Nature*. 377:348 (1995).
66. L.A. Tartaglia, T.M. Ayres, G.H. Wong, and D.V. Goeddel, A novel domain within the 55 kd TNF receptor signals cell death. *Cell*. 74:845 (1993).
67. N. Itoh, and S. Nagata, A novel protein domain required for apoptosis, Mutational analysis of human Fas antigen. *J. Biol. Chem.* 268:10932 (1993).
68. B.Z. Stanger, P. Leder, T.H. Lee, E. Kim, and B. Seed, RIP: a novel protein containing a death domain that interacts with Fas/APO-1 (CD95) in yeast and causes cell death. *Cell*. 81:513 (1995).
69. A.M. Chinnaiyan, K. O'Rourke, M. Tewari, and V.M. Dixit, FADD, a novel death domain-containing protein, interacts with the death domain of Fas and initiates apoptosis. *Cell*. 81:505 (1995).
70. M.P. Boldin, E.E. Varfolomeev, Z. Pancer, I.L. Mett, J.H. Camonis, and D.Wallach. A novel protein that interacts with the death domain of Fas/APO1 contains a sequence motif related to the death domain. *J. Biol. Chem.* 270:7795 (1995).

71. M.P. Boldin, I.L. Mett, E.E. Varfolomeev, I. Chumakov, Y. Shemer-Avni, J.H. Camonis, and D. Wallach, Self-association of the "death domains" of the p55 tumor necrosis factor (TNF) receptor and Fas/APO1 prompts signaling for TNF and Fas/APO1 effects. *J. Biol. Chem.* 270:387 (1995).
72. H. Hsu, J. Xiong, and D.V. Goeddel, The TNF receptor 1-associated protein TRADD signals cell death and NF-kappa B activation. *Cell.* 81:495 (1995).
73. G.H.W. Wong, and D. Goeddel, Fas antigen and p55 TNF receptor signal apoptosis through distinct pathways. *J. Immunol.* 152:1751, (1994).
74. K. Schulze-Osthoff, P.H. Krammer, and W. Droge, Divergent signalling via APO-1/Fas and the TNF receptor, two homologous molecules involved in physiological cell death. *EMBO J.* 13:4587 (1994).
75. L.A. Tartaglia, M. Rothe, Y-F. Hu, and D.V. Goeddel, Tumor necrosis factor's cytotoxic activity is signaled by the p55 TNF receptor. *Cell* 73:213 (1993).
76. A. Sarin, M. Conan-Cibotti, and P.A. Henkart, Cytotoxic effect of TNF and lymphotoxin on T lymphoblasts. *J. Immunol.* 155:3716 (1995).
77. R.A. Heller, K. Song, and N. Fan. Cytotoxicity by tumor necrosis factor is mediated by both p55 and p70 receptors. *Cell* 73:213 (1993).
78. K.C.F. Sheehan, J.K. Pinckard, C.D. Arthur, L.P. Dehner, D.V. Goeddel, and R.D. Schreiber, Monoclonal antibodies specific for murine p55 and p75 tumor necrosis factor receptors: identification of a novel *in vivo* role for p75. *J. Exp. Med.* 181:607 (1995).
79. J. Rothe, W. Lesslauer, H. Loetscher, Y. Lang, P. Koebel, F. Kontgen, A. Althage, R. Zinkernagel, M. Steinmetz, and H. Bluethmann, Mice lacking the tumour necrosis factor receptor 1 are resistant to TNF-mediated toxicity but highly susceptible to infection by *Listeria monocytogenes. Nature (Lond)* 364:798 (1993).
80. K. Pfeffer, T. Matsuyama, T.M. Kündig, A. Wakeham, K. Kishihara, A. Shahinian, K. Wiegmann, P.S. Ohashi, M. Krönke, and T.W. Mak, Mice deficient for the 55 kd tumor necrosis factor receptor are resistant to endotoxic shock, yet succumb to *L. monocytogenes* infection. *Cell* 73:457 (1993).
81. S.L. Erickson, F.J. de Sauvage, K. Kikly, K. Carver-Moore, S. Pitts-Meek, N. Gillett, K.C.F. Sheehan, R.D. Schreiber, D.V. Goeddel, and M.W. Moore, Decreased sensitivity to tumour-necrosis factor but normal T-cell development in TNF receptor-2-deficient mice. *Nature* 372:560 (1994).
82. P. Vassalli, The pathophysiology of tumor necrosis factors. *Annu. Rev. Immunol.* 10:411 (1992).
83. C.O. Jacob and H.O. McDevitt, Tumour necrosis factor-alpha in murine autoimmune 'lupus' nephritis. *Nature* 331:356 (1988).
84. X.D. Yang, R. Tisch, S.M. Singer, Z.A. Cao, R.S. Liblau, R.D. Schreiber, H.O. McDevitt, Effect of tumor necrosis factor alpha on insulin-dependent diabetes mellitus in NOD mice. I. The early development of autoimmunity and the diabetogenic process. *J. Exp. Med.* 180:995 (1994).
85. S. Gerder, D.E. Picarella, P.S. Linsley, R.A. Flavell, Costimulator B7–1 confers antigen-presenting-cell function to parenchymal tissue and in conjunction with tumor necrosis factor alpha leads to autoimmunity in transgenic mice. *Proc. Natl. Acad. Sci., USA* 91:5138 (1994).
86. A. Schattner, Lymphokines in autoimmunity--a critical review. *Clin. Immunol. Immunopathol.* 70:177 (1994).
87. H. Ishida, T. Muchamuel, S. Sakaguchi, S. Andrade, S. Menon, and M. Howard, Continuous administration of anti-interleukin 10 antibodies delays onset of autoimmunity in NZB/W F1 mice. *J. Exp. Med.* 179:305 (1994).
88. C.O. Jacob, Studies on the role of tumor necrosis factor in murine and human autoimmunity. *J. Autoimmun.* 5:133 (1992).
89. P.W. Gray, K. Barrett, D. Chantry, M. Turner, and M. Feldmann, Cloning of human tumor necrosis factor (TNF) receptor cDNA and expression of recombinant soluble TNF-binding protein. *Proc. Natl. Acad. Sci., USA* 87:7380 (1990).
90. F.M. Brennan, D.L. Gibbons, A.P. Cope, P. Katsikis, R.N. Maini, M. Feldmann, TNF inhibitors are produced spontaneously by rheumatoid and osteoarthritic synovial joint cell cultures: evidence of feedback control of TNF action. *Scand. J. Immunol.* 42:158, (1995).
91. A. Ashkenazi, S.A. Marsters, D.J. Capon, S.M. Chamow, I.S. Figari, D. Pennica, D.V. Goeddel, M.A. Palladino, and D.H. Smith, Protection against endotoxic shock by a tumor necrosis factor receptor immunoadhesin. *Proc. Natl. Acad. Sci., USA* 88:10535, (1991).
92. M. Feldmann, F.M. Brennan, R.O. Williams, A.P. Cope, D.L. Gibbons, P.D. Katsikis, R.N. Maini, Evaluation of the role of cytokines in autoimmune disease: the importance of TNF alpha in rheumatoid arthritis. *Prog. Growth Factor Res.* 4:247, (1992).
93. C.O. Jacob, Tumor necrosis factor alpha in autoimmunity: pretty girl or old witch? *Immunol Today* 13:122 (1992).

INDEX